The Oral History of Adm. Stanley R. Arthur, USN (Ret.)

Interviewed By
Edward J. Marolda

U.S. Naval Institute • Annapolis, Maryland

Preface

Stanley R. Arthur served in the U.S. Navy during a time of war and crisis in Asia, global confrontation with the Soviet Union, and major social changes in the U.S. military establishment. He flew more than 500 combat missions in attack aircraft during the Vietnam War. He describes in detail not only the professional and operational successes but the restrictions, frustrations, and failures that accompanied carrier operations in Southeast Asia. His postwar experience included naval aviation commands and staff assignments. Arthur relates the Pacific Fleet's involvement in Operation Desert One, the failed hostage rescue attempt in Iran, U.S. naval actions off Central America in the 1980s, and his work as the first Commander U.S. Naval Forces, Central Command. Arthur, who served successive tours in the Pentagon during the revolutionary Maritime Strategy and 600-ship era of Navy Secretary John Lehman, made key decisions on naval aviation and budgetary and logistical matters.

An important element throughout the oral history is Admiral Arthur's connection to the men and women he worked with who inspired him, shared the dangers of combat with him, and provided him with sage advice. In many cases these individuals became lifelong friends and confidants. Equally influential and supportive was his wife Jennie Lou, who hailed from their hometown of Jackson, Ohio. They raised four children: Erin, Brian, Derek, and Brad. Like any service family, the Arthurs had to adapt to numerous moves and school changes during his career, but the experience only strengthened their familial bond.

As Commander Seventh Fleet, Vice Admiral Arthur led U.S. and multinational naval forces during Operation Desert Storm, one of the most successful military campaigns in modern history. He describes his interactions with Defense Secretary Richard Cheney, Joint Chiefs of Staff (JCS) Chairman Colin Powell, and Chiefs of Naval Operations Frank Kelso and Jeremy Boorda during Arthur's tenure as Vice Chief of Naval Operations from 1992 to 1995. Perhaps the most significant aspects of Arthur's time as VCNO related to the Tailhook sexual harassment scandal and the status of women and other minorities in the service. He relates his handling of the Rebecca Hansen case and his early retirement from the Navy. Arthur was also a key figure in the base realignment and closure (BRAC) process, the development of the F/A-18E/F Super Hornet and the V-22 Osprey, and major Navy acquisition programs. In short, this oral history documents the service of one of the Navy's most significant warriors and leaders of the late 20th century.

Edward J. Marolda, Ph.D., interviewer

June 2016

The U.S. Naval Institute Oral History Program

Researchers and authors have been drawing the Naval Institute's Oral History Program since 1969, the year it was established by Dr. John T. Mason Jr. He and his successor, author and historian Paul Stillwell, sought to capture, preserve, and disseminate a permanent record of the stories of significant figures in naval history. Under the leadership of Vice Adm. Peter H. Daly, U.S. Navy (Ret.), CEO of the Institute, the program has expanded, with increasing numbers of historians conducting more interviews.

These oral histories are carefully fact-checked and reviewed by both historians and interview subject before they are made available. The Naval Institute is known for this high level of editorial intervention and polishing. The reader is reminded, as with all oral history interviews, that this is a record of the spoken word.

The Naval Institute wishes to acknowledge the many donors who make this program possible, in particular the generous support of the Tawani Foundation of Chicago and Jack C. Taylor of St. Louis.

Admiral Stanley R. Arthur, USN (Ret.)

San Diego, California, native Stanley R. Arthur attended Miami University (Oxford, Ohio), on a Navy ROTC scholarship. Graduating in 1957, he entered flight training and in 1958 earned his wings as a naval aviator. During the Vietnam War, he completed 514 combat missions, many as commanding officer (CO) of Attack Squadron 164 (VA-164) on board the aircraft carrier *Hancock* (CVA-19). He commanded the combat stores ship *San Jose* (AFS-7) and the aircraft carrier *Coral Sea* (CVA-43) and served as Commander Carrier Group Seven. Admiral Arthur's staff tours included assignments with Commander in Chief, Pacific Fleet in Hawaii; the Rapid Deployment Joint Task Force (RDJTF); Commander in Chief, Central Command (twice); and the Bureau of Naval Personnel. He served as the Deputy Chief of Naval Operations for Logistics before being selected to command the U.S. Seventh Fleet based in Yokosuka, Japan. Admiral Arthur capped his 38-year career in the Navy as the Vice Chief of Naval Operations.

Upon retirement in 1995, he entered the private sector and served consecutively as Vice President for Naval Systems with Loral Corporation, Vice President for Washington Operations with Lockheed Martin Corporation's Electronics Sector, and President of Lockheed Martin Missiles and Fire Control in Orlando, Florida, retiring from Lockheed in 2005. He holds a B.A. degree from Miami University, a B.S. in aeronautical engineering from the U.S. Naval Postgraduate School, a master's degree in administration from George Washington University, and honorary doctorate degrees from Miami University and the Citadel. In 1996 the Navy League honored him with the Admiral Arleigh Burke Leadership Award. The Deputy Chief of Naval Operations for Fleet Readiness and Logistics established the Stanley R. Arthur Award for Logistics Excellence in honor of this distinguished naval leader.

Deed of Gift

The U.S. Naval Institute is hereby authorized to make available in any format it chooses, from bound-book hard copy to electronic/digital Internet access, the audio recordings, transcripts, and videorecordings of the oral-history interview series conducted concerning the life and career of the undersigned. Disposition, repositories, and access shall be at the discretion of the Naval Institute.

The undersigned does hereby release and assign to the U.S. Naval Institute the rights and title to these interviews, with the exception that the undersigned and heirs retain the right to use the material for personal, noncommercial purposes. The copyright in the oral, transcribed, and videorecorded versions shall be held by the U.S. Naval Institute. All recordings, transcriptions, and videorecordings of the interviews shall remain the property of the U.S. Naval Institute.

Signed and sealed this 23 day of June 2015.

Signed name SR Arthur

ADM Stanley R. Arthur, U.S. Navy (Retired)

Interview Number 1 with Adm. Stanley R. Arthur, USN (Ret.)
Date: 23 June 2015

Ed Marolda (EM): This interview is being conducted on the 23rd of June 2015, with Admiral Stanley R. Arthur. Interviewer is Edward J. Marolda of the U.S. Naval Institute. Admiral, it's a pleasure to be here this morning.

Stanley Arthur (SA): Ed, nice to have you here.

EM: I'm enjoying Florida. I haven't been here in a few years, so it's nice to return to get some of the beautiful sights, although the weather is the same as in D.C. [Laughs]

SA: The weather never changes in summertime.

EM: No. Well, this is the first of hopefully several interviews, probably four or five, who knows. It depends on how much you want to say. And we have plenty of digital electricity, so we can do all that. Today I'd like to get through your early life, and if we have time today and tomorrow, get into your Vietnam experience. I've got hopefully most of your particulars from various sources, the internet and who knows what. I mean, it's all out there.

SA: Yes.

EM: We're all open to the world these days. If they can get your address, they can get everything. So, for the record, you were born on the 27th of September 1935 in San Diego. I know you were only there a few years, but do you have any memories of San Diego?

SA: I do. We lived at 1333 Highland Avenue, and that address has never been lost in my memory for some reason, I guess just because it was one of the first things that I remember. But we had a Navy pilot that lived the street behind us on the corner and I used to see him drive off to work. My dad told me, yeah, he's a Navy pilot.

EM: That impressed you?

SA: And that impressed me. Of course, the carriers were based out of North Island [Naval Air Station] and you'd see the airplanes fly in, the old F4B-4s with the bi-wings, with the yellow wings and the blue stripes on the wings. I'd see them up there.

EM: So this was when you were like five years old?

SA: Yes, yes. So I was really fascinated with airplanes from that moment on. So that's one of my memories. Our house was up a bank from the street, and we were on a hill that went down, and that's where I first learned to ride a bike. And I can remember my frustration and my dad's frustration at me trying to get myself balanced on a bicycle. Then when I finally did it and was able to go racing down the hill [laughter], he was a little worried whether or not I'd stop by the time I got down to the bottom. But I do have some glimpses of things that happened in those early years.

EM: You left there and you went to Hawaii in what, '41?

SA: In 1941 we went off to Hawaii. Dad was on a repair ship, the *Altair* [AD-11, actually a destroyer tender], which had gone out to Hawaii, and so once they got out there he sent for us, and Mom, my sister, and I jumped onto the *Lurline,* the Matson Line. Off we went to Hawaii, and that was really interesting. Some memories of the *Lurline*—my sister is two years younger than I am, Sandra, so by that time I was coming up on my sixth birthday, and so she was four. Mom had got us all settled one of the nights on the voyage and we were in the stateroom. Mom was going to go to the adult dinner. We knew we were going to stay in the room. Everything was fine; well, after she went down for her dinner, we decided, we had been around the ship enough we knew where to go and how to stay safe. We decided we'd go out and do our own little excursion around the ship.

EM: Adventure.

SA: Eventually, somebody caught us and said, what are you two doing out here? We said, "Oh, we're just out here walking around." Back to the room we went, and they got Mom. We were in a little bit of trouble, blaming each other for the excursion. The arrival in Honolulu, in those days, you know the pictures you always see of the streamers, they gave you rolls, little paper rolls, that you could throw them over, you know, colored streamers. So that was our first impression of Hawaii.

EM: Did you get leis as well?

SA: And leis. In fact, I've got a picture of my sister and I right after we got off the ship there. So Hawaii was a great adventure for us.

EM: How long were you there?

SA: We were there—I'm thinking that we left in January to go, and we left the last weekend in October of 1941. I had my birthday. Then I had appendicitis. They took me to the hospital there, and while I was in the hospital they took my appendix out, and by this time, during the summer of '41, Dad's ship was put on alert that they were going to move to the East, to the Atlantic at some point in time.

Each day he went off to the ship, we never knew whether we were going to get a phone call and he would say, "We're getting under way and I won't be back tonight." That finally happened just shortly after my birthday in September. Off he went. They're gone. I go in the hospital for an appendix operation, and Mom gets the notification that we are no longer "sponsored dependents" in the territory of Hawaii and we've got to leave.

EM: That's pretty harsh, jeez.

SA: Yeah. And so we were put on the *Wharton* [AP-7], a troop ship, sometime right at the end of October, and came to the West Coast.

EM: You went back to San Diego?

SA: No, the house was rented by then. Mom decided that we would go back to Ohio and stay there. We went back to Ohio. At that time, her folks lived on a farm outside of this little town of Jackson. We stayed with Grandma and Grandpa Keller out on the farm. I had just started first grade in Hawaii in a Baptist church school there. There was no public school near the house that we rented, so I had started school there, so then we came back. Still at home with my grandparents were their last four children. My Uncle Dan, who enlisted in the Marines shortly after we got home. There were the twin brothers, Jack and Pat, who both enlisted probably a year later, I guess into the Army, and the youngest girl, Joan.

EM: Your mother was one of 12 children?

SA: One of 12, yes. Joan was the youngest. Joan is just five years older than I am, so she was like a big sister to me. The school that I was in out there in the country was a one room schoolhouse. She would drag me up to the bus and off we would go to school. As a young kid, I had nosebleeds all the time. She was always called to come out and clean me up when I had one of my nosebleeds. She's still alive and we still stay in touch.

EM: One room schoolhouse, so she was in the same schoolhouse with you? Small town!

SA: It was, it was really small. Very few families out there on the farms. It was quite a change for me from the Hawaii school into this one room schoolhouse where you had older kids.

EM: It was different in Hawaii?

SA: In Hawaii, it was a school like a small campus. So the first grade, we had our own room. We didn't really interface with any of the other students. One of the requirements to come to school, you had to bring our own tatami mat, little rice mat.

EM: For nap time?

SA: Nap time. That's where I learned my first major lesson about school. I liked to socialize during nap time, and the teacher was not having any of that. So one day she said, "I've told you that you have to be quiet, and so now I'm going to make you be quiet." She got a roll of that old white adhesive tape, about two inches wide, and she just taped my mouth shut.

EM: You couldn't do that these days.

SA: But it was a great lesson. From that moment on, when I was in school, I was in school listening.

EM: You were embarrassed by that?

SA: I was, but it kept me quiet. I never had a problem. As soon as I walked into a classroom and the teacher showed up, almost to the degree that I would hardly ever ask a question unless I really had a burning question. I just stayed quiet.

EM: That memory was strong with you?

SA: Yes it was. I think now, yeah, like you say, my gosh, it could never happen. But I thought it ended up doing me pretty good.

EM: That's what discipline is all about, and learning, you learn from your experience hopefully. Your father was one of five children.

SA: He was one of five. He was the youngest. He had two sisters and two brothers older than him. His older brother was in the Navy for a while during the war, Olen, and Lena lived in Xenia, Ohio. She had a son, Jack, who was in the Navy, and he was not much different in age than my dad. He had served on the original aircraft carrier.

EM: *Langley* [CV-1].

SA: On the *Langley,* and so that was another Navy tie-in. The next sister, she married a farmer up in London, Ohio; Naomi was her name, and Uncle Ed. Every time we went back to Ohio, we always went up there because Dad used to go there on his summers. He used to go up and live with them and work on the farm in the summertime with them. They were such a sweet, sweet couple. Then his brother, I call him Uncle Dick, his name was really Arbella. Uncle Dick worked for the railroads and ended up coming back to Jackson late in life. But he followed the railroads up into Michigan and spent most of his working years up in Michigan with the railroads.

EM: Your dad's dad owned a grocery store?

SA: Yeah. He owned a grocery store, and when we came back from Hawaii—they lived in town, and so we spent quite a bit of time with them initially. But he died rather suddenly in June, so just a few months before he died. Dad was not able to come home. By that time the ship was down in Trinidad. That's where the *Altair* came to Bermuda, through the canal to Bermuda, and then down to Trinidad. The Navy, they just couldn't let him come home. So he was not able to come home for the funeral. I guess Granddad Arthur died of a heart attack as best as I can recall.

EM: You didn't know him that long or that well?

SA: I didn't know him that well at all. He was a very quiet, quiet guy, very pleasant. My Grandpa Keller was quite a character. He was a different—

EM: Can you tell me about your father's Navy experience?

SA: Dad enlisted in the Navy right out of high school and went to Great Lakes [Naval Training Center]. That's where his recruit training was, and then to San Diego. He was a machinist's mate. He was a second class machinist's mate when I was born. By the time we got to Hawaii he was a chief petty officer. He loved the Navy. He just absolutely loved it. He was very proud to be a "snipe," as they called it, engineers, you know. He used to tell me, he said, "The one thing we were most proud of is that when we went ashore, we had the best uniforms of anybody on the ship. We made sure, even though we were always dirty and greasy in our work on board ship, when we went ashore there was"—it was special to him. I mean Dad was a dresser. He always dressed as spic and span as possibly could be. He used to tell me, "Boy, when we went ashore, we looked good."

EM: They were not the "black" gang there.

SA: No. He ended up staying 23 years. He retired. He was selected for lieutenant commander when he retired. He was just turning 40, and he said, "It's time to do something else." He was no longer qualified to go to sea.

EM: How did he become an officer?

SA: He, when he came off of, I guess, I'm trying to think when he was actually promoted, because he went away. He went away and he was a chief, and he came home and he was a lieutenant. It happened as he came off of the repair ship. Whether he got the commission while he was still there or he got the commission en route, because he came back to Little Creek [Naval Amphibious Base] in Virginia in '43, late '43, as best I can recall, and went through commando training en route to go out to the Pacific.

EM: Commando training?

SA: Yeah. And while he was there, he pulled his Achilles tendon and they repaired it. Dad could never stand on steel decks again from that moment on. It would cripple him up right away. But he went out and ended up on Guam with a ship repair facility. As soon as they were able to get a safe spot there in Guam, they went ashore and set up a machine shop and ship repair facility that ended up being the naval shipyard out there in Guam built in. And that's where he ended the war. Then they came back. He went to Lake Forest, Illinois, up there at Great Lakes as an instructor.

EM: So do you think he was someone recommended—why don't you become an officer, this is wartime, you have a chance to get up?

SA: I think it was a straight—I think it was wartime.

EM: Battlefield commission, they call it in the Army.

SA: Right, I think it was the same thing. Because when he retired, he had to revert back to chief until he completed 30 years, and then when he completed 30 years in the fleet reserve, he was restored back to lieutenant.

EM: So he did that?

SA: Yes.

EM: So his memories of the Navy; his service in the Navy was positive?

SA: Very positive, he had great stories. Dad kept a log book up until the time he was married. You know the old green ledger book that he had covered in canvas and his name and all this stuff. And in there were beer bottle labels, theater tickets, dance hall tickets, receipts for when he and one of his shipmates wrecked a car going up to Oregon—one fender for $2.50, one headlight for $1.00. He had the repair bills. It was a great thing to leaf through. Then he closed it off when he got married. He says, "Well, dear diary, this is the end of the diary. You know I'm now married, and I'm off into a different life."

EM: Right. [Laughter] I'm sure your mother had something to say about that.

SA: I'm sure, I'm sure. He loved the Navy and he loved the people. He met a lot of great people.

EM: What did he want to do in post-retirement?

SA: This was another learning experience for me. He wanted to come back to the old hometown because his mother was still alive and, of course, my mother's family were all still there. By that time Mom's folks had moved into town—and right on the outskirts of town—from the farm. They had a plot of land, and dad said, "Well, sell me the plot that you are not using for your house and barn and all this stuff. I want to build a house." And so they sold him that piece of land. He and Mom had sent away for house plans, and they modified these house plans to be the house that they wanted. Dad came back to Jackson on leave and subcontracted to a guy to pour the foundation and get that all started, and then subcontracted the contract for the building of the house to a guy by the name of Don King, who was one of the builders in town and owned a big lumber yard.

Dad was a very meticulous planner. I mean, I always accused him of being a communist because he always had a five-year plan, and he had all these notes, and had it all mapped out for five years, and each year he would revise it. He did the same thing with the house. He kept a daily summary of where he thought things should be at the house, so while he's in Philadelphia, he's sending back, and calling back to the contractor, saying okay, you should be here at this stage, and I want you to do this, this, and this. And make sure that this and that, and you know, he had all the little things he wanted done.

The school year ends in Philadelphia, and Mom, my sister, and I come back and move in with Grandma and Grandpa next door to where the house is being built until Dad can come back and—I think it was July he came back. He had his retirement there in Philadelphia on Friday, and on Saturday he hopped in the car and drove back.

EM: What year was this? '43?

SA: It was '49. He comes back Sunday afternoon, and Don King, the contractor for the house, shows up and says, "I expect you at work at 7:00 tomorrow morning."

Dad says, "What are you talking about?" He says, "I'm retired, I'm not going to work."

Don King, who was a pretty aggressive guy, he said, "Oh no, you're not retiring. You are coming to work for me, and I want you there."

EM: I think he liked his organized way of doing things.

SA: He absolutely did. He said, "I'm going to teach you. I'm going to teach you the business, and you're going to enjoy it."

So Dad went to work for Don, and they had a great time. Don, he let Dad go for a couple of years. I never quite understood this, but Dad ran for the office of mayor of the town while I'm still in high school, and so Don says, "Okay, but you know, when your time is up playing politician, you are coming back to work for me."

And so he did. Dad was a two-term mayor of the town, and again, the project was the town. We had two blast furnaces in town that made [silvery] pig iron, and every summer it takes a lot of water to cool those things for the molds and all. Every summer we would go dry, and—

EM: Really?

SA: Yeah, and so then the blast furnaces would have to shut down for a few days until we had a good rain or something. Dad, his first project was to build a lake, a reservoir for the summer months. This turned out to be a real godsend, because what happened was the blast furnaces failed later on. But in the meantime, after they got that, they didn't have any problems with water problems. What the city and county did was, then turned that reservoir into a recreational area. People built cabins out there and had boats and fished and all that sort of stuff.

EM: Is that called Arthur Lake now?

SA: No. [Laughter] They didn't name it after him. He was very proud of the fact that he was able to sort of drive that one to completion. Once it was complete, he said, "Okay, I've done what I've really wanted to do as mayor." And so during the latter part of his second term, he resigned and went back to work for Don King. He did some major construction projects around the state as the head guy for Don on those projects.

EM: What influences did your parents have on your development?

SA: Well Mom, she really, really believed in school, you know, education. She was valedictorian of her class. Of course, her class was, I think, probably seven people or something like that. But she was very smart. Mom was always attentive to making sure that when we were in school, we did well. She watched our grades, she made sure we didn't miss school or get in trouble or play hooky. I did one time, and we got a letter. Mom was sort of fearless, I mean she was a fearless person. She was very happy. She always had a big smile, and there was just nothing that she didn't think she could do. She was a great role model as far as, hey, if you think you want to do that, go for it! She was always sort of—I was very small. I didn't grow until actually—until really after I graduated high school. So I was usually the smallest guy in the class, and she was always a little apprehensive about me getting hurt. But she finally got over that.

EM: Did you have any boyhood scrapes?

SA: Oh, I, yeah, a little bit, I guess, but nothing really—no, I didn't break any bones. I finally got my nose broken in high school in football, but I didn't have any of the disasters. My sister, on the other hand, who was really—she had her mom's spirit. I was a little more on the cautious side and figured things out and took a little time making sure I knew all the risks. My sister was one that could just jump into something. So while we there on the farm, she jumped up on the tractor and pulled back one of the handles that had the ratchets, and of course she had her hand down and she cut off one of her pinkies.

EM: Pinkie?

SA: Yeah. We had to get her into town and get it sewed up. But for me, I was pretty lucky.

EM: How about your dad? I mean you've talked a bit about your dad, but in the general sense, other than clearly you were impressed by his being in the Navy and enjoying it.

SA: Yeah. Dad was—his machinist mate training—he was always tinkering. He loved to tinker with things and like I say, he always made his lists of projects that he had to do. Dad was a car nut, and I became one. I became one worse than he was; his cars were always cleaned and polished, and you keep your hands off of them and don't get them dirty. So I inherited that from him. While I was in high school, I hated to mow the lawn. Dad, for some reason, and I guess it was from time being at sea, he just loved it. I said, "Hey, we are going to—let's make an arrangement here. Why don't you do the yard work and I'll take care of the cars? I'll keep them washed, and I'll keep them clean and polish and wax them." So we ended up making that agreement while I was in high school.

EM: So you liked things mechanical and tinkering with equipment?

SA: Yeah. And that was Dad. When he retired he went to Sears, and he bought a whole woodworking shop, I mean a whole shop down in the new house that we built there in Jackson. He had a table saw, he had a jigsaw, he had a planer; he, I mean, he had it all set up to do all sorts of woodwork. He did a little bit of this and that, but he never got into it like he wanted to, because he got busy with Don. But he could fix anything. He was always tinkering around the house. And he was very precise about everything. There was just no wishy-washy, it was always pretty straight.

EM: Sounds like both of your parents were inclined that way.

SA: Yes.

EM: They knew what was right and what was wrong.

SA: Yeah. Mom was a lighter spirit than Dad. Dad had a little more serious note to him, but Dad also had a little mischievous side to him. You could always tell when he'd get a little twinkle in his eye when he was thinking that maybe there was another way to skin the cat or he knows something. He was very even tempered. I think I learned a lot from him, watching him. I used to work in construction in the summertime with Don King as well. A couple times I would end up on projects where Dad was the overseer of that particular thing, and I could see him come in and see that things were wrong. He was very good about not flying off at the handle, but he would stop it right then and there. I mean, he was very—"Okay, let's stop it right now."

EM: He didn't lose his cool?

SA: But he didn't lose his cool. I tried to learn that, but I didn't learn it as well. [Laughter]

EM: Can you tell me about your sister and your relationship with her?

SA: My sister, Sandra, she was a great kid. I was always happy to have her around, and then years later, in fact, just a few years ago, she said, "You know, the thing that I remember most was that you never shut me off from being around you or around your friends. You always made me feel that I was free to come and go." And again, she was a lighter spirit than I am. She has a little bit of a gypsy about her. She didn't go to college.

She married right out of high school. Raised a great family. Her husband ended up working at Lawrence Livermore Labs out in California, and for a while they ran a karate school. She can do karate, and the thing that—the only disappointment I have with my sister is that she's still the same size as she was when she was 18 years old. [Laughter]

EM: Elaborate on that; I'm not sure, in height or weight?

SA: Height and weight, everything. I mean she grew, she looked like she stopped at 16 or 17, and she's never changed. She's never changed.

EM: Well, you can't help that.

SA: No, I know, I wish I had inherited that.

EM: She lives in California?

SA: Yes, and most of her family lives around close by her, and she's managed well for—I think Bub, her husband, has now been gone almost 20 years, and she's still by herself and lives by herself. While we were in high school, I thought she took it up [shooting] because she really liked to shoot. She confessed to me a couple of years ago that she took up shooting because the instructor was a cute guy. I didn't [laughs] know that at the time. But she was quite a good marksman. She shot competitively for a while.

EM: Oh, she did?

SA: I told her that part of doing this whole history is that I go back and research and try to get the timing down on everything that I'll finally write up. Then I want her to add in because she's got such a different set of eyes than mine. She says, "Well, you know I remember everything through you."

I said, "No—once I put it all down, you'll start remembering pieces that are unique that you had seen, that I didn't see. I want to get your view."

Because my sister is somebody that can—she writes like she talks. In other words, if she writes you a letter, it's just like she's talking to you. Which I find is a very unusual skill, and my mom used to say the same thing. Sandra sends me these letters and it's just like I'm talking to her.

EM: Are there any particular events or developments in your youth that really stand out, something that guided you in your future?

SA: We bounced around at so many schools, so I really got a perspective. A different perspective about education, and also got a much different perspective of how different this country is relative to the school systems. Like when we went to Great Lakes, the war had just ended so there were still no places to live. Dad found a Scottish lady who had just recently been windowed who was amenable to renting out the second floor of her home to us, Mrs. McCray. We lived upstairs, and she let us share part of the kitchen. On Sundays she had a beau who came and played his bagpipes. [Laughter] Just drove my sister crazy; she hated those bagpipes. The school that I was in—we were in Lake Forest, which was a very wealthy suburb north of Chicago. Big mansions and all this stuff. I go off to the fifth grade and here are my classmates being driven to school in limousines, and I'm really intrigued by this, you know, people are riding in limousines. I managed to get in fights there just like every place else where you were always the outsider. They had a very nice course in astrology for fifth grade, and I just loved it. I thought that was the most wonderful thing, learning about the stars. They took us down to Chicago to the big museum. And I remembered that place; one, the difference in—you know, even the rich kids had to go to school, so that was fun, and really it was a very, very nice school.

Philadelphia was a real eye opener for me. I had gone to school for a very short time down in Norfolk [Naval Operating Base, today Naval Station Norfolk] when Dad was there for that commando training, and I could see the white-black issue. There was none in southern Ohio. I mean, virtually we didn't have—we only had two or three black families, and they were just part of the fabric, and nobody—they were just there.

EM: Exactly like my hometown in Connecticut, we had one or two black families, and it wasn't until you went to Hartford you'd see black people.

SA: And they were just part of the fabric, and they were great kids at school. But Philadelphia was a different story. I can remember, I think it was my second school there. Yeah, it was the second school that I was in, and I was walking to school one day, and I'd seen these two kids, or three, two guys beating up on this smaller kid. So I jumped into the middle of it and said, "Hey, what are you guys doing? Get off of him."

And they turned around at me and they said, "It's none of your business."

EM: Were these white guys beating up on a black guy?

SA: No, this first incident was all three white guys. I'm saying, "Well hey, you just can't do that."

"Well, we're going to do it because he's a Jew."

I said, "What's that mean?"

That was my first encounter with—and so anyhow, I was able to sort of break that up. But it so happened that they then decided that I was their next target. [Laughter] At the schoolyard, within a couple of days or something, they ambushed me in the schoolyard and started beating up on me. And a black kid that was in our class, he did what I did for the other. He came to me and he was able to get the other guys off of me so I could take care of the main culprit. He took care of the other guys and then he was—we were in class, but we didn't really know each other. We sort of—

EM: After that you formed a bond with him?

SA: It was my first real awareness that there was all this tension going on between religions.

EM: Race?

SA: Race. And so Philadelphia was a tough, tough place.

EM: The Navy yard is in South Philly, so that's even tougher than most of Philadelphia.

SA: But we lived—in all these moves, we never lived in Navy housing. And we were always away from—

EM: Why was that?

SA: It was just something that Dad said. We were going to live on the economy.

EM: Because he got an allowance, right? Housing allowance?

SA: Yes, he had an allowance, and so we always did. And that was unique. Most of the schools I went to were not overly populated with military kids. In Philadelphia, where I finally ended up in seventh and eighth grade, we were out in East Lansdowne, which had very few ties, really, to the shipyard or anything down in South Philly. Although Philly was a tough town as I look at it today, I had a lot of freedom. I had a paper route. I passed the *Philadelphia Bulletin,* so I had a paper route at the end of every day. I went out and delivered papers and then collected on Saturdays and paid my bill, and then went to the movies and then got up early Sunday morning and passed the Sunday paper. Usually fixed breakfast for the family and stopped by the bakery that was just opening and brought hot donuts home. So I had my own money. The folks would let me go down to the ballpark and watch baseball. I could go down and for 25 cents, sit down in the bleachers, and watch the Philadelphia A's [Athletics], which was my team, and three or four of us would ride the El downtown to the stadium and come back. Today I don't think a young kid could do that. Sunday afternoons, I would get on my bike and ride down to South Philly to the airport where the Philadelphia airport is, because when Lockheed built the Super Connies [Constellation], one of the first Super Connies—they came in and had it on display down at the airport, and I rode down to see that.

EM: So even then, even though your father was surface Navy from the earliest time, you were looking at aviation that really got your interest?

SA: Aviation was always my thing.

EM: Did your father encourage that, or say "Oh, those flyboys"?

SA: He used to try to, you know, those guys, the real Navy is down there in the—so I had a love affair for the engineering, in fact all the times I was in squadrons on the ship I would volunteer to be—they always needed somebody to grade their engineering exercises and stuff. I would always do that, and they would say, "Why are you here?"

I said, "Well, my dad lived in these spaces, and he told me it was a good place to be, so I'm down here looking to see what you guys are doing."

I learned a lot about people and a lot about—I could see the things that still trouble this country. The one impression that I had that has stuck with me all these years, with those bicycle rides on Sundays down to the airport, I would ride my bicycle all alone right through the heart of the black community. Nobody ever bothered me. Nobody ever said anything but waved to me, or hi. The thing that I remember on Sunday afternoons was the dads out polishing their five- and six-year-old Cadillacs that had been retired from driving the wealthy around town, and they had them, and they were shining and they were clean. Everybody in the family was dressed to the nines. I mean it was like an Easter parade every Sunday afternoon. I was just the—you know, I said boy, they really have a nice sense of family.

EM: Community.

SA: And just sort of all blew away someplace. It really—

EM: Commentary on the entire country at the period. I mean they were segregated, yet, I don't know, seems like the problems came in during the '60s.

SA: Yeah.

EM: For many reasons. All right, your hometown, if you want to talk a little bit about that. Was it Anglo-Saxon, Protestant mostly, or were there ethnics in town?

SA: The community is very Welsh and German. The town, as I can reconstruct how it came into being, is the soft coal fields sort of play-out in southern Ohio. In other words, the soft coal in Virginia and West Virginia comes across the Ohio River; those soft coal fields, they just sort of play out south of Columbus, which is in the middle of the state. Jackson is in the southeast, and if you draw a line where Kentucky and West Virginia meet, from that point come up 30 miles, and that's where Jackson is. The area is called the tri-state area. So a lot of the old families were coal miners. They followed the soft coal fields up, and then they ran out there, and they all became farmers. The two blast furnaces that came in brought the industrial piece of the pie, so the railroads came through there to haul the iron. There were a lot of fire brick furnaces around the area because of the cheap fuel from the soft coal and stuff. In school, every springtime we had what we called an *eisteddfod.*

EM: Say again.

SA: Eisteddfod. Yeah. I think the real pronunciation is "ess-te-fudd," but I'm way off—it's a Welsh term, and it's a singing contest. D. Merrell Davis was the musician in the school system, and he ran these things. So every spring, you were either going to sing a solo or a duet or play a piano, or something, and you get ribbons for this thing. But you had to participate. It was one of those things that it was the thing to do and they still do it in the elementary school. I don't think they do it in the high school anymore. But when I was there, we did it all the way from first grade to high school. So there's a very musical community from the Welsh influence. A lot of the coal miners were German that settled—large German settlements down in the southwest corner of Virginia. These folks were all sort of related that came up through West Virginia into southern Ohio. That was the German influence.

EM: The Arthurs, were they Welsh?

SA: No, we're Irish. Well some say we're Irish, some say we're Scottish, that they were related.

EM: Maybe Scotch-Irish?

SA: Yeah, we're really a MacArthur, everyone dropped the Mac and we're Arthurs, so there is some confusion.

EM: What about on your mother's side?

SA: My mother's side was German, Keller. Oh, yeah—*so* German. So that was the makeup of the town, and it was a population of about 5,000 people.

EM: Any inspirational teachers from that time period?

SA: D. Merrell Davis, who was the choir guy, was also the choir director at our church. I was going to the Methodist church in those days and sang in the choir. The second year that we were back, when I was in the high school, between my freshman and sophomore year, D. Merrell Davis had been appointed.

The rich family in town that owned the blast furnaces, the Joneses, very wealthy family, he bought a property out on the lake and built a camp out there for kids and for the people in Jackson. They would issue 10 or 15 scholarships every summer for kids to come out that wouldn't be able to afford the camp, to come out and be part of the camp. My second year there, I got one of those scholarships. D. Merrell was the camp director. Very, very nice guy. He sang, he led glee clubs for different 4-H conventions, well known around the country as a matter of fact—and so he was the camp director. His two daughters, one was a year behind me and one was two years, I think, or three years behind me, and I think one of them was in Jennie Lou's class.*

They lived in a little bungalow out there for the summertime, and he directed the activities of this camp. It was a very, very good experience, because they had kids from Chicago and Cleveland, New York and elsewhere, and some locals from the state as well. It was a good mixing bowl and lots of skill sets. We went canoeing, we learned to fly fish, we learned to navigate over land, you know, a compass and a map, and those kinds of things.

EM: Did you ever think of going into music?

SA: I did. I started piano in Philadelphia and then gave it up. I think my teacher thought I had more talent than I did, because she got me playing classical music early on, and I just couldn't keep up with it. But then I switched to trombone, and then I played the trumpet, and so for the first two years I was back in Jackson, I was in the band. I played in the high school band.

EM: And you played football as well?

SA: I didn't do that because Mom said I was too small, and finally in my junior year in high school I said, "That's it. I'm not going to play in the band anymore. I'm going to play football." So my junior and senior year I played football. I've always had a flair or a feel for music. I go back again to the school, the difference in the school systems. In East Lansdowne, when I got there for the seventh and eighth grade, I had a teacher who was an absolute tyrant. Just, I mean, his face would get red, and he would yell and scream and terrorize the class. One day he was really upset and yelling and screaming, and all of a sudden he points me out and he says, "See him? He's not having any trouble with this. Why are you all having trouble with this? That kid has been bounced around from school to school to school, and he doesn't even hardly know where he's at, yet he knows—"

---

* Jennie Lou Mapes would become Admiral Arthur's wife.

I said oh, don't put that on me. And what he was talking about was, I think, he had a class where we studied the derivatives, so how a word was formed, and what the front end of it meant and what the back end of it meant.

EM: Right, the roots.

SA: The roots, and so some of that made sense to me, because the languages were a little bit different as I moved around, so you had to have an ear for picking up the nuances of what they really were meaning. He said, "See, he can figure this out."

I thought, maybe bouncing around all this time has not been so bad after all.

EM: Yeah, a broadening experience.

SA: Yes. What he did, he was also a singer or actor with the Philadelphia Opera. He would take us on field trips in the afternoon for the operettas down at the Metropolitan Opera House, which was just a completely out of the blue—

EM: Beautiful.

SA: And it was.

EM: Do you have a liking for classical music today?

SA: Yes.

EM: Because of that experience?

SA: Yes. I started listening more because of that, and then I found that I had to pick and choose pretty carefully on the classical side. I cannot be anywhere without music. I've got it on my computer, I always have music playing through the computer.

EM: What do you listen to routinely?

SA: Usually just easy listening anymore, but I use Pandora so I can select artists to play. I've got an XM radio out in the garage. If I'm in the car, I've got XM radio playing all the time, so I've always got music going.

EM: I guess we are going back a little bit, but what is your remembrance of World War II? I mean in a general sense, because you had four or five years of that experience. Looking at the larger world events, I mean how did they impact you?

SA: When we were in Hawaii, of course the fact that Dad's ship was going to have to go to the Atlantic at some point in time—the question was, "Why do you have to go to the Atlantic?"

"Well, we're a repair ship, and ships are being sunk in the Atlantic from the war in Europe," so we knew about the war in Europe.

We had a big avocado tree in the front yard of this house that we rented from this—my folks called her a White Russian, I guess, that was a term used in those days, a White Russian lady, and she rented us her house. I'm a big avocado fan from those days. My mom taught me how to eat avocados and I like them. These two Japanese guys would come down in the morning and take the avocados that had fallen on the ground. I was taking a little umbrage at that because I didn't think they ought to be taking those.

Part of the dialogue associated with that was at night, people were reporting Morse code coming, bleeding through on our radios, and they were talking about the Japanese sending messages to Japan about ship movements and what all was going on in Pearl Harbor. This was already coming into my mind that they were potential enemies. The night we got on the troop ship to head out of Honolulu and come back to the West Coast, sometime in the late evening the ship went to General Quarters. Now, whether it was a practice drill or what, I was pretty well tuned in.

EM: It got your attention?

SA: Yes, and the dialogue that went along with it was that they had come across a ship that was running without lights. They had challenged the ship, the ship didn't answer. They assumed it was a Japanese ship and went to General Quarters. The fact that there was a war coming was—

EM: You felt that?

SA: There was no doubt in my mind, and I always said the reason they enforced the rules so rapidly on us as being non-accompanied dependents was to get us out of there. So we got back to the farm in Ohio and five weeks, less than five weeks' time, Pearl Harbor. It was not a surprise.

EM: But were you shocked at the severity of it?

SA: Yes, because, you know, we'd been down there.

EM: Were there any folks that your parents knew in Hawaii who got involved?

SA: No, not that I know of, because most of the folks they knew were tied to the ship.

EM: Was your mother worried about your father being in the Atlantic? Because there were ships, we were losing ships in '41, even though we weren't at war yet.*

SA: Yeah, she was, but then, as he told her, "We're a repair ship, and basically we're pretty much pierside and they bring the stuff to us."

It was a real shock after we were back there and then Pearl Harbor because, gee, we were just there. We wondered about our old house, but it was way down toward downtown, so there was no problem there.

---

* A German U-boat sank the destroyer *Reuben James* (DD-245) in the Atlantic in October 1941, a little more than one month before Germany declared war on the United States in December.

EM: During the war, did you, like a lot of kids have a chart with all the different aircraft profiles?

SA: I had all the those cards, for the civil air patrol, civilian or whatever the air defense was, I had all those cards, all memorized. I had little miniature models of the airplanes. I built airplanes all the time. When I was in the hospital with my appendix in Hawaii, Mom brought me an aircraft carrier with airplanes on it, and it had a catapult that I could launch the airplanes off of. [Laughter]

I could listen for them, of course [in Ohio]. Wright Patterson Air Force Base over in Dayton was about 80, 90 miles away, but a lot of times the planes would fly over down in our part of the state. I could hear them and say, "Oh, that's not an airliner, that's a warplane." I followed it fairly closely. I never worried too much about Dad. I don't know why. I always felt like he was going to be okay, and everything was all right. I really didn't know what he was doing out in the Pacific after he went out that way until—

EM: So you really didn't see him for a couple of years?

SA: No, we didn't. First time we saw him after he left Hawaii was in '43 when he came to Norfolk for the commando training, and then he went away. Next time we really, really saw him I think he was—don't think we saw him until right at the end of the war, when he came back to go to Great Lakes. So that would have been '45, so from'41 to '45, we probably saw him twice.

EM: Do you recall your feelings toward the end of the war: The war is over?

SA: Yeah, the first part of that was when [President Franklin D.] Roosevelt died. I thought well—

EM: That was traumatic to a lot of folks?

SA: For a lot of people; particularly my mom's side of the family were die-hard Democrats. My granddad [Keller] was, and he had been in politics a little bit too. Dad's side of the family were Republicans.

EM: Interesting.

SA: Mom was a die-hard Democrat until [Richard M.] Nixon became president. She became a Nixon fan of all things. So she ended up being a die-hard Republican until the day she died. But Granddad [Keller], oh God, Granddad was a Democrat through and through. When Roosevelt died, there was lots of sadness there. For some reason I thought it was a good thing.

EM: From your mother?

SA: So, you know, I was sort of overjoyed for some reason. I thought, "Well, this is a good thing, and the war's going to end soon."

Then all the guys started coming home. My uncles started coming home.

One of the stories about when Dad married Mom, he came back to Jackson and they really hadn't known each other. They got hooked up somehow, because Mom was working at a beauty parlor up near the Washington Courthouse, but somehow or other they met. They ran away and got married. Well, Granddad just had a fit. Just had a fit. Then Dad said, "Look, I'm taking her away, but I promise you I will bring her back to Ohio every year. We will never miss a year of coming back here."

And he held onto that until the war came, and that's part of why we moved back at the end also. In the meantime, he gets out to California with Mom, and Mom was quite a looker. And his shipmate, Victor Abherve, who was the son of a French chef from Hollywood and had a big falling out with his dad and ran away and joined the Navy, he and Dad became big buddies. Vic says, "Are there any more like her back there?"

Dad says, "She's got five sisters."

"Okay, next time you go back, I'm going back with you." He came back the next year and married my mom's next sister.

EM: Is that right?

SA: Aunt Ann. So Granddad knew—he was really miffed now [laughter], two daughters off with sailors.

EM: With sailors, yeah, watch those sailors.

SA: Uncle Vic and Aunt Ann—Uncle Victor, was, you talk about a character. He was an absolute wild man, they used to say. They'd have him come to the house in San Diego on payday, and by the time he got there all his pay was gone. He'd start drinking and realize he was supposed to come to the house for dinner. And whatever loose change he had left, he couldn't figure it out to give to the bus driver so he would just empty his pockets.

EM: Take it all.

SA: Then he was a boiler tender, so they were both snipes. Uncle Vic, during the war, was out in the Pacific almost the whole time. He was on a couple ships that were hit by kamikazes. He got out just like Dad did. Then he got recalled for the Korean War to be an instructor up in Bainbridge [Naval Training Center], up in Maryland, for the engineering guys up there. He had been offered a commission and wouldn't take one. Dad—they weren't interested in having Dad back, but they were interested in Uncle Vic going back, and so he went back for a while. Then the war was over, the Korean War was over, and he went to work for International Harvester.

EM: During World War II, did you follow the carrier operations out in the Pacific through the newspaper? Air fights?

SA: Yeah, through the newspaper and newsreels, you know, the Saturday matinees and the newsreels with the—

EM: Was it Pathtone?

SA: Or Path something [Pathé News]. They were great. I was always intrigued with those.

EM: Your time in high school, you played football, you were in the glee club. Were there other activities?

SA: The band for the two years, Hi-Y for the guys. I was a chaplain for the Hi-Y association.

EM: Were you particularly religious? In fact, was your family particularly religious?

SA: We were medium religious. In high school I went to church most of the time.

EM: What faith are you?

SA: Methodist. When I was in the eighth grade, I had, in Philadelphia, I had sort of a religious epiphany and started to get serious about going to church.

EM: What brought that on?

SA: I'm trying to remember. I think it was a little camping expedition with a faith group, a church group that I went on and they—.

EM: Inspiring?

SA: Born-again type thing. I remember I surprised my folks one night at some function when I said, "I'm born again." And they sort of looked at me and wondered. They were off and on churchgoers. Mom—well both of them—there was always a Bible around the house, and Mom sometimes would sit and read. I read the Bible a lot. I drifted away during the Vietnam War when the churches became active in anti-war. I couldn't put the equation together. I can see it now; I couldn't see it then. So I feel very religious, but I don't practice as much as I should.

EM: That's probably true of most Americans.

SA: I guess, and it's sad, I think.

EM: I wanted to get this on the record, your high school yearbook was the *Osky Wow?*

SA: *Osky Wow.* [Laughter]

EM: I really got a kick out of that.

SA: There's a little ditty that goes along with it, but I don't know it all. But it's "Osky wow wow, osky wee wee, something," and it goes on from there.

EM: Do you know where that came from?

SA: I don't. It's still used.

EM: Interesting. All right, you graduated high school in 1953, and of course that's the last year of the Korean War. Do you remember your thoughts about the Korean War? Did it affect you at all?

SA: Of course there were a lot of guys from our hometown who went to Korea and left school early. The guys ahead of, you know, the '49, or '48–'49 guys, some of them left school and went off. One of my aunts was dating a guy who went off to the Korean War and got killed. That was sort of tragic. Then Uncle Vic getting called back to Bainbridge, Maryland. I was still following the airplanes. I was watching what the airplanes were doing and all of that.

EM: Did you see the movie *The Bridges at Toko-Ri?*

SA: My favorite all-time movie. In fact, on the message that I sent out on the night that we kicked off Desert Storm, I used part of that line from *Bridges,* and I wrote [James A.] Michener and told him.

EM: What did you say?

SA: I said something like, "I know you are prepared and I know you're wondering what's going to be next. Some of it's going to be—you're going to be frightened. I know that when the"—I'll have to dig it out to get it all right. I put the line, "Where do we find such men and women?"

EM: If you could dig that out, I'd like to recount exactly what you said.

SA: Yeah, I'll try to see if I can dig that out.

EM: So that movie impressed you?

SA: It did. It really did.

EM: Had good stars?

SA: Yeah, it had good stars and good flying, and you know Tom Hudner—*

EM: Oh, I know Tom Hudner.

---

* Thomas J. Hudner Jr., then a Navy lieutenant (junior grade) and later promoted to captain, earned the Medal of Honor in Korea for his courageous, albeit unsuccessful, effort to save the life of his wingman, Ensign Jesse L. Brown, one of the first African American naval aviators.

SA: And of course we had become acquainted over the years since.

EM: He's a good guy.

SA: Yeah, really good guy. Yeah, that movie, that was right down my alley.

EM: You know, James Michener, he was with the Naval Historical, whatever they called it back then, division. They sent him out to record World War II in the South Pacific. He didn't record much of the war. He wrote his book. [Laughter] I've seen reports on him back at the archives. You didn't do what you were supposed to do.

You graduated high school '53 and you went off to Miami University, Miami of Ohio, and why did you pick this school?

SA: It was in the erstwhile business of the Navy. They picked it for me. I had listed it as number two on my NROTC [Naval Reserve Officers Training Corps] application as a school because it was in state.

EM: So you had applied for NROTC first, before you knew what school you were going to?

SA: Yes. Let me tell you the story about how I got into the NROTC. Football season ends Thanksgiving Day 1952. I'm looking around saying, "I'm not learning that much at school. I know this stuff and I'm itching to get on with life." I'm thinking that, you know, I've got to get started doing something. I made up my mind that I wanted to go into the Navy. I thought that I didn't know for how long, but I thought it would be a good thing to do. I come home one day and I approached Dad first. I approached him before I did Mom, because I thought I'd have an ally, because I knew Mom would push back. And, so I said, "Dad, I'm going to go down and talk to the Navy recruiter. I'm gonna quit school and join the Navy."

He said, "No, no. You're not going to quit school."

And I said, "Well, I can take the GED [General Educational Development Test] and get my high school diploma."

He says nope. "No," and he said, "Now that that's decided [laughter]. You ought to be applying for the NROTC scholarship."

I said, "What's that?"

He saids, "Well, I don't know that much about it, but I do know that the Navy has an ROTC program and they offer scholarships. You ought to take the exam for it. And oh, by the way, the exam comes up in December. And so you better get busy figuring out how you're going to take the exam."

EM: He was determined you were going to be an officer. He didn't say you've got to be an enlisted guy like he was to start?

SA: No. He just said this is what you are going to do. And so I said okay, I guess. I took the exam. Part of the exam was you had to have an interview. The interview was by somebody from CNET [Naval Education and Training Command], I guess. I had to go up to Columbus, Port Columbus, which had a little naval air facility up there. I sat down with this lieutenant and he says, "Why aren't you applying for the Naval Academy?"

I said, "Oh, I don't want to go to the Naval Academy."

He said, "Why not?" He was a Naval Academy graduate.

I said that my social skills weren't that good. [Laughter] I said, "I think I need to go someplace where there's boys and girls and I learn a little bit more how to get along with the other people."

He said, "Well, you ought to think about the Naval Academy."

I said, "I'm just not sure that's for me, but I know I want to be in the Navy." He said okay.

When I came back, Dad said, "Well, what did you talk about?"

I said the guy wanted me to go to the Naval Academy and I told him I didn't want to go. He says, "What?" [Laughter]

We were sitting around waiting for this thing, and so in the process, I decided—well I looked at the schools, where they the NROTC units were. I went up to Ohio State and talked to the NROTC unit up there, and I said, "I've taken the exam, but I don't know how I've done. I'm planning on coming to Ohio State. I'd like to be in the NROTC unit."

They said fine. "If you register for Ohio State, you're in. We've got your name down." Then I get the notification I'd been accepted as an alternate for the NROTC scholarship and where would you like, what are your choices? I put Ohio State University because I wanted to get aero engineering. Then I looked, and we had a hometown friend of the family was on the admin staff down at Miami University and a couple people from Jackson had gone to Miami. It was a very good, very good school. So I said that would be my second choice. And I think I put Purdue as number three. Well, the thing that happened was—right away they said, "You've got the full scholarship and you're going to Miami University."

I said I'd been accepted at Ohio State in their NROTC program. They said, "No, you're going to Miami University."

EM: The Navy said this?

SA: Yeah.

EM: I wonder why; what was their rationale?

SA: I guess to just fill in the quota. [Laughter] Actually, it was a blessing. Smaller school. We didn't have any calculus or advanced algebra or anything in the [Jackson] school. I would have probably survived, but Miami did for me what I had in my mind. It socialized me a lot. It was a very good school. The NROTC unit was a great unit. That worked, that ended up working really good.

EM: You majored in aeronautical engineering?

SA: They didn't have an engineering program, so I majored in the School of Arts and Science. They had an aeronautics course to really teach you basic aeronautics and then the business side of running an airport. It was meant for people who had business interests in aviation more than people who wanted to do engineering in aviation. But it gave me the basic aeronautics and so that worked. Then I took all of the physics, calculus, and all the hard math and sciences, chemistry. That played into when I went to Monterey [Naval Postgraduate School] and got my aeronautics engineering out of Monterey.

EM: Any inspirational professors there at Miami, either in the NROTC or in general?

SA: Yeah, Ara Parseghian was our football coach at Miami. I didn't play football at Miami. I was still too small.

EM: And this was before he went to Notre Dame?

SA: Yes. Johnny Pont was an assistant at Miami who went on to coach later at other bigger schools. We had mandatory PE [physical education] at least for the first two years. I'm not sure if we had to do it all four, but I think just the first two years. My PE instructors were Ara Parseghian and John Pont.

EM: Is that right?

SA: The head coaches, I mean the head coach and assistant coach of football.

EM: Even then, did you know that these guys were special?

SA: Yeah. I knew they were really good. We had a really good football team when I was there.

EM: You didn't play football though?

SA: No, I didn't play football. I was pretty well acquainted with most of them because when I went to Miami, I needed to work, so I got hired on as one of the busboys, whatever you call them, for the dining rooms, which were the jobs they gave to all the athletes on scholarships. I was bumming around with a lot of the guys who played football. What struck my mind was, here are these guys were doing really—their job was to make the football team a great football team, and yet they were still teaching PE. They didn't seem to be—it didn't seem to be a burden to them. In other words, they seemed to be enjoying it. They were engaged.

A good friend of mine and I were both in the NROTC unit, and we stayed close all the way through flight training, all the way through our careers, and until his career tapped out at commander. We still stayed in touch, and then he died here just a few years ago. He was from a small town close to Jackson, Athens. Dick [Barker] and I, we used to, we were both car nuts, so we always were talking about cars and stuff.

EM: Did you own a car at that point?

SA: I did. At that time I had a 1950 Plymouth Business Coupe, little two-door Business Coupe.

EM: Was that your first car?

SA: It was my second car. My first was a 1935 Chevrolet two-door sedan. [Laughter] But in Miami you couldn't have a car on campus. I had my car parked in a farmer's field just outside of town.

But Dick and I got into an argument in PE class, and I think Pont was the instructor that day. We were wrestling, and I thought he did something dirty, and I lost my temper and went after him. Pont said, "Okay guys, I'm not going to break it up. Just keep going until you get tired of it." We just pounded on each other until we finally said that's enough. [Laughter]

EM: Get it out of your system.

SA: Yeah. I thought back about that and I thought that was pretty smart of him, because we didn't walk away.

EM: Let the steam blow off.

SA: Yeah, let it go.

EM: All right, during this NROTC experience you took some midshipman cruises?

SA: Oh yes.

EM: Tell me about them.

SA: Those were great. First cruise was on board the USS *Missouri* [BB-63]. Mom and Dad and my sister drove me down to Norfolk. They went on to Florida to have a vacation, and I jumped aboard the battleship *Missouri.* The midshipmen were a mixed group from many NROTC units and Naval Academy. Predominantly football team Naval Academy guys, at least in the division that I was with. The first event of that cruise was we were getting under way from Norfolk. I'm aboard ship, find my compartment. And I'm in first division on the deck, and we are at quarters for getting under way. We're out there on the teak deck all in our whites looking spiffy, and the ship's about ready to get under way and the afterbrow is still down, and you know there is some screwing around. We're still waiting for somebody, we're still waiting. And down the pier comes this bright red Cadillac. It comes whistling down the pier, screeching to a halt down by the afterbrow and this little, I mean he must have been about a five-foot-three, or five-foot-four sailor, black, tiny, real tiny guy, he jumps out of this Cadillac and he comes hustling aboard the ship. He's the leading seaman of the deck department.

EM: Little bit late to the job.

SA: But he's there, and we get under way. I'm thinking, "Hey, this is really something. This guy must really be something." Well, he was something. I mean he was the leading seaman and I mean, everybody deferred to this little guy, and it was just fun to watch all these big guys, you know, boy he ran things with a—

EM: So he had a little deference there?

SA: Yeah. Many years later we go to this up-and-out policy, and in my mind I always flash back to this guy among several others that I met much, much like him. But he was the man for the deck, and everybody knew it. And they knew if you wanted something done, he could get it done for you.

EM: Yeah.

SA: And I've thought, you know in today's Navy, we can't have that guy. He'd be kicked out 'cause he—

EM: Couldn't follow every rule?

SA: Yeah, and he couldn't make grade, he couldn't pass the exam to be a third-class.

EM: Could you elaborate on that? I mean, so there's a difference from that time?

SA: In that time a guy could go, he could spend 20 years in the Navy and never go past seaman or leading seaman. Today, if you can't make third-class petty officer on time, or within a certain limit of time, you're gone. So it's an up-or-out policy. I ran into this years later, with a lot of times when you get first-class petty officers who top out at first class and can't really do the job of a chief, but they can certainly do the job of a first-class petty officer, and we would throw them away.

It was just one of those instances in time that sort of stuck with me all the way through, so that was my first day at sea. The second part of that day was once we got under way, all four battleships, all four of the *Iowa*-class battleships were under way that day, and we all went out and steamed in formation.

EM: That must have been impressive.

SA: I've got some black and white photos someplace—

EM: You steamed up out of what port?

SA: Norfolk. So all four of the *Iowa* class were out there doing—*

EM: This was what, '54? '55?

SA: This was the summer of '54. So we did diamond formations and line, and then we went off, and our first port of call was Lisbon, Portugal. The Texas Bar in Lisbon, Portugal, was the place to go. First time a bunch of fresh midshipmen got to see real sailors in action, and it was a sight to behold. We walked into this place, and of course we got some strange looks. They weren't very happy to see some middies floating around in their midst, but we tried not to be too obvious. We were standing at the bar, and all of a sudden I hear this *whooo, whooo, whooo, whooo, thunk.* Somebody threw a bottle and hit one of the guys.

EM: Who else was in there, were there Portuguese? Or were there other sailors, marines?

SA: It was all sailors and marines. A true sailor's bar. And so the riot starts. We're scrambling, we're getting back. By the time the riot was in full swing, some of the girls that were there in the bar had already made their marks, they weren't going to lose their marks. They were there in the middle of the brawl, trying to drag their sailor out so they

* Included in this class were *Iowa* (BB-61), *New Jersey* (BB-62), *Missouri* (BB-63), and *Wisconsin* (BB-64).

can continue their adventure. And it was just one of those—never forgot the name, the Texas Bar.

EM: Did the shore police show up?

SA: Yes, they finally showed up and we all disappeared.

EM: So that was your first stop out of the United States? Well, you had been to Hawaii.

SA: Yes. Lisbon was a great place. We had some nice tours there and went to a bullfight. Bought some, still got some of that nice Portuguese lace. I think it's still in the original wrapping that I got for Jennie Lou.

EM: On the ship, what division were you in?

SA: I was in the deck division, first division, and then we rotated through the different jobs. We rotated through being on watch up on the bridge and then lookouts, and then we had a period where we were in engineering. We did the engineering watches. It was a great experience.

EM: Did you get seasick at all? I mean on a battleship it's kind of hard to feel it?

SA: No, not on a battleship. I didn't get seasick. Luckily my berthing compartment was on the main deck, so we had the portholes that we could keep open, we didn't have to close at night. We had a lot of fun. We went from Lisbon to Cherbourg [France]. From Cherbourg we took a tour into Paris, which was a lot of fun.

EM: On the train?

SA: On the train. Yeah. And Cherbourg was very interesting, because there were still a lot of the World War II damage and relics and stuff, that was good. Let's see, trying to think.

EM: Then you were in [Naval Air Station] Corpus Christi [Texas]?

SA: Yeah, the second year, the second cruise we went to Corpus Christi and Little Creek. We did amphibious training at Little Creek, and then we went down to Corpus Christi for aviation indoctrination. Got to fly, and that's the first time I got to fly.

EM: Even commercial, you were never on a commercial airline?

SA: I never had been on a commercial airline or anything until the Navy flew us in a C-54 to Corpus, and then we got to fly in the T-28 when we down there. I flew home from there on a United [Airlines] flight.

EM: What was your reaction to being in a T-28?

SA: I was so excited I could hardly stand it. We had NROTC instructors. They went on, so the guy I was flying with wasn't a regular instructor. And so he tried to do a loop, and he stalled out in the top of the loop and we fell out of the top of the loop.

EM: It didn't scare you off?

SA: No, it was all exciting.

EM: So he stalled out and then he restarted the engine?

SA: Yeah, the engine was still running but it coughed, and he also was upside down a little too long.

EM: You didn't get airsick either?

SA: I didn't get airsick. I'm surprised, because I don't ride very good in the back of an airplane. Flying it doesn't bother me at all, but riding in it is not good.

EM: Interesting.

SA: I'm scared of heights. I've always had this fear of heights. [Laughter]

EM: That's great. Let me tell you a little story; I hate to use up time, but I was in Colorado at the Air Force Academy conference, and we went up in Pikes Peak, and one of the guys, Ken Coskey, was in Vietnam, got—he was in prison and all that—so there were four of us coming down Pikes Peak road, where they told us, "Don't come down it 'cause you're next to the precipice."

I'm a little nervous, 'cause the guy driving was my boss, smoking a pipe, one hand on the wheel and we're hitting all these ruts, and I look in the rearview mirror and there's Ken Coskey, white as a ghost. I mean here's a former pilot. [Laughter]

So you were afraid of heights?

SA: Yeah, well, the story with that, when we were in Philadelphia Dad took us out one weekend for a trip to Gettysburg [Civil War battlefield]. I always had this desire to get to the highest point there is, but getting there is the problem. We get to this watchtower, we're going to go up to this watchtower, and Dad starts walking up and he isn't paying any attention, and all of a sudden he realizes I'm not beside him. Now this is when I'm in seventh grade, I think, so I'm probably 11, something like that? Eleven or 12. And he looks back and I'm crawling up the steps. [Laughter] He says, "What are you doing?"

I said, "I'm coming, but I hate to look down, and by doing it this way I know I'm covering a whole lot of ground and I'm not going to fall through anything." He just lets me meander my way up there, and then I very gingerly have to get to the front and have to look over the side. That stayed with me.

EM: Being in the back of a plane, like in a COD [carrier onboard delivery] flight, would you get a little queasy?

SA: In a COD flight, yeah, I don't like—

EM: All the windows are closed and you're in the cranial [protective headgear]?

SA: Yeah.

EM: I had the same reaction, though I'm not an aviator. Interesting. And the last midshipman cruise was in a destroyer?

SA: Destroyer *Gearing* [DD-710]. We went off to Barcelona, which was a great place. I loved Barcelona. Then we went up to Belfast, [Northern] Ireland, and we had a great time up there. We still hadn't figured out the Irish problem in those days, so here we are in Northern Ireland, we're in a bar singing the only Irish songs we know, which are—

EM: Yeah, southern Irish. [Laughter]

SA: Once we started seeing the very hard stares, we realized we were a little off base and cleaned up our act. My good friend that I talked about, Dick Barker, he decided hey, if we're in Ireland, we've gotta play golf. We've got to play on one of these golf courses because these are gorgeous golf courses. Anybody that's a golfer—

EM: Everything is green there, right?

SA: They all wanted to. I said I don't know how to play golf. He says, "Just watch me."

First thing that I realized was he didn't know that much about golf anyhow; he was pretending he did. Because when we went in to get balls for the—we needed to buy some balls, he was surprised to find out that their golf ball was a different size than ours.

I guess today they're all the same, but in those days they were different, they were a little bit smaller. So we buy the balls, then he says, "Okay, just watch me."

Of course all the Irish guys have come out of the clubhouse, and they're standing out there on the porch watching us. Dick gets up there and he swings, and he hits the ball, and *fisssssssst,* big slice off over the hedgerows. And he's got a real temper. Dick's cussing and swearing and throwing his golf club. [Laughter]

I get up there, and the only thing I'd ever swung is a baseball bat, I had played some baseball. So I just backed away as far as I could with the club and swung it like a bat at a low ball and I hit the ball and boy, I was really surprised. I *really* hit that ball. I said, "Boy, did I hit that thing?" I'm looking out there to see the ball and I can't see it. It's so far out there I can't even see it. And the guy behind me is falling down laughing.

EM: He said, "Look down"?

SA: Well, I had laid my golf bag up at the side of the tee, and it had gone right in the golf bag. And so we're both laughing and carrying on, and Dick is just beside himself now. Not only did he embarrass himself, but now we've embarrassed the whole tribe. [Laughter]

EM: I'm familiar with that. I'm a sort-of-golfer.

SA: So I never took up the game of golf.

EM: The fact that you went through ROTC and you didn't go to the Naval Academy, did that affect you throughout your career? Did people hold that up to you?

SA: It popped up in different places, in strange places. On the midshipman cruises there was always a little bit of rivalry between the NROTC guys and the Naval Academy guys, although great friendships were developed. There was always a little pull and tug. One of the things I noted that differentiated a lot of us was the fact that when we were ashore and testing our newfound ability to be out in the big world and having a drink or two—

EM: Attract females?

SA: Yeah. The Naval Academy guys had—the first they could think of was going to find a drink. Now those of us who had gone up in the university system, even though Miami was dry—we were a dry school, there was no hard liquor at Miami; in Ohio you had the dry counties and the wet counties, so it [Miami] was in a dry county. All we could have was 3.2 beer, which was dangerous enough in quantity. I noticed right away that the Naval Academy guys had more adjustment to make on how to take care of their drinking. That it was a new thing that they hadn't mastered quite as well as the NROTC guys did.

In flight training, when I went to basic training, basic training in those days had T-28s, and they were just starting to phase out the SNJs. I went to Corry Field [Naval Auxiliary Air Station], and Corry—Whiting already had all T-28s. Corry was still a mixed base, with SNJs and T-28s. Just out of happenstance, a guy says, goes down the names, "Okay, you're going to be over here flying SNJs," and the guy beside me, "Okay, you're going to be here." Or, "You're Naval Academy? You fly T-28s."

I said, "Wait a minute, what's the difference?"

"He's a regular."

I said, "Well, my commission says I'm a regular too."

"Yeah, but you're really not." [Laughter] It just showed up. When I was on the flag selection list, that list, I think there were four of us—I believe, I'm trying to remember now—but I think there were four or five of us that were all in ROTC. First time that been that kind of number. It has faded away completely.

EM: Were there times when you thought it might have been to your detriment to not be a Naval Academy graduate as you were working up the ladder?

SA: There were times where I thought it would have been, yeah. I would have had a better chance if I had been, but as it turned out, it didn't hold me back.

EM: You graduated and were commissioned in 1957, and you got married that same year. How did you meet Jennie Lou?

SA: Jennie Lou. We met my senior year in high school. Well, I knew her. She lived right across the alley from a very close friend of mine who was a classmate. He had a sister who was a year younger than us, or two, I guess two years younger than us, who one of my other friends was dating. I saw Jennie Lou across the alley, so I knew who she was. She was dating a cousin of mine. Lots of cousins in small towns . [Laughter] Right at the end of the senior year, there was this hayride scheduled to go, and I don't have a date. My friend that's dating the girl on the other side of the alley says, "Well, you know Jennie Lou, that daggum Danny Coll that she's been dating asked some other girl to go on a date and she doesn't have a date. You ought to ask her out."

Well yeah, yeah, why not?

EM: You had your eye on her?

SA: She was a cute girl, so I said yeah. I invited her, and off we went on a hayride. Then I'm gone, and she's got three years of high school left. But we dated all the way through college, we dated—

EM: Jackson's not that far from Miami?

SA: It's about a three hour and 15 minute drive.

EM: Did you go back home or did she come down to school?

SA: She came down senior year. She came down for a couple of functions. I would come home quite a few weekends. Sometimes when I had jobs back there, or when I would come back and work for Saturday, and back and forth. We dated pretty steadily all the way through school. When she graduated I still had a year to go, so she hired on as a secretary at the high school. When we got married I went off to flight training, and she still had another month or two of her contract to fulfill. I went down to flight training in July, right after we got back from our honeymoon. She came down, I think it was September.

EM: Well this is in the mid-'50s so the Cold War is raging, and you were a young person as I was, and I'm sure the larger world was kind of out there. Did you have an impression of what was happening in the Cold War or our competition with the Soviets, and did that impact you in any way?

SA: It was there, I'm not sure—I really didn't think that much about it until after the flight training ended.

EM: What about communism? Was it important to you from the start?

SA: Yeah, the thought of it was, I couldn't figure it out, because all the images you saw belied the verbiage. In other words, the hard light of the camera and the hard light of an eyeball told you that it didn't match the rhetoric that went on with all the oratory. There was something wrong in that equation. Of course, I guess today you'd say the same thing about our malfunctioning system. The threat was there, but quite honestly, at that stage we were so excited about flying airplanes. And you said, "Well, we will worry about that when the time comes. In the meantime, let's go out and fly."

EM: You're off to Pensacola [Naval Air Station] June of '57 for your flight training. Was it what you expected when you got there?

SA: It was. They had the program pretty well in place. In other words, they were moving you through. It was a machine. You could tell that you were on a conveyor belt, and you were going to go through these many stations where people would pick and probe at you and do things, and away you went. The competition, there was a lot of competition. There was a change in the equation that happened right away, just about the time we got through our ground school, which was about six weeks. Then we went to our primary base, Saufley [Naval Air Station Saufley Field], for training for our first flights in the old T-34B. While we were there, there were seven of us from Miami that all came into the aviation program. We were all in the same class. While we were there, they came out with a new regulation that if you stayed and got your wings, you incurred an extra obligation to your contract. The idea that your four-year contract was now going to be five and half, six years, maybe even seven years, depending on how long it took you to get through flight training—five of the guys that we went down there with quit. They just said, "No, we're not going to continue in the aviation program. We are going to go out and be fleet sailors. Drive ships. And when our obligated surface is up, if we want to stay, we'll stay. If not, we'll go home after four years." Dick Barker—

EM: It didn't deter you though?

SA: No, the question didn't even enter my mind. My friend Dick Barker stayed. Now Dick had not been—he knew he didn't want to drive ships, so I talked him into going aviation. He's the guy that we had the wrestling with. I talked him into aviation. Well, Dick was going to go into aviation. He wanted to fly multi-engine and I wanted to fly fighters. We signed this contract for extended time. Our serial numbers were, I'm 564262 and Dick was 564264, so we are lined up A/B all the way through. At assignment coming out of primary, I'm going to multi-engine and Dick's going to fighters. He says, "I don't want to go to fighters. I want to go to multi-engines." I said, "I want to go to fighters."

"No, you guys can't do that."

Well, what were the criteria? In other words, there were really no criteria. "We sort of just went down the list because you're all the same grades," et cetera, et cetera.

"Well, how do I get into this other program?"

This guy says, "If you go over and flunk your code exam, you'll be in the next class and maybe you can." [Laughter]

So I went over and flunked the code exam. He said, "No, no, you can't do that. You've got to come back and take a retest."

So off we went. I went to the multi-engines side of the house and Dick went to the fighter side of the house. That was a disappointment. I told Jennie Lou, I said, "I've got to make a decision." I loved to drive cars by myself, and so I thought flying airplanes with other people was just not going to work. I said, "I think I'll just go off and be a destroyer sailor. But this flying sure is a lot of fun. I guess I'll stay."

So I stayed. Again, one of those fortuitous things, the SNJ. I learned more about flying from the SNJ—because I ended up flying the T-28 later on in life and I had a lot of fun with it—but as an airplane, that [SNJ] taught you that the minute you start to turn over the prop, you had better be paying attention. Because the SNJ was a squirrelly airplane, and you could ground loop in the chocks if you weren't careful. If you had a cross wind or if you hit the brake too hard, you would just whip around. Anytime that thing was running, you were eyeballs. You were actually thinking you were flying, because there was always something going on that you needed to watch for. The T-28 was like your car. I mean you got it started, tricycle gear, and away you go. It [the SNJ] was a great teacher to tell you you've got to be ready to fly the minute you jump in the cockpit.

EM: Serious business.

SA: I was glad that I had that opportunity, especially once I was able to transition later on in life.

EM: How did you handle the academics there? Other than failing the one test.

SA: The academics were, of course with the aeronautics at Miami, I had all the basic courses. I had the navigation, I had the weather and all this stuff.

EM: It was old stuff.

SA: It was old stuff to me, and so for me the academics were easy, although they always had trick questions so you couldn't go to sleep. You still had to read the question. That part didn't bother me, and I was able to help the other guys that you could help who were having trouble with some of the stuff that was sort of second nature.

EM: How much flying did you do there?

SA: I think we got about 30-some hours in the T-34B in primary, and then I think we got 60 or 70 hours in the SNJ and basic. Then we had instrument flying out at Barron Field [an old Army airstrip], and then we were supposed to get our first carrier work out at Barron also. We went out to Barron to fly in the Beechcraft, the twin-engine Beechcraft with an instructor, and fly navigation and instrument training routes. You got your basic instrument qualification, and then we were supposed to come back into the SNJ and do the field carrier landing practice to go out to the carrier for our first—I think they gave them something like six traps, or something like this. At that time the SNJ had had a couple of failures with our tailhooks, and so they grounded the SNJs from any further carrier operations, since we were sort of the class that were flying SNJs. We didn't get to do the field carrier work or the carrier landing, so I missed out on that altogether.

EM: Do you remember any remarkable instructors or professors on the academic side?

SA: My primary instructor was a young JG [junior grade] I think he was a JG, maybe a lieutenant, but I think he was a JG. Real quiet guy, very nice guy, really, really easy to fly with. He didn't put the fear of God into you. He taught you well. His name came across later with the *Oriskany* [CVA-34] fire. He was the ordnance handling officer on the *Oriskany* when they had the flare fire [in 1966], and he took a professional hit because it was his guys.

EM: Flare into the locker?

SA: Yeah. His name was [R. L.] Zerwas. Z-e-r-w-a-s. Really nice, really nice guy. In the SNJ with the basic instruments, I had this instructor that, we had to sit in the back there with the hood over us, and I was really having trouble with the recovery from unusual attitudes. They would put you in an unusual attitude, and you would have to recover it and get your wings level and nose level. I was always overcorrecting, getting nose low, and had to cycle it out. He was getting frustrated with me. [Laughter] So he got me in a real high nose unusual attitude, and I basically just put us into a spin coming out of that thing. He's up there and he says, he starts cussing at me and he says, "And I'm not touching this thing until we hit the ground." [Laughter]

I immediately went wings low, nose level. He says, "Okay, now I got your attention." [Laughter]

I never forgot that. I believed him. He said, "I'm not touching it until we hit the ground."

EM: Do you remember your first solo flight?

SA: You know, I don't. I just remember that it happened and that I was—the thing that worried me was would I get back into the break pattern, into the landing pattern the correct way and not on the opposite side of the circle. But as much as I had built up to that thing, it disappeared from my memory. I remember the first solo in the SNJ more than I remember the first solo in the T-34.

EM: Maybe you were comfortable with it at that point.

SA: I guess. It's sort of surprising that I don't because, gosh, it had been years in the making.

EM: Anything else that sticks in your mind from Pensacola?

SA: Two things, and they happened about the same time. Dad and Mom came down to Pensacola, and I was in formation stage in the SNJ, and I said, "Dad since you're an old retired Navy guy I bet they would let you fly in the back of the airplane."

He said, "You think?

I said, "Let me go check."

So I went to the ops boss in the squadron, and I said, "My dad's retired Navy. Would he be allowed to fly in the back of the instructor's airplane on one of my flights?"

The guy says, "Yeah."

So we are on this four-plane formation flight, of course, these airplanes being all new to Dad, so I don't know what in the world he's thinking other than it looks like a big cluster out there. You know, airplanes going every which direction. The guy that was scheduled to be number 4 in our flight, I think I was scheduled to be number 2 on the takeoff—no, I was scheduled to be number 4 and he was scheduled number 3. And number 2 was a pretty squirrelly guy who wasn't doing too well. Number 3 didn't want to fly on number 2's wing. We launch, and on the rendezvous turn coming off the runway, number 3 cuts number 2 off [laughter] and takes his place, and then I've got to watch for now 2 is 3, who I've got to fly on. The flight sort of went that way. It got all jumbled up, and I got back and I asked Dad, I said, "How did you like your flight?"

He says, "Well, it was sort of exciting, but I couldn't figure out where everybody was going." [Laughter]

EM: You couldn't either.

SA: I said, "You weren't alone." [Laughter]

Then we had a snow in Pensacola, and when you're flying formation in the SNJ, over the years they had learned that they had too many wingtips hitting canopies in formation. The canopy was a birdcage, lots of bars, and it would jam the canopies. You couldn't get the canopy open and bail out if you had to. They made us fly formation with the canopies open. So here we are. We have to go out to the airplanes and sweep the snow off the airplanes, get it all off the wings and everything else, and then get in the airplane and then fly it with the canopy open. Of course you are up there in all that cold air and stuff. Your hands are frozen to the stick and the throttle. What a world this is to be up here motoring around—

EM: Snow in Pensacola. It's a historic place, having been there, the barracks, the forts there.

SA: Yes, it's an interesting place. We get back over there almost once a year, sometimes twice a year for gatherings and stuff.

EM: Your dad must have been pretty proud of you when you got your wings.

SA: Yeah.

EM: Before that, when you got commissioned?

SA: Dad was really proud of me. It's too bad he didn't live long enough—when I had the *San Jose,* he would've really loved it because I had a lot of flexibility there. I could have brought him out on a couple of cruises.

EM: When did your dad pass away?

SA: He passed away in '75. In fact, the *Navy Times,* sometime there in the fall of '75, had my name on the captain's list and his name was in the obituary list. At that time they had a column for the old sailors, so our names were in the same issue of the *Navy Times.*

EM: You finished up at Pensacola and you're off to Kingsville [Naval Air Station, Texas] in March of '58?

SA: Yeah, '58. We go to Kingsville. We first get out to Corpus Christi and they do some of the physical training. I think we did the water swim again, the oxygen chamber, the high-altitude chamber again, and then they gave you your final destination. Well when we got there, the guy in charge, the student control officer, says, "All you guys are going to seaplanes."

EM: From multi-engine to seaplanes?

SA: Oh, no. So, I said, "Can I come and see you?"

He says, "Yeah."

I said, "Look, I'm a single-engine guy. I've always wanted single-engines, but I ended up in the wrong pipeline. I'm here, and you're still training in Spads, the AD." I said, "I want to fly the AD."

He looks at my record. He says, "Okay, I think we can do that." He said, "But there's a lot of push on. We've got a big, big gap in the VP [patrol squadron] pipeline, especially in the seaplanes, and single-engine guys are all stacked up. But let me see what I can do. I'm pretty sure I can get you AD."

So I come in the next morning and he said, "I'm sorry, I couldn't do it. I couldn't get you switched." He said, "However, I got you—you're the last guy going to the S-2s up in Kingsville.

I said, "Carriers, okay, I'll go. That sounds good."

EM: S-2 Tracker?

SA: Yeah. He says, "Wish I could've done more, but that's the best I could do."

I said, "Well, I appreciate it." And the rest of my guys all went to seaplanes, and I went up to S-2s.

EM: You weren't too unhappy then?

SA: No, the thought of getting—at least I was going to get to a carrier, fly off the carrier. I was happy. One of my guys—three of us really palled around a lot in Pensacola. Our wives would always get together on Saturday night and play canasta and watch *Gunsmoke* and Paladin and all those things.

EM: *Have Gun Will Travel?*

SA: *Have Gun Will Travel.* One went to seaplanes, so he stayed in Corpus. Dick Barker, who had been in the single-engine jet pipeline, went to Beeville [Naval Air Station Chase

Field, Beeville, Texas], and I went to Kingsville. Then we rotated while we out in Texas. We all rotated from weekends at Kingsville, Beeville, or Corpus.

EM: Close enough together.

SA: Yeah. That was good.

EM: Anything else that you remember from Kingsville?

SA: We had some Italian students come through. They were going to fly the S-2 in Italy off their carrier. I was paired up with one of the Italian students. Not politically correct today to say typical, but he was a typical Italian. He'd get excited and the English would falter, and off he would go into Italian. We're doing our instrument training in the S-2. When you're doing that, you've got the instructor up in the right seat, you've got one of the students in the left seat, and the other student is in the back of the plane. Back there where the radar and all that stuff, which has been deactivated, but you're still in this little—you'd have to just look around to see what was going on up front. He's getting ready to make an approach, an instrument approach, and you're supposed to have the instrument approach plate out on your kneeboard, all ready to go to execute all the turns.

He casually reaches down into his NAV bag to pull out the chart, the approach chart, puts it on his kneeboard, and then he looks down and sees he's got the wrong one. He just sort of takes that one, reaches in, puts another one on, it's the wrong one. Well, this one gets flipped back. Reaches down and picks up the third one, and it's the wrong one. Now both hands, the hand comes off the yoke and he's down, and he's throwing stuff like this [gestures], and now he's in a spiral turn, descending, and he's just lost all contact with the outside world. He's just buried into this bag. [Laughter] And I sat back there, and of course I can't help him. I'm trying to reach in and get his attention. Finally the instructor takes the airplane and says, "I've got it." [Laughter]

I asked him after, he says, "Once I couldn't find that thing I just got so excited I just had to have it."

EM: There are people like that though, they lose it.

SA: Yeah. As it turned out, several of the guys who were in different classes, that had been there earlier, then came later—a lot of us, like Dick [Richard] Gentz, I don't know if that name rings a bell.

EM: Yes, as a matter of fact I've met him. He's a retired captain?

SA: He's a retired rear admiral. He was a vice admiral at NavAir [Naval Air Systems Command].

EM: Yeah, I have met him.

SA: He does a lot of work with the [Smithsonian] Air and Space Museum.

EM: The Naval Historical Foundation, I think that's where I've met him.

SA: Dick's a good guy. Dick was a car guy too. At this time I had a '56 Plymouth Fury and he had a '57 Ford Fairlane, I think. I told him that he needed to drive my car someday to drive a real car. We're at happy hour and it's one of those things, the S-2s had just gone down again, so I wasn't going to get to go. Neither one of us were going to get to CQ [carrier qualification] and finish our flight training for an extended period of time, unless they wrote off the carrier, which they finally did. I'm one of the rare guys that never got a carrier landing all the way through flight training. I got my wings without any carrier landings, even though the VP guys got it. Dick and I were drowning our sorrow at an early happy hour and decided on the way home that we would trade cars. Since we had been talking about it for so long, and now it really made sense. So we get home and stop both cars out in the middle of the street and go in our houses, and our wives look out and say, "Why are your cars out in the middle of the street?" [Laughter]

"Well, I guess we'd better fix that." [Laughter]

EM: This is now September '58 and you're in Antisubmarine Squadron 21 on *Kearsarge* [CVS-33].

SA: Yeah. At that time VS-21 had just come back off of a deployment and they'd been on the *Philippine Sea* [CVA-47], a straight-deck carrier. There were, I forget how many ensigns, a whole bunch of us checked in about the same time. We hadn't been assigned to *Kearsarge* yet. Our assignment, I think at that time, was we weren't going to go back on *Philippine Sea,* but we might go to *Princeton* [CVS-37]. The squadron was just getting reorganized. In those days you had a complete flush-out—everybody that was going to not make the next cruise would flush out—and all the new people would check in. The big body of us new guys. I ended up having to go to the hospital to have an operation. I had a pilonidal cyst, and the flight surgeon said the number of hours you're going to be sitting on these hard seats, you'd better have that thing taken out now, before it gives you some problems. So I missed about six weeks there.

We started working up to go to sea on *Princeton,* which was a straight deck. Now I'm getting my first opportunity to fly a FCLP [field carrier landing practice] pass on paddles with the old LSOs [landing signal officers]. This is my first time to really see what a carrier approach looks like. It's fun and scary at the same time because, for the old paddles passes, you're flying just a little bit above the stall speed, and you still have to make the turn and all. I'm getting used to that.

The job, I'll talk about the job first I think, just because it was something that stuck with me the rest of my career. I was assigned to be the material officer, which is sort of cats and dogs job.

EM: Supply guy?

SA: Yes, like the supply guy. The guy that had the job at an earlier time in the previous cruise had been a guy by the name of M. Staser Holcomb. Staser became vice admiral. He was brilliant guy and was looked at as sort of a godlike figure. I thought, holy mackerel, because all I heard was Staser Holcomb could do anything, everything, and here I am now learning this job. I'm an early riser and I never liked to be late for anything, so I would always be to work really early.

There's this first class petty officer in the maintenance control shack one morning, he says, "Ensign Arthur," I say, "yes," he says, "I noticed that you're here early every morning."

I say, "Yes, I like getting up early and I like to be at work early."

He says, "What are you learning while you are here early?"

I say, "Well, I'm trying to get my job done."

He says, "I'm going to give you some homework."

EM: This is a first class?

SA: Yes, first class. He had been a commissioned officer at the end of World War II and had been reverted back. He says, "I'm going to give you some homework." He says, "You're going to start reading every instruction we got down there, and I'm going to give you a quiz." I said okay. That's what we did. He'd lay out an instruction. The next morning I'd come in. He'd give me a quiz on it. We'd talk about it.

EM: An instruction on antisubmarine warfare technique, or—

SA: No, this was all sort of the maintenance instructions, how-to, or policies and squadron instruction, policies and the duties. I thought, as time went on, why did this guy take the time to take care of me? Because that's really what he was doing. He says, "Here's a guy wasting a lot of time, comes in early, but he doesn't know what to do. I'm going to teach him what to do."

I thought, "Boy, what a learning experience."

EM: Selfless.

SA: Yes. I never forgot that you just never know how much talent resides down there that sometimes you don't see. You don't get a chance to see that, you've got to look for these guys because they are all out there.

EM: Well a lot of enlisted guys have master's degrees, PhDs, they are pretty sharp.

SA: Well, the funny thing—sad thing—much later, as I'm leaving the second squadron, we stayed in the same air wing, second squadron. He was moved to that squadron too. He became a chief—

EM: That was VS-29?

SA: Twenty-nine. One of the last jobs I had, the skipper assigned me a summary court-martial on a chief petty officer, and who was the chief petty officer?

EM: Is that right?

SA: Something had gone wrong at home and he bailed out. He took off, went to Las Vegas, got drunk, I guess, and stayed there for a week, lost it, and then came home. The skipper assigned me—I'm a lieutenant by this time, getting ready to leave the squadron. So he says, "You're the summary court."

I say, "Skipper I can't do that. I know this guy too well."

He says, "Why do you think I'm giving you the court-martial?" I worked out what I thought was a good punishment and all that. I think justice was served, and he didn't get hurt too bad. Of course, it's a terminal thing on his record. He wasn't going to go any further, but he wasn't going to go any further anyhow. It was, again, sort of bookends of a learning experience of one kind and another learning experience of another kind.

EM: He had screwed up, but he still had value?

SA: Still had value. How do you not recognize that? It just turned out to be one of those things with the two pieces of it that served me well for my whole career.

EM: On board *Kearsarge,* were you involved in the Taiwan Strait Crisis of '58? The ship was there?*

SA: Yeah.

EM: Do you remember that?

SA: Yes, I do. We were out there. We flew around-the-clock for, I forget now, a long period of time where we just didn't have any breaks.

EM: What were you flying?

SA: We were flying S-2s, and there was nothing—it was just presence more than anything. We were out there looking for stuff, but we didn't know what was out there.

EM: You were told the Chinese may come out? Did you have nuclear-capable aircraft on board?

SA: We did.

EM: Guarded by Marines?

SA: On the S-2s we had nuclear depth charges. So we weren't part of the SIOP [single integrated operational plan] as such, you know, for the year I started flying A-4s and we were on SIOP. But we had the nuclear depth charges, so we had to go through the nuclear training and go through the exams and flights with the shapes, and what we had was a

* In late 1958, President Dwight D. Eisenhower deployed the U.S. Seventh Fleet to the vicinity of Taiwan, governed by the Republic of China, to deter aggressive actions by the People's Republic of China.

dash 2 model of the S-2 that had a pregnant bomb bay. It had a little bulge on the side of the bomb bay that would hold the nuclear depth charges.

EM: What was your thought about dropping one of those? Did you think you'd be toast?

SA: None of us were convinced that we could get away.

EM: It may go in the water, but it spouts up in the air.

SA: They kept telling us, "Oh no you guys, this thing's not quite that fast." [Laughter]

EM: Well, this is your first actual operational assignment out on *Kearsarge* with both squadrons. What was your impression? I'm here now, I love this stuff?

SA: Yeah, it was exciting. There was a lot to learn. ASW [antisubmarine warfare] is a chess game. To be good at it, you have to think about it. You've got to be able to think and fly at the same time. You've got to get your crew organized so that people are clear on what their part of the equation is. There's a lot of satisfaction in one getting to understand the problem and how to develop a plan and also how to execute. You take advantage of everybody's strengths, and you got the right people doing the right thing. A lot of boredom. The flying is not that exciting, until all of a sudden it is.

EM: What about your carrier landings and launches?

SA: For the S-2s, we could either catapult or deck launch. When I first started flying, we did night deck launches. The accident rate was extraordinarily high in those days. I think about the transition that happened from the Second World War, and the transition was really starting just as I started flying the S-2. One of my skippers who came along later, I think had 33 carrier landings and he's a Navy Cross winner in World War II. By the end of our first cruise, all of us had 100 landings or more. A lot of guys and a lot of the senior people didn't like to fly at night. Us young guys didn't know any better, so we loved to fly at night. The mission was low and slow at night, a prescription for any momentary—

EM: Did you ever lose your night vision?

SA: Yeah, you had to be careful. We had the big searchlight out there, so if you're going to run in on a target to identify it with the searchlight, the guy that was flying the airplane really had to get down under the cowl and make sure he didn't get trapped by that 70-million candle-power searchlight. One of the things we also learned, we always had flashlights to pre-flight the airplane. Go out pre-flight. The World War II mentality of no lights on deck at night was still in place. We'd go out to pre-flight our airplanes at night with a red lens on our flashlight.

Well, the biggest problems we had were hydraulic leaks where you can't see a hydraulic leak, red fluid and red light, you can't see it at all. Some of us decided it was a lot safer to just crawl in the airplane and not be out there on the deck with the chains and the props turning and all that stuff, and just get in your airplane and say the plane captain must have done a good job. On the cowl, if you weren't thinking, the first thing you would do is you'd get into the cockpit, get the battery switch on, and then you'd have lights. You'd lay your flashlight up on the cowl. If you left it up there and were going to take a cat [catapult] shot, you'd be wearing the flashlight. [Laughter] Lots of guys did. We always had to have that on our check-off list to make sure we put the flashlights in the bag beside us and not on the cowl in front of us.

EM: Did you have many bolters [waved-off landings] when you came in?

SA: No, not too many. Enough to get your attention. The S-2 was a very, very, very nice carrier airplane. The ability to have a bolter on the angle deck on *Kearsarge* and not catch a wire or barricade on the straight decks, what a difference that made. A bolter is nothing. You just get to go up and try again. On the straight decks, if you didn't catch a wire, you were into the barricade. That was not fun at all.

EM: Did that ever happen?

SA: No, not for me, no.

EM: Did you lose any squadron mates, wing mates?

SA: Let's see, yes. In the S-2 we did. We lost some in our helo squadron too. We had a couple of accidents on the ship where we didn't lose anybody, we got some people hurt. We were up off of Alaska, not on a cruise, but we'd been sent to sea on a Soviet submarine alert. We chased this guy all the way up into the Gulf of Alaska. We got into 60-foot swells out there, and we were still launching airplanes. Crazy. We didn't have enough gas to get to a field, so it was either get back on board the ship or else. We had one of our guys get lost, and by the time he got back to the ship he was really low on fuel. When you have a deck moving that much, it's a matter of just getting into the cycle. You've got to get in cycle. You've got a small period of time where the deck is sort of level and you just happen to be there.

Couldn't get him in cycle with the deck for several passes. Now he's really low on fuel, and he comes around, and the deck's down but not too far down, but it's starting back up. He makes a play for the deck at the same time, so he increases his rate of descent. The deck's now coming up, so that doubles the equation, and he flattened all the gear. They got hurt getting out of the airplane. They came out the door thinking that the door was still in the air, but the door was at deck level, so when they came out they didn't realize that they were now on the level with the wing. The flaps were down and they cut their heads on the trailing edge of the flaps.

EM: But they survived?

SA: They survived, yeah. But we had guys fly into the water and of course you lose all of them on that.

EM: Do you remember any of your COs on the ship or your group commander?

SA: The CO of the ship was [Captain Robert L.] Townsend. He ended up as a vice admiral, I think. I think he was AirLant [Commander, Naval Air Force Atlantic] at one time, if I'm not mistaken. Very distinguished guy. My first cruise, we didn't have an air wing commander. The second cruise we had an air wing commander, and I was just trying—his name all of a sudden escapes me—I know it will come back. He was our first CAG [carrier air group commander] for the VS [antisubmarine aircraft] community, and he was really an interesting guy. Towards the end of my VS-29 tour, he became sort of the chief inspector for all the VS squadrons on the West Coast. Also for the reserve squadrons who were activated because of the Berlin crisis in '62 or '62 timeframe.

EM: From '59 to '63, there was a crisis like every year.

SA: So we went around. He formed an inspection team that would go around and inspect all of the squadrons and give them an up or a down, whether they were ready. I was one of the lieutenants on the team. He had rules, and I can't remember them all. Later on in life, his son was my catapult officer on *Coral Sea,* and I asked him to get his dad to give me the full set of rules, but he was never able to get them.

If you were up doing your part of the debrief, which always got confrontational—"We saw you guys didn't do this"; "Oh, yes we did"—he had these rules numbered. Now all of a sudden if the conversation got too heated he would say, "Rule 1." Rule 1 was don't get in a pissing contest with a donkey. [Laughter] Rule 2 was never stand in front of a man who is about to barf. [Laughter]

EM: Sound advice.

SA: They went along that line. They were wonderful rules. Some of them stayed with me, some of them didn't. But it was very clever that he had these things, and he briefed us on them, and he would just sit there. Unlike a lot of the bosses who wanted to do everything themselves, he let you go as far you could go before he would step in and bail you out, you know when Rule 1 got displayed.

EM: You were a plank owner of VS-29. What did that entail?

SA: Well, what happened was in, let's see, that was '60, they decided to take the VS squadrons and split them in half, and then bring an air wing commander in charge. So they added an air wing commander. Now you had two VS squadrons, and you had an HS [helicopter squadron], and you had your early warning squadrons. So now the air wing commander had four squadrons do deal with. VS-29 was going to be the new squadron. We cut VS-21 in half. The COs must have had a lottery, and what they did was they divvied up the talent, and half stayed with 21 and half went with 29. I went with 29 and so we were plank owners, you know, the first inhabitants of the squadron. Which was really, really fun. A lot of work, because we didn't have anything of our own. We had to make all of the instructions and everything had to be changed.

EM: You couldn't use VS-21 instructions and adapt them?

SA: We could, but we wanted to be our own squadron, and so those of us who came over decided that we wanted to write things the way—in VS-21, we had instructions that did not match the way we were doing business. Our challenge to ourselves, those of us who had to write a bunch of the instructions, was we were going to write the instructions so that if you read them, that was the way we did business. So the dialogue and instructions matched.

EM: Did you get any flak from your superiors?

SA: No, they embraced it and turned us loose, and away we went. We were sort of proud of ourselves. We thought we did a good thing. We did. The squadron had—when you have something like that happen—you add a little bit of competition into the game with a known quantity, because you were there. Hey, we think we know a better way to do this, so let's see if we can do it. It was an exciting time, and some of the drudge work of doing instructions became a challenge that we all embraced.

EM: You were JG at this point, right?

SA: JG and lieutenant shortly thereafter.

EM: Were you treated a little differently from when you were an ensign?

SA: Yes. Lieutenant is still the best rank of all.

EM: Why?

SA: Your superiors think that you're pretty well adjusted to the system by now. They give you a lot of responsibility. But if you screw it up, they say well, he's just a lieutenant. [Laughter] And so you had this opportunity, I mean the things they let me do as a lieutenant were just phenomenal, and yet I knew that there was a fallback position, that if I really screwed it up they would just write it off as, he's just a lieutenant. I still think lieutenant is just absolutely the best rank of all time.

EM: You were about 25 years old then? In '60?

SA: Yeah. They sent me off to do all sorts of things. One of the last things that I did in the squadron was, the skipper called me in one day and said, "Hey, I've got to send a crew up to China Lake [Naval Ordnance Test Station, California]. They need an S-2 up there to drop some new torpedoes that have been developed. I want you to take a plane and your crew up there and drop the torpedoes for them. I don't know what they're going to do or what they want you to do. All I know is they want an airplane that can drop a brand new torpedo."

I said okay, I will go up to China Lake. It's tonight or tomorrow night or tomorrow, I forget; it was really a short fuse thing. We get loaded up and fly up to China Lake and get up there at night. I landed at China Lake, go in the next morning. Point of contact is a civilian tech director, and he says here's what your job is. You are going to drop torpedoes.

I say, "Good. Out in China Lake, right?"

He says, "You are at China Lake."

I say, "Yeah, well, we don't have any water. Where am I going to drop a torpedo?"

He says, "You're going to drop a torpedo in the sandbox."

I say, "What?" [Laughter]

He says, "What we want to do is, we are going to measure the angle of entry, and what you do is fly your normal torpedo pass," 100 feet to 120 knots or whatever it was, "and you pickle this thing off, and you've got to put it in the sandbox so we can measure what the angle was as it entered the sand." [Laughter]

It was a great time. I thought, just one of those crazy things you get to do. I was awe-inspired to go up there and drop something in a lake. [Laughter]

EM: No lakes around there, unless you go to—that's not too far from Las Vegas, is it, China Lake?

SA: Yeah, it's not that far.

EM: While you were on *Kearsarge,* did you get any WestPac [Western Pacific] liberties? Do you recall any ports of call?

SA: Oh yeah. We went lots of places. Our first cruise on VS-21, the first time we came to Japan, they had the big typhoon that had struck Nagoya. I'm a co-pilot, I'm one of the new guys, and my lieutenant is plane captain. We're launched to go in and fly into Nagoya Bay to check to see if the carrier can get in there amongst all the debris that was supposed to have been there, bodies and boats and stuff all floating out there in the bay. So we go in and we take a look and say, "Yeah, there's all that, but there's plenty of room to get the carrier in here—if the water depth—but yeah, there's a clear path."

We come back to ship with that report. Ship goes in and we anchor out there, and then we provide relief for two or three days, helo squadron picking people off of rooftops and picking up people, dead bodies floating by the ship in the water, and all that sort of stuff. Of course, in Yokosuka [U.S. Naval Base, Japan] we got liberty there. We were in Hong Kong, Singapore.

EM: Manila?

SA: Manila.

EM: Subic Bay [U.S. Naval Base, Philippines]?

SA: Subic Bay. I had my seaplane friend that was in—he was now stationed at Sangley Point [Naval Air Station, Philippines] down there in Manila Bay, so I spent Thanksgiving with them down there. We had good liberty. The old hands, the lieutenants in the squadron from the previous cruise, had been to most of the ports of call, and they all had tours and shopping places for us all lined up. We had instant tour guides to get us around.

EM: What was your impression of Asia?

SA: I loved it. Yeah, I really did. I was surprised by Japan. I think I was—

EM: What did you expect?

SA: I didn't know what to expect. I didn't know how they—what they thought of us and what we would think of them. I just didn't—I thought it would be sort of a strained environment. But it was anything but. They were so gracious, and people always wanted—The second cruise over there, the funny incident was, like I said, one of my skippers, Buck [Delmer A.] Schatz, was a Distinguished Flying Cross winner from World War II.

EM: Buck Schatz?

SA: He didn't want to go ashore. I forget now what port we were in. It was not Yokosuka, one of the smaller ports. It was going to be a going-away party for a couple of guys in the squadron. I said, "Skipper, you've got to present the plaques. If you don't feel comfortable, myself and one of the other lieutenants will come back to the ship and get you when it's time. Take a cab to the hotel where we are having the party. You can present the plaques and we will bring you back to the ship."

EM: He still had residual animosity toward the Japanese?

SA: Yeah. He said, "I just don't want go." We finally, we kept working on him, kept working on him. Finally he says okay, okay, I'll go. "Anytime you want to leave, just let us know." Well, he gets there and he has a few drinks. The party breaks up, and he goes into the bar. He's sitting there, Japanese fellow sits next to him, "American pilot?"

"Yes, Japanese pilot?"

"Really? Where were you?" They ended up having flown against each other.

EM: Oh, jeez.

SA: The next day I get a call from the duty officer, he gets a call to come to the quarterdeck. He goes to the quarterdeck. This newfound friend of the skipper has just delivered four cases of beer to the quarterdeck.

EM: All is forgiven.

SA: All is forgiven. [Laughter]

EM: Okay, we are on to the Naval Postgraduate School, where you began your instruction in July of '62. Why this education at this stage in your career?

SA: Well, back in that day your first shore duty was sort of an open-ended, "What would you like to do in your first shore duty?" Because I hadn't got an aeronautical engineering degree when I was in [the] NROTC program, and that's what I always wanted to have, I thought going to PG School [Naval Postgraduate School] would be a good way to do that.

EM: Did you ever think that you might want to be an aeronautical engineer? Design, develop aircraft?

SA: In the early days of my life, that's where I thought I would end up. If the Navy wasn't going to be my career, being an aeronautical engineer was what I really wanted to do. Probably—we can talk about this later when I talk about life after the Navy, but after I got to manage a bunch of engineers here with Lockheed Martin in my afterlife, I decided that, boy, did I make the right decision to stay Navy. [Laughter]

EM: Why is that?

SA: I worked with some of the most talented engineers that I think there are in the world in my time with Lockheed Martin, really, really extraordinary people. They had great careers, they developed wonderful systems, but yet when it came time for their retirements—we had retirement parties, and it was always a grand event—I would find out that here one of my vice presidents, and one of my younger vice presidents, he retired and he had never moved any further than from one end of this house to the other end of this house in his whole career.

EM: Missed out on a lot?

SA: And how, you know, one, how did he maintain his sanity. I mean, part of the adventure in the Navy was, if you finally got fed up with one job, you got rid of that job and they assigned you someplace else. It was something completely new and different. You went off and found a new challenge. My admiration for these folks who spent all these years, 20 and 30 years, sometimes 40 years in the same place, yeah they had different responsibilities and different products that they worked on, but man, how did they maintain their sanity?

EM: I worked in the same place for 37 years. [Laughter] I hear what you are saying, though.

SA: I thought that's what I would do. Fortunately I found enough to keep me excited in the Navy, and then I got to do the engineering part at the end.

EM: How did you put your education to use in your subsequent career?

SA: Well, part of the requirement was if you went to Postgraduate School, you owed the sponsor of the program a payback tour. While I was there at Monterey, the options were stay three years and get a master's degree, which was—at that time the PG School hadn't figured out how to do this. They do now. One of my classmates was—had a bachelor's degree in mechanical engineering from Penn State University—really a good engineering school. No, it was in aeronautical engineering. He had his bachelor's degree from Penn State, very high academic credentials, and to get a master's he had to stay at Monterey for three years. Now, me, a bachelor of arts graduate from Miami, had to spend the same three years to get the same master's degree. I thought, holy mackerel, there's something wrong with the system there.

The more I thought about staying three years to get a master's in aeronautical engineering and the opportunity to get a payback tour on the second half of my shore duty tour, flying airplanes and paying off my education appealed to me. I was in the two-year program, and when I left Monterey then I went to VX-1, a developmental squadron.

EM: Did you think it would have been too much of a diversion from your Navy career and set you back with your contemporaries?

SA: Well, yeah, there was that. Three years ashore is—even though the tours, you know, two to four years—anytime you're away from the operational guys is not good time.

EM: Not good?

SA: Not usually. The fact that my sub-specialty now was aero engineering and that VX [test squadrons] were flying projects with new weapons systems and stuff like this, I was back sort of working with the fleet. I would go out to the carriers and travel around the country to Point Mugu [Naval Air Station, California] and places like that and Pax River [Naval Air Station Patuxent River, Maryland]. I was still sort of close to the operational side of the fence. That really appealed to me, and it was a great payback for the education. I learned some stuff at the school that I was able to put into practice right away.

EM: That was my next question. What was your impression of the curriculum at PG School?

SA: Well, at the time I thought it was pretty archaic. One of our major programs was diagramming the air flow over a Spitfire [British World War II fighter plane] wing. Why the elliptical wing was such an interesting development for the Spitfire, and why it was so successful as an airplane. I thought, well, you know, we were past the Spitfire [laughter] right now, although it was a beautiful airplane and I would love to fly one. I didn't understand that I was learning that much. I thought I could learn something—the same lesson on a more modern tack. However, one of the projects that I had to develop a program for was a submarine-launched antiaircraft missile. Which, as time went on, that became something of interest, the one that the Soviets developed, something like that. We thought about doing something like that with ours. That sort of brought me into the spectrum of, yeah, what could the future do with submarines that had an antiair missile that could—once they knew you were in the vicinity, could launch something submerged, and it could float to the surface. The next time you flew over that sonobuoy [acoustic buoy], it would—*psssst*—pop you out of the sky.

EM: Did that weapons system go anywhere?

SA: No, didn't go.

EM: Did we have anything comparable to that?

SA: You know, I think the Soviets do. I don't know whether they still have it, but at one point in time, I thought I saw something where they had an antiair missile.

EM: You mean the Soviets or the Russians?

SA: Well, the Russians now, I guess.

EM: They are always Soviets! Do you remember any of the instructors you had there at Monterey?

SA: I'm trying to remember the name. Our aero instructor himself was a Brit. Fairly young, I mean he was probably mid-40s. Very entertaining and very professional.

EM: He must have been the Spitfire aficionado?

SA: I think that's where the Spitfire came from. I can't remember the Brit's name, but he was good. He was a good instructor. We had an older gent who was a materials guy. Taught us all the stuff about metallurgy and tension and torsion and all that stuff we used mostly. The Russians, of course, were the experts in that field, and most of the textbooks were English versions of Russian manuals. He gave us a lot of practical time in the lab where we broke things and instrumented stuff for stress and strain—you know, stress and strain gauges, so that we got a feel for not only what the formulas told you you could do, but how, when you put devices on the stuff and then put them into a frame where you could actually twist them and pull them and push them, to see that the formulas were good were good predictors of where you could go. You had to know where to put them relative to where the forces were being applied. One of the first things that happened after

I got out, we had to instrument a P-3, one of the new P-3s, for max G. We were going to do maneuvers with the P-3 at max G allowable loadings, which the fleet normally doesn't do. First thing we came across was we needed to know where to put the sensor, because the length of the fuselage of the P-3, depending on where you wanted to measure the G force, it was different at each place. We wanted to make sure that we were measuring the G force that the wings felt, not necessarily the G force that the tail felt. There was a good payback relative to that.

EM: How about classmates, anyone stick out?

SA: We had great classmates. In the class ahead of me was Gene [Eugene A.] Cernan. In my class was Ron Evans who was an astronaut and then was killed in one of the accidents.* Bob [Robert H.] Shumaker was in our aero class. He was in the class ahead of us. He had also been looked at for the astronaut corps, and they found that he had a—from an early injury as a teenager, I think he had a split sternum and that disqualified him for astronaut corps, but it didn't disqualify him from being a POW [prisoner of war].†

EM: No. [Laughter] He gave a talk not too long ago at the museum.

SA: He's just a wonderful guy. I thought that of him when we were out there, even though he was a year ahead of me. I just thought he was just such a prince of a guy, and he's always been that way, really.

EM: At this point, let's see, Erin, Brian, and Derek was already born?

SA: Yes.

EM: Your children.

---

* Captain Ronald Ellwin Evans Jr., U.S. Navy (Ret.), actually died a natural death in 1990.

† Robert Shumaker was the second naval aviator taken prisoner after being shot down in an attack on North Vietnam in February 1965. He spent the next eight years in captivity in Hanoi.

SA: Yes.

EM: You were at sea when each of them was born?

SA: I was at sea for—Erin was born before I went to cruise, Brian was born while I was on cruise; he's a Fourth of July baby, so I was on cruise. Derek was born in Monterey. Then Brad, our youngest one, was born while I was at sea. Ten-year span between the daughter, who is the first child, and Brad, who is the last child.

EM: The social life at Monterey, did you have enough time, or were you studying so hard you couldn't get out there and have fun?

SA: Jennie Lou had a great time in Monterey. I sort of hated it. [Laughter] I didn't like the weather. The damp weather, and it seemed like the damp weather was always around.

EM: Yeah, they get a lot of fog there, don't they, right off the sea.

SA: I was wrapped up in the program much more than I should have—I think I should have taken a more relaxed view of the world. We had some friends up in the San Francisco Bay area that we'd go up and visit, and they would come down and visit there. The folks that lived in La Mesa, the government housing, they had a very active social group there. We lived up the road out of town, north of Seaside in Marina.

We had a Marine who lived next door who was in the aero program with me. One of my squadron mates from VS-21 lived up the street about five houses, he and his family. Another squadron mate from VS-21 lived a block over; and then the Army classmate, was in high school a year ahead of me—and his wife and Jennie Lou were classmates in high school. Helen was maid of honor for Jennie Lou when we got married. He was at the Defense Language School, and they lived on the street right behind us. In the neighborhood we had a little Ohio clan and old VS-21 clan. I had two VS-29ers with whom I'd been in the squadron who lived in town, so between all that we had a social life, but it wasn't real, real active. The guys and I would get together and tinker on cars. Two of them were car nuts like me, so we would find a way to tinker on cars on weekends.

EM: During the time you were there, this was during the Cuban Missile Crisis. Do you recall anything, in fact, let's see JFK's [President John F. Kennedy] death was in '63.

SA: Yes, JFK, that happened.

EM: Well, what happened during the Missile Crisis? Were you put on notice you guys had to get ready?

SA: No, we were wondering what was going to happen next and whether things were going to change, but basically we didn't change our status there at all. In those days we still did proficiency flying, so they had a fleet of airplanes there. We still went out and flew around the countryside, not too far from Monterey.

EM: What kind of planes were they?

SA: We had Beechcraft, twin-engine Beechcrafts; we had T-28s, the single-engine T-28s; and we had T-2s, a little two-seat jet. There were three types of airplanes there.

EM: Was this the first time you'd flown a jet?

SA: I was flying the T-28 for the most part there. I did fly the jet a couple of times—we had a mini test pilot routine, where we had to go out to determine the dynamics of the airplanes, whether it was longitudinally stable or unstable. They would rotate us through the airplanes, and we would go out and fly a performance card.

EM: What I meant, though, was that the first time you had flown a jet?

SA: Yes.

EM: Was there a period of adjustment? How different did it seem?

SA: No, the jets were much easier to fly than a prop plane. I mean you push the go button and it goes.

EM: You didn't have to worry about throttles?

SA: And then the rest of the controls are the same, except you don't have as many. You don't have to worry about the prop and the mixture and all that stuff. So it was push-pull on the throttle, and same motions on the stick. It's just everything happens a little faster.

EM: I forget, I think it was Jerry [Gerald E.] Miller was telling me that during Korea with the new jets, the Panther [F9F], there were a hell of a lot of accidents because they were too fast for the carriers. There was a period of adjustment there, it took a while.

SA: The problem there was the jets were heavier, they landed faster, and you had to be very careful about overstressing the arresting gear. The other thing with the jets, especially in that time, was the engine lag. On the piston-powered airplanes, it's just a like car. You'd tap the accelerator, you got instant power. You let off of it, and it's instant off. On the jets, when you push the throttle forward, there's a lag time depending on the design of that particular jet engine, but it's significantly more than a prop. Around the carrier, guys that were used to flying piston-driven airplanes around the carrier, you've got absolute positive control of your power. If you need a little bit of power, you can get a little bit of power.

With a jet, if you wanted a little bit of power, you had to add more power than you wanted, and as soon as it started to take effect, then you had to pull some of that power back, or it would be too much. You had to play with that lead-lag situation, which got you to the end of the—

Coming over the ramp, if you just really were not solid on airspeed and sink rate, well then, bad things could happen. They went through a long period of where the accident rate was really out of sight. Most of it was the fact that you didn't have as much leeway with your airspeed. You had to be more precise on your airspeed, and being able to control that airspeed with power was much more difficult than it was on a piston airplane.

EM: What was your reaction to John Kennedy's death? The assassination?

SA: That was a real shock, and the memory of that day was—

Like I said, the folks from our hometown who lived a street behind us—Jennie Lou I think was down at the Army base at the time—and Helen came around. She had just heard it on the news, and she was really distraught. I opened the door and there was Helen, and tears are rolling down, and I had not heard yet. She said, "The president's been killed." It was a shock.

EM: Everybody that age remembers what they were doing.

SA: Yes.

EM: I was in the library cramming for an exam. It was the end of that, because they shut the library down. Of course, Vietnam in the early '60s, we started to do things there. Wasn't huge on the radar screen, but it was certainly there at that point. Do you have any recollections of what you thought about Vietnam?

SA: Well, because the class ahead, like Bob Shumaker and those guys, they were headed, they thought they were going to be out there soon, because everyone was going back to the fleet. My thoughts were—I'm wondering when I can make my next play to get back into single-seat and get into a jet. Some of the jet pilots who were in our class were clamoring to go. Get ready, because something is going to happen out there, and we're going to go out there.

EM: That was the general feeling? Things are going to start really ramping up?

SA: Yeah, we thought for sure.

EM: So you're off to VX-1 June of '64? Where was VX-1? Pax River?

SA: VX-1 in those days was in Key West [Naval Air Station], Florida, at Boca Chica, the air station there that was just north or east as it curls around, but on the road to Miami. It's just one bridge away. We left Monterey and we drove to Ohio. My wife and the kids were going to stay there with her mother while I went down to Key West to find us a place to live. Jennie Lou was not happy about going to Key West. She had heard all sorts of terrible stories about how wild and woolly Key West was, and it was awful.

EM: Even then, huh? [Laughter]

SA: Yeah. "It's no place to take a family," and all this. I had orders when I dropped them off there in Ohio and drove down to Key West: "I'm not coming to Key West until you've got a place to live." I said okay.

I drove down to Key West and I arrived on Friday, early Friday, and so I got into the BOQ [bachelor officer quarters], I guess it was, and got into uniform and went over to the squadron to check in. They said, "Well, you're just here—"

And there were some people I knew in this squadron, who had been in VS-21—youngbloods—in an earlier day. "You got here just in time. We're going to have happy hour over at so-and-so's house. We're all going to gather over there a couple of hours from now."

I said, "No, I need to go out and find a place to live."

"Well, there's some guys that are going to be there that are leaving Key West. Maybe they've got some ideas for you, so you ought to come and meet everybody."

So I said okay. I get the directions, get in there. And everybody I see, I ask, "Do you know where there's a place for rent? Do you know where there's a place to rent?"

This one guy walks in and says, "Hey, I just got notified by housing that they've got quarters available for me."

I say, "Are you renting?" He says yes.

I say, "I want your house."

He says, "Well, it's right down the street. You can see it from the door here."

I look out there and I say, "I'll take it."

EM: I'll take it? [Laughter]

SA: He said, "We will have to talk to the guy who owns it." It was another Navy guy, a commander, surface guy, who didn't live in Key West. He said, "We will have to call him and check and see."

I said, "Okay, well I want it. Tell him that I want it." By that afternoon I had the place rented. I called Jennie Lou and said, "Well, I understood the orders and I got us a place to live." [Laughter]

EM: That's great. So she came down there after a bit?

SA: She did. It turned out that she had a great time. But the trouble was I was gone most of the time. I was on projects. I went to Paris, I went to Marseille.

EM: To test this equipment?

SA: Yeah. Fly with the French a little bit and learn how to fly one of their missiles from Nord Aviation and how to control one of their missiles; went to their factory training. Was out at Point Mugu for a couple of months and ended up at Pax River for three or four months. I was at White Sands [Test Facility], New Mexico. I was at Fort Benning [Army Base, Columbus, Georgia]. I was all over the place.

EM: This was a follow-on from your training, your education at Monterey that you were expected to get another shore duty related to your training?

SA: Yes. Usually that happened, the next shore thing, but since I still had time on my original shore tour, I went there and got right into payback in the same period of time.

EM: What systems did you test?

SA: My job was to test weapons. We were sort of divided up into guys that tested the sensor systems and computer systems and guys that tested weapons. I was in the weapons business. I was involved in mines, torpedoes, Bullpup missiles, Sparrow missiles, and the Nord Aviation AS-12 wire-guided missile. Let's see, the Mark 46 torpedo was probably my biggest project that I had.

EM: Did the Mark 46 make it? Mark 48 is the standard torpedo, isn't it?

SA: Yes, the Mark 46 was the lightweight for the air, and it came online. We had some really interesting—I flew down at St. Croix [U.S. Virgin Islands]—we had an acoustic range where we launched torpedoes down there and tested there. We were testing for interference with the helicopters, with their dipping sonars. I had one torpedo come out of the water and try to—I almost sank a helo with a torpedo. [Laughter] The Mark 46 was a fast, a very fast torpedo. We were checking to see if the sonar dome on the helos would attract it. It did. The 46, in a search pattern, it had gone down to quite a deep depth, and it was on its way back when it acquired that signal from the helo.

EM: Did the pilot see it coming and head up?

SA: No, no. Well, they could hear it, but they thought it was chasing in circles. What it had done, it had recognized it and it came up vertically. And it ended up coming out of the water up in front of the helo, and went up above where the rotor blades were spinning. [Laughter] Then flopped over and came down in the water and broke up. Well, that helo guy—he wasn't going to go back out there again until we figured out that interference problem. It was lucky that it didn't come up through his rotors.

EM: So you worked on Sparrow as well?

SA: Yeah. We were looking to equip the P-3s with Sparrows or Bullpups so that they would have a forward-firing weapon. What we were looking at was for torpedo boats.

EM: The Osa?

SA: The [Soviet] Osa class, and we needed something that could stay out of the range of—I think they had 37-millimeters.* We needed a forward-firing weapon that would keep pretty much an unarmored airplane, like the P-3 or the P-2, out of their range. But they could still kill the platform. We were evaluating those, and for the S-2, which couldn't carry that big a weapon at the time, we were looking at the AS-12, the wire-guided missile from the French. It was an air version of their SS-11 antitank wire-guided missile that they used in the field for their infantry. Very good system.

The thing that I got to see, because I was working both the Bullpup and the AS-12, was the difference in training. Of course I hadn't been in an attack squadron yet, so I didn't know what the Navy was doing with the Bullpup. I went to France to learn how to fire the AS-12 from the French. They had a very sophisticated training program. They had a psychologist who sat in all of our training sessions and watched each of us go through the training. They could predict which one of us was going to be the best shooter.

EM: Based on what characteristics?

SA: They would never fully disclose what they saw. In basic terms, what I got from when I talked to my closest French buddy there, who would usually tell me most everything, was that it was sort of personality-dependent. How much concentration somebody could develop to the task without getting—

EM: Mental diversion?

---

* These were actually twin 30-millimeter guns.

SA: Diversion, exactly. Although sometimes you would sit in the classroom, and what we did was, we drove a little spot of light around. They had a computer that drove us a spot of light that played like the missile. Basically, you took that spot and drove it out through a picture, and you were supposed to hit something on that picture. You mentally had the time, the time of flight, because there was no way of knowing, because we didn't know exactly the distance, other than what we could see in the picture.

So you said, you know, that's sort of the max range of this, and the max range is probably 30 seconds of flight time. Okay. Instead of putting the pipper right on the target from the start, they taught you a method where you took the light on the back of the missile, below the target or above the target. Got it stabilized, and then started a descent or ascent, depending on where you put it, so that it would arrive on the target just at the time of impact, so that you weren't—you didn't get so focused on holding that spot of light on that point, and you got to over-control it.

They taught you this very methodical way of getting it captured, getting it under control, measuring the time of flight in your mind, and just letting it slowly drift down or drift up and arrive at the target at the time. It was completely different than the Navy with the Bullpup; they tried to teach people, get it off the airplane and try to drive it in there immediately. Of course, people had terrible luck with the Bullpup. When I got hold of the Bullpup with the French training, it was a whole different world.

EM: It was an effective weapon for you?

SA: Yes.

EM: Because that's exactly what the—you always hear the Bullpup was a dog.

SA: Yeah, well, I can see why. When I got to Point Mugu and we were going to shoot at boats, I kept hitting the boats and they said, "We've never seen that happen before."

I said, "It's because of the French training. It's so much different than what we've trained." So when we did fire Bullpups in combat, I loved to get ahold of them, because I was pretty comfortable with them. It is a much harder missile to control because it is so fast. The French were very clever. We tend to think of circular error probable, CEP, like a 10-foot CEP or a 5-foot CEP. The French talked only in hit/miss.

EM: Either you hit it or you missed it.

SA: It was horseshoes, you either hit it or you missed it. It wasn't, "Oh, why, you're at five feet." No, it was, "Did you hit it or miss it?"

EM: Well, if you've got an A-bomb, I mean that's going. [Laughter]

SA: That's a different thing.

EM: So you mentioned Sparrow and the AS-12, was it? What other systems did you work with? And the Bullpup?

SA: And the Bullpup and then mines, we did 2,000-pound mines. We were dropping them off of Mark 52s, I believe. They brought out the heavyweight P-3 for mining. It had the beefier wing stations for carrying mines. They hadn't cleared the P-3 for those. The first drops, I had to have an A-3 [Skywarrior] come down so we could drop them from an A-3 and see what the aerodynamics were. When we got our first heavyweight P-3s in, we took them out to sea to see what the drag did to the airplane as far as fuel consumption. They were blunt-nose, big, high drag. That's when we started talking about—if we were really going to use the P-3s for a lot of heavy mining, then the P-3 was going to have to get an in-flight refueling system. The range was too short to carry that thing out on the wing stations. The bomb bay was fine—

EM: Of course, when we finally did drop the mines it was A-7s [Corsairs] and A-6s [Intruders]?

SA: A-7s and A-6s, yeah.

EM: Any other weapon system that comes to mind that either panned out or didn't?

SA: Let's see. While we were there, the new version of the P-3—Ed Waller was the NavAir guy at the time, and he was bringing out the new P-3; we were going to get one of the early ones to start the systems test on. I got a chance to look at it when I was up at Pax River, because we did a lot of our weapons work out of Pax River. I got a chance to look at what Ed was doing on that. The P-2 was just going out; basically the P-3s were doing it. My flying partner for the P-2 there in VX-1 was a Royal Air Force squadron leader. He and I did most of the work up at Pax River with the P-2 up there. He had flown the P2V-5 version with the Royal Air Force, and we had one of the 7s down there.

EM: So you had number of contacts with Brits, French, other NATO allies? Aviators?

SA: In our squadron we had a Royal Air Force pilot, I think two. For our helos and S-2, we had a Canadian pilot who was with us. The French—we didn't have anyone in VX-1, but the project we worked with Nord Aviation which was the interface between us and the French Navy. Then the rest was all U.S. Army. We did a lot of work with the Army because of their SS-11, the antitank missile. We worked with them to see what their training and experience was with the SS-11, so we could carry that over to the AS-12. Then White Sands, New Mexico, we did a lot of the land-based firings of the AS-12 out at White Sands under—there was both a Navy and Army detachment out there that we worked with.

EM: At this point in your career, you're really focused on antisubmarine warfare? That was your thing.

SA: Yes, that was my thing.

EM: Were you happy with that? Or you wanted to get into attack?

SA: I still wanted to get into attack, but I enjoyed the ASW. It's a thinking man's game. The movement in the sensors was starting to take place, so we were really getting some sophisticated automated systems.

EM: Your primary opponent would have been the Soviet submarine force.

SA: Soviet submarines, yes.

EM: Did you read up on Soviet capabilities?

SA: Oh, yeah.

EM: Intelligence kept you up to speed?

SA: Intel kept us up, and every now and then we'd get something new they would ask us about. They would see something new and ask us to take a look and see if we could handle it.

EM: What was your opinion, and those in your squadron, about the Soviet Navy's submarine capabilities?

SA: We thought they were pretty good. We thought tactically they were good. They had a good awareness of the situation. They did a lot of maneuvering and some very sophisticated maneuvering tactics. They had some good decoys; we spent some time trying to sort out how we could identify the decoy before we got led astray. We thought they were very capable. We didn't think they were as good as our guys, but we thought they were very capable.

EM: Did you actually go out and track any Soviet subs?

SA: The guys who really did the systems work did. They went with the fleet as well. Sometimes they went with the fleet guys, and sometimes they took the new stuff out and tried it.

EM: Any American squadron mates, folks you saw later on in your career? And who was your CO?

SA: Paul [L.] Ruehrmund was my first CO down there, and I'm trying to think of my second. They were both captains. Paul had been an ex-S-2 guy—Jack Laney, I think, was the second one. He was a VP guy. Paul liked to fly the P-3 every now and then. I remember we were pretty casual. He was flying P-3 for—I forget what kind of flight we had, but we were coming back into Key West, and Paul was known to fly the P-3 like he flew an S-2. Like he was coming aboard the carrier, which just drove the VP guys berserk, because it was not the same pass that they were used to flying.

He's coming back in, and there must have been eight of us up there on the flight deck—the pilot, the flight engineer, the co-pilot. There was me. I think Dick [Richard]) Mieldazis and a couple of other guys. So there were seven or eight of us up there on the flight deck. We're watching Paul coming into the runway there at Key West, and Paul's a little high and a little fast for the VP guys, and for us carrier guys he was really high and fast. [Laughter] All of a sudden he sort of realizes that he's got to do something drastic if he's going to get that airplane down on the deck and not halfway down the runway. He takes a whole bunch of power off and drops the nose. Well, that cleared the flight deck. [Laughter] We all went back, we were jumping into seats wherever we could find them. He ended up getting it on the ground, but it was a little—a little stiff landing. We always—after that we were cautious when we went with Paul in the P-3.

In the S-2 he flew—I went out with him one Friday. The squadron had whatever was the forerunner to the DEA [Drug Enforcement Agency], a fishing boat that they had acquisitioned from the drug lords someplace. So we had this really fancy fishing boat that belonged to the squadron. The skipper was an avid fisherman. He wanted to launch every Friday morning in an S-2 and go out and see where fish are. I'm the duty guy to go with him this Friday morning, and we're out there, and the next thing we see is way out—you know, at Key West there are certain areas out there that you can be miles from land and be walking, still walking in, and we see this little trail of muddy water. At the end of the trail was this guy walking. Then we look around, and we see about half a mile away his boat floating away. [Laughter] We take a closer look, and out around him, about a 100 yards out are sharks. [Laughter]

EM: What if the tide goes up too? He's out of luck.

SA: Paul calls back to the squadron and has them launch one of our helos to come out and pick up this guy.

EM: Did he ever get his boat?

SA: Oh, yeah. We got him to his boat. They really had a lot of fun with that Friday morning fishing expedition. Everybody would go out and fish. They tried to get me to go fishing for the whole time I was there, and I wouldn't go. I've had some fish stories from my youth that tell me that I'm not a fisherman.

EM: Had no interest in it?

SA: No interest. They said, "You've just never been at the right place. I mean, Key West is a fisherman's paradise."

I said, "Look guys, I just—you don't want me to go." They said yes we do. So finally, just before I left the squadron, we went out on a fishing expedition. Didn't catch a single fish. [Laughter] They came in, they were madder than hornets. They were cussing at me. And, I said, "Look guys, I have been telling you all this time, you know the year and a half I've been down here in this squadron, I told you I didn't want to go fishing, and what would happen if I did. And it did, and now you guys are complaining?"

EM: You warned them. This is ’64 through ’66, so Vietnam is really heated up at this point. What were you hearing about Rolling Thunder [bombing campaign against North Vietnam] and how it was being conducted?

SA: The first thing that we heard down there was we had a group of guys, P-2 pilots, that were in the squadron. All of a sudden they were getting calls about forming up a new squadron.

EM: For coastal patrol?

SA: For Vietnam, special mission. One of the guys—he had just left or was just leaving—had headed up the weapons group that I was with, and he got assigned to the standup group for the VO [observation] squadron.

EM: VO-67 [Observation Squadron 67]?*

SA: Yes. That was our first inkling that things were really—they were going to do some, sort of different stuff over there. We didn’t know anything about the sensors at the time. For us, we didn’t have any work going on those kinds of sensors. We were all intrigued about what that was. Those guys had a tough time.

EM: They lost a number of crews and planes?

SA: Yeah.

EM: It was an interim operation, too. The Air Force, I guess, was coming on board with some aircraft, but they didn’t have them yet?

* The highly secretive Aerial Observation Squadron (VO-67) lost 20 aircrewmen killed during only two months of sensor emplacement missions over North Vietnam and Laos during 1968. For additional information on the “Ghost Squadron,” see Richard A. Mobley and Edward J. Marolda, *Knowing the Enemy: Naval Intelligence in Southeast Asia* (Washington: NHHC/NHF, 2015), 32.

SA: Yeah, they didn't have them yet.

EM: So they paid a price, flying into mountains? It was tough there in Laos.

SA: P-2 [Neptune] was a wonderful airplane. All my VP buddies they were in love with the P-2s. They didn't know how they were going to feel about the P-3s. I flew the P-3 first, before I ever flew the P-2. I thought the P-2 was really ancient history, but I had to admit, it was a really smooth-flying airplane. One of the smoothest airplanes I think I've ever flown in. We were hearing—of course, with me going around White Sands and the Army guys—I was already starting to fly with guys who had been flying in Vietnam, especially in the helos.

EM: What were they telling you?

SA: They told me it was wild.

EM: Tough?

SA: Really tough. When I got to work and started looking at the Komars [Soviet fast attack missile craft] for something to shoot at those, I picked their brains about what you could expect to see. I knew then that you didn't want to—we didn't want to get our airplanes over land to do anything. If we were going to tangle with something like the Komar, we had to be pretty careful not to wander into his realm without being prepared. Those guns could really stitch an airplane in half.

EM: This is jumping way ahead, but your experience with VX-1 about attacking Soviet-made vessels. We shot up a whole bunch of Iraqi vessels of the same type. Did that somehow inform that operation later on?

SA: It brought back to me the problem that we were starting to see there. We had the fixed-wing guys who were starting to figure out how to put something on the fixed wing to handle the threat. We talked a little bit about the old hunter-killer concept that we had initially with the ASW, with the old guardians, the AFs, where we had the AF with the radar and the AF with weapons, so that the radar guy could vector the guy in and give him the best range information.

Then the thought was going, you know, what are we going to do with helos? Well, nobody really wanted to talk about arming the helos, although when I got over, all the Army guys were doing was shooting. They wanted to shoot all time, and I thought they were all just going and dropping people off and picking people up. But no, they were shooting. And they were getting in trouble. We come up to Desert Storm, and here we are, Uncle Sam's Navy, still doesn't have anything to fire off of a daggum helo other than stick a gun out the side. The Brits had their Lynx with their Sea Skua [British air-to-surface missile], and so turned it over to Ron [Ronald J.] Zlatoper and his gang and said, marry our guys up with the Brits and see if you can go out there and do something. Of course, they did.*

EM: Very effective!

SA: Yeah. I put that in the lessons learned, and that's still sort of wandering around. People were very nervous about arming helos. [Laughter]

EM: Some things never die.

SA: Never die is right.

EM: Well, back to the VX-1, and you're heading out to *Bennington* [CVS-20] for service back to sea. Was *Bennington* slated for Vietnam service? You're looking at WestPac?

SA: Yes.

---

* For a detailed description of the employment of U.S. and British combat helicopters in the Persian Gulf War, see Edward J. Marolda and Robert J. Schneller, *Shield and Sword: The United States Navy and the Persian Gulf War* (Annapolis, MD/Washington, D.C.: Naval Institute Press/Naval History and Heritage Command, 2001), 231–33.

EM: Did you take any pre-Vietnam training while you were still at VX-1?

SA: No, I didn't have anything. I went directly to *Bennington.* She was home-ported in Long Beach [California]. We rented a house in Garden Grove. I went down to the ship to check in, and the ship was on "fast cruise." They had been in yard period, and they were doing their two- or three-day where you play like you're at sea, but you are alongside the pier. Nobody can come on or off the ship.

EM: What's the point of that?

SA: Well, it's to test all your gear, get everything all hot, and you're testing everything but you're not at sea. You got everybody to their stations, and you're getting all the organizations established and make sure the communications all work, all the sound-powered telephones and all that. It's a good exercise. If you haven't been at sea for a long time, to have a couple days where you pretend like you are at sea, but if all of a sudden everything goes dark [laughter] you can still—

EM: Have another three days of training.

SA: You got time to bring somebody on board to fix it. I can't get aboard the ship. [Laughter] Finally they get word to me. Well, you don't need to get aboard ship, you need to get to San Diego [Naval Base] to go through OOD [officer of the deck] training, one of the mandatory schools for being the assistant navigator. I went down there and ended up in class with Phil [Philip M.] Quast, who was a brother of Harry [Harold] Quast, who was one of my Miami ROTC guys. Phil and I were in the same OOD class. Phil was a surface guy, then came back up to the ship and we started getting ready. We were going to go on cruise pretty fast. I forgot what date we left on cruise—we headed to WestPac.

EM: So you've been promoted in February to lieutenant commander, right?

SA: Right.

EM: Now you're looking at a different approach to your naval career. It's not so much aviation, you're doing deck duty, ship driving and that sort of thing. That was programmed into every officer's career?

SA: Yeah. That was one of the things, if you were a carrier aviator, or even VP aviator, they expected you to get to sea sometime as the ship's company. I was happy to get the job as assistant navigator. I didn't realize how much fun it was really going to be. Navigation was always something—I like navigation. It was interesting.

EM: Did you have any training in it, though?

SA: Well, you had training in NROTC courses. You had training in navigation.

EM: Celestial?

SA: Some celestial and so you had to figure out stars and all that sort of stuff, and I had, along with the OOD school, taken another course in celestial nav. I was interested in navigation, and when you're assistant navigator, you're right behind the bridge all the time. You get a lot of time on the bridge. I got my OOD qualification real fast. You're up there for every sea and anchor detail. You are up there for entering and exiting port. You're right in the center of where things are happening. It was a good experience for me. The quartermasters, they were pretty smart sailors. Easy to be their division officer.

EM: Did you have any memorable events while you were acting as assistant navigator?

SA: Oh, yeah. Lots of fun things. My first skipper on *Bennington* was Captain Wiley B. Howell.

EM: Gee, that name rings a bell.

SA: Wiley B. Howell went to England to fight with the RAF [Royal Air Force] in 1940, or something like that. Got a Distinguished Flying Cross for sinking one of the U-boats [*U-502* in the Bay of Biscay on July 6, 1942]. The U.S. enters the war, puts out a call to all the Yanks who are over there flying with the Brits, saying, "Come on home, we've got a need for you guys. Where do you want to go?"

Well, the couple guys Wiley was flying with—this is his story now—they said, "Oh, we want to go to the Army Air Corps." Okay Majors, you're in the Army Air Corps.

Wiley B. Howell says, "I want to fly with the Navy." Okay, Ensign [laughter], you can come fly with the Navy. He comes back, and they send him to Grumman Aircraft Company to fly acceptance flights on new airplanes, the F4s and F6Fs and stuff. Wiley's up there flying these test flights for a while and realizes he's not getting any flight pay. He goes into the system and says, where's my flight pay? And they say, well, you're not a designated naval aviator. And he says, wait a minute—I'm flying airplanes for the Navy. Why aren't I a designated naval aviator? They say, you've not been through Pensacola.

EM: He says, "I've been flying in combat." [Laughter]

SA: I'm sure the story is true. It really tickled my fancy when he was telling me this. He says, "What I had to do was, I took a brand new F6F from the factory. I flew down to Pensacola. They put me in the backseat of an SNJ and took me out and said fly this, fly this, show me this, show me that, show me this—then I was a designated naval aviator."

EM: He could get his pay. [Laughter]

SA: Wiley was a real character. Obviously from the background of a guy just taking off and going to England and joining up—I mean he was a true character. He had a really uncontrollable temper. Something would set him off. The old sound-powered telephones up there in the bridge, you know, all that those nice coiled cords. The OOD would do something that he didn't like and—

EM: He'd rip it out of the wall?

SA: No, he'd throw it. [Laughter] One night he was up there, and he got mad at the OOD and he threw that thing, and it snapped back and I just happened to be in between. I grabbed it, or it was going to catch him right in the head. [Laughter] He was a wonderful guy, a fun to work for.

Pete [Evan Peter] Aurand was our flag officer. He was a very decorated World War II fighter pilot. Now he's ASW, becoming the ASW czar. He went from us to Third Fleet, as a matter fact, to be the head ASW guy out in the Pacific. Pete had all of these ideas about the flag—he didn't like the way the ensign was being flown while we were at sea. He thought it should be more visible than it was on the gaff, where you normally fly it. So he said, "No, I want a flag, one, when we are in port, that is very light and the least breeze would make it flutter." He had the supply officer find somebody who could make a flag out of silk. Then he said, while under way, "We need someplace more visible. Now, I want the O7 [deck]." There was an old gun tub back there, and we navigators used to go out there for the sextant, to do our—to get the stars at night and stuff like this. Then at vulture's row, when flight ops were on nights, everybody would gather back there to watch airplanes. He says, "I want a flag on a mast that comes up from the aft end of that gun tub back there, instead of flying it from the gaff."

Well, the skipper said, "Over my dead body."

EM: Wouldn't that interfere with flight operations?

SA: Not really, because it would be off to the side. It was an unusual place to be looking for the ensign, because it should be way up there on the gaff. The skipper was saying, no, no, I'm not going to do this.

I said—of course I'm the action officer, a duty of assistant navigator—"Well, Skipper, the admiral says it's got to be there, he says he wants it, and unless you tell me that I can't do it, I'm going to have to do what the admiral says." He says all right—grumble, grumble, grumble. We get the welding crew up there and we put up a mast, and it makes the admiral happy. At Wiley B.'s going-away party, he's talking about his new house. He's got a swimming pool, and so I had the guys cut down that pole and deliver it to his house and make him a flagpole out by his pool. [Laughter]

EM: Momentous events.

SA: Yeah.

EM: What did *Bennington* do? You were out there in WestPac. Were you looking for submarines, or were you getting into the air war?

SA: We provided the armed helos for rescues, so for the North SAR [search and rescue] rescue destroyer. We flew every morning, I think it was HC-7, the armored helo up off of our—we got it ready to go—flew it up—*

EM: SH-3s?

SA: SH-3s, and we had body armor on the seats and body armor on some of the sides and underneath, and they had machine guns on it. Then they would go up and spend the day on the North SAR Station, and then in the evening when the sun went down, they would fly back. It was my job—the CO made me the CDO [command duty officer] for launching the SAR helo every morning and my other lieutenant commander Air Department buddy was the air boss for getting the SAR helo launched every morning. That was our job about 4:30 in the morning, we would get up there and get things ready

* Throughout the air war against North Vietnam, the Navy maintained units in the Gulf of Tonkin to retrieve downed fliers from the sea and from North Vietnam and Laos. Normally, two destroyers were deployed to the forward, North SAR Station and another two to the South SAR Station. To carry out rescues in North Vietnam's lethal environment, one UH-2 Sea Sprite helicopter was nested on board a ship at each station. Another four SH-3A Sea Kings were based in one of the aircraft carriers in the gulf. During major air operations, one or two SH-3As orbited over the destroyers. Under the overall control of Commander Task Force 77, the SAR coordinator directed the Navy's effort from a North SAR Station destroyer.

and get it launched, headed up to the North SAR Station. The VS squadrons, during the day, flew maritime patrol in the Tonkin Gulf and tracked WBLCs [water-borne logistics craft], fishing boats and merchant men that were coming up into Hanoi. Off the Vinh, they would get the hull numbers and see who it was.

EM: Loaded with trucks and missile bodies.

SA: SAMs [surface-to-air missiles] and all that stuff. The interesting thing was that from the navigation standpoint, although the area was covered with LORAN [long-range navigational system], the LORAN was very unreliable. We were there in a period where we didn't have many clear days, so we didn't always get a good sun line, and a lot of times we didn't get the stars at night. We were finding ourselves—when we would launch the S-2s and they would go in to pick up their routes and some of them were right off of the coast—they were telling us, hey, you know, we're mispositioned by 10, 15 miles. Much to my chief quartermaster's chagrin, I found an old VP buddy of mine and we started experimenting with a bubble sextant, which you don't need a horizon for. You have the horizon as artificial with the bubble that shows that the thing is absolutely level. So you don't need a horizon to measure the azimuth of the star. You've got the bubble to give you the right azimuth.

EM: Why did the quartermaster not like that, because it's not traditional?

SA: Not a traditional sextant. A traditional sextant is, you've got to get the horizon, and then you've got to get the star.

And I said, well, all this means is you've got to get the bubble in the middle, and you get the star—

"Oh, no, no, no, none of that fancy stuff for me, you daggone airdales."

Back there on the famous platform, I built another platform for the bubble sextant, so you could put it in. That helped. We got the radar guys, and we started mapping the mountain range on the western side of Vietnam into Laos, where you started to get into the karst ridges and the mountain. So we started mapping those on a day-to-day and night-to-night basis, and we finally were able to take the radar and change the azimuth, on the radar until we could just get a couple of well-known peaks that we could almost always find. We started getting some radar ranges off of those and getting better as far as not getting in closer to shore than we thought we should be.

EM: What about the waters, were there shoals everywhere?

SA: No. For the most part, out in the middle of the gulf, we didn't have to worry at all. If you wanted to get in close to ports, a couple of times we didn't go into port, but we went in close enough to do some transfers, like the Bob Hope Troupe show, so that it wouldn't be too far for them to come and stuff like that. You had to be careful, but out in the normal operating area, we had very few hazards.

EM: Well, northeast of Haiphong of course there are all kinds of islands and rough stuff there.

SA: Yes, but we never went up that far.

EM: Right. What about Hainan [territory of the People's Republic of China], which is off to your right?

SA: Yeah. We had to be very careful with Hainan, because they were sensitive to that area. We took the normal boundaries that the system had sort of said stay out of, and then we added some cushion there, because we didn't want to get tied up with Hainan. You had to watch that. You had to watch coming out of the gulf and turning and going to the Philippines, because then you got into the Spratly [Islands] and some of that stuff. You had some hazards there that you had to worry about.*

---

* In April 1965, the People's Republic of China broadcast that a U.S. F-4 Phantom had shot down one of their MiG-7s over Hainan. That aircraft was apparently one crewed by Navy Lieutenant (jg) Terence Murphy and Ensign Ronald Fegan. The last communication ever from Murphy was that his plane had expended its air-to-air missiles. Later during the war, Chinese MiGs destroyed five Navy and two Air Force aircraft operating over or near Hainan. Navy Lieutenant Robert Flynn of Attack Squadron 196 was the only survivor of these

EM: This ties in with the Tonkin Gulf incidents of '64 about ghost radar emissions. Did you see a lot of that?

SA: Yes, there's no doubt that there were environmental issues with the radar. We saw that time and time again.

EM: It amazed me, because I saw the operational reports from the Tonkin Gulf episodes and fire-control radars were locking on to a supposed target.

SA: Yeah, now that's pretty amazing there, because that frequency is such a—but with the normal search radars, it was not uncommon to have ghosts all over the place out there. I think that's part of the reason LORAN was so squirrely out there, too. It was just totally unreliable.

EM: The water quality, was there a lot of effervescence?

SA: There was.

EM: In the wakes?

SA: Yes. And a lot of sea snakes.

EM: Really?

SA: Oh, gee eww. They couldn't get on the ship, but I remembered it when I started flying a lot—

---

encounters. The Chinese government did not release him from captivity until the end of the war in 1973.

EM: You didn't want to go down.

SA: I didn't want to be swimming with them. [Laughter]

EM: I've got here that you flew one combat mission during this period?

SA: I did, with one of the VS squadrons. They said hey, why don't you go fly with us one day? I said hey, yeah. I'd like to get some flight time. We launched—it was a search pattern that was right off the coast, and I'm in the co-pilot's seat, trying to keep track of where we are going. We're in the S-2Es at that time, so we had a pretty good navigational system and radar. Radar says we are going to get pretty close to land here in a little bit. And the guy says, "It's a little too early. Do we need to make our turnout now?"

"No, I think a little bit longer, but yeah, pretty soon we're going to have to turn out to sea."

He says, "Well this is a little short, a little quicker than I thought."

"Yeah, but I can see the land up there quite clearly."

It was what we called—I later learned but didn't know at the time, but all the aviators called it the hourglass. It's where there's two rivers that come down and they sort of bounce out and look like an hourglass.

EM: I've seen it on the map, yeah.

SA: And that was hot country, and it's not a place you wanted to fly over in an S-2. We vectored out and went out. Other than that it was a routine flight, but I remember that and then—just about a year and a half later, I'm flying into—I said, I've been here before. [Laughter]

EM: Any last thoughts on the *Bennington* experience?

SA: Well, we had the Bob Hope Show on there, which was great. We had a lot of fun with that. We had one of the destroyers, the *Kirk* [DE-1087] I think it was, that when they were on the gun line took some rounds. Four sailors were injured, two of them died. They were brought in over to us on *Bennington* on Christmas Eve of '66.

Dave [David] Kasselbaum, who was in the Air Department, my lieutenant commander counterpart, he and I flew the ship C-1 into Danang with the two bodies of the guys. The helos took the wounded guys over to the hospital ship. That was sort of a downer. It made the war seem more real. The Bob Hope Show, they came aboard, I think it was a couple of days after that and we had a good time with them. Bob Hope was a real trip. We had Anita Bryant, she came and—

EM: Martha Rae?

SA: Martha Rae wasn't with this group. Joey Heatherton was. Joey, she was not a happy camper. Vic Damone was with this group. What a gentleman he was. I always thought he was a wonderful singer, but he was just such a nice guy, really, I mean, no pretense about him. Just as good looking as his picture, I mean he was just a poster boy.

EM: My father liked him, he liked his music.

VSA: I was amazed at how friendly and easygoing he was.

EM: What was Joey Heatherton upset about? Too long out there?

SA: "It's too much, it's too hard, I don't know if I like this food, I don't know—jeez, how do you people live on a thing like this?" She was sort of just being a brat. Anita Bryant, she was good, very nice.

EM: Well, I think we will break off for the day.

SA: Okay.

EM: Good stuff, and tomorrow we will pick up with Kingsville and Lemoore [Naval Air Station, California] and onto *Hancock.* We will do the Vietnam story.

SA: Okay.

[End of 23 June 2015 interview]

Interview Number 2 with Adm. Stanley R. Arthur, USN (Ret.)
Date: 24 June 2015

EM: All right, sure, we hadn't finished the *Bennington.* Do you want to add a few things, Admiral?

SA: There was one period on *Bennington* that I found interesting, and it sort of carries into the things that have happened today. In May of '66, we were designated a ship for a ship evaluation on the Ling-Temco-Vought XC142A. It was the tilt-wing: it had four engines, but the engines were rigid to the wing and the wing tilted. It so happened that the test pilot was Lieutenant Roger Rich, later Rear Admiral Rich, helicopter pilot. He and I had been in basic training together at Corry Field when we were getting our wings, and so he brought the airplane out. That was '66, and of course the argument was, do you have tilt-wing? Or do you have tilt-rotors? Like the V-22 now. It was sort of the predecessor to that concept of a vertical-lift airplane that you could transition to horizontal flying.

EM: That particular model didn't go forward?

SA: It didn't, but it was really interesting to watch it fly. It looked like it flew great. We had a great time.

EM: What torpedoed the project?

SA: I think they still hadn't really decided which the better way was, either tilt the wing or tilt the rotors. I think in the long run it was much easier to go ahead and do the tilt-rotor with a fixed-wing instead of trying—

EM: More strength to the wing that way.

SA: But it was an interesting concept. It was fun and very different than anything we had seen onboard ship before.

EM: You didn't get to fly it? [Laughter]

SA: Oh, no. Sadly, no.

EM: Anything more on the *Bennington* experience? We talked about your one combat mission.

SA: The *Bennington* was really a turning point in my career, I thought. I learned so much on that tour and I was given so much responsibility. I was the only lieutenant commander that the skipper let become a CDO under way, at night. I would be up there at night for the OODs, in case they got in trouble. I got to see a lot, got to watch a lot of different styles of leadership, and you got a feel for what was going on more so than when you're in the air wing.

EM: Part of the ship's company.

SA: So you got to see a lot more of the interplay and what was going on outside the ship, and the demands of why the ship had to be at a certain place, at a certain time, and how when you're flying, you say, why in the world is the ship trying to go that way when the wind's going that way? So it was a great tour.

EM: That's a good convention to have, aviators go toward the surface, and vice versa.

SA: I think so. It was a great learning experience for me.

EM: And you were off to the Naval Air Auxiliary Station, Kingsville, Texas.

SA: Yes, for jet transition. When I had got to *Bennington* shortly after, I finally went through the schools and got aboard ship. We got under way from Long Beach and we went down to San Diego for refresher training—I believe that's what we were down there for.

But we pulled into San Diego and so I had been on the ship probably three to four months, and this guy, a lieutenant, came up on the quarterdeck and said to the OOD that he wanted to see Lieutenant Commander Arthur. So the quarterdeck called and I came down and the guy said, "I'm so and so—I'm your relief."

And I said, "I just got here." [Laughter]

He said, "Well, I have just been told that I'm getting orders to come and be the assistant navigator."

I said, "That's very interesting. Let me show you the ship."

Then as soon as he departed, I ran down to the XO [executive officer] and I said, "XO, do you know something I don't know? There's a guy on, that was just here who says he's got orders to come be the assistant navigator."

He said, "What?"

"That's sort of that I said. But if that's the case, I really want to transition to jets. I understand that with the war going on, there's some opportunities to do that."

He said, "Well we wouldn't let you go in less than a year on board. You've got to get your OOD qualifications and all that sort of stuff."

"Yes, I understand that and I want to."

"But," he said, "if that's the case, then we'll certainly endorse you for that."

So it turned out that wasn't it; it wasn't the same guy who ended up coming to relieve me. I still don't know what happened there, but it set the ball rolling so that basically we went on deployment, I was on the ship 13 months, and I left the ship early. We had finished all of our operational stuff and the ship was headed to Australia for a port visit, which I hated to miss. I got orders to Kingsville to jet transition.

EM: So did you fly off to Subic [U.S. Naval Base, Philippines]?

SA: Yeah. Got off the ship at Subic and then flew home from there. The family was still in Garden Grove there up in the LA area, and so I went on to Kingsville. The regular training squadron was there, well, they rotated us through; they had three different squadrons there in Kingsville at the time that were doing jet, you know, training for pilots to get wings—student aviators to get their wings. There were quite a number, not a big number, but I think there were probably five or six of us who were all in the transition stage, lieutenants or lieutenant commanders, who were down there to get checked out in jets.

EM: Could you elaborate on why you wanted to go to jets? I mean, you had been antisubmarine warfare from the beginning.

SA: Yeah, it was just one of those things that when I had started to fly, I wanted to fly single-engine airplanes. I thought this was a good opportunity to see if I could make the try one more time. It was just something that I had always wanted to do. I said, "I'll do it."

Everyone told me not to do it. Even the captain of the ship when I was leaving, he said, "I think you're really making a bad mistake."

EM: Why?

SA: He said, "You've already established yourself in the ASW world. You've had two deployments in ASW, and you were at a VX squadron that specialized in the new stuff that's coming along, and now you've been on a CVS [antisubmarine aircraft carrier] as an assistant navigator so you know all the routine about the ships, the staffs, and how to work with them. You know the community and the community knows you." He said, "You're going to go into another community, and you're a lieutenant commander. You're going to be competing for department head jobs with no experience on the airplane. I think that as far as a career move, this is not a good move."

EM: You weren't deterred by that?

SA: No. [Laughter]

EM: Well the war was going on, too; did you want to get in, drop bombs?

SA: I wanted to get in dropping bombs. Being a part of it.

EM: So, do you want to go into some of your training at Kingsville?

SA: Kingsville was good. They were set up pretty well for us old guys coming in. They threw us right in with the young guys, and we had an abbreviated syllabus, but we got all the parts that they did, just not as much. We got rockets and instruments and carrier qualifications. We went out to the carrier with the young guys. It was fast-paced—I think we were probably there two months, two and a half months.

EM: It was accelerated because of the war?

SA: It was accelerated for us, just because some of the basics we really didn't need. Once we got to the point where we really understood the airplane, some of the repetitive stuff that they do with the normal student aviator coming through, they shortened it because we didn't need quite that much repetition.

EM: The basics?

SA: Yeah. But made a good friend there, Jack [John F.] Calhoun. Jack and I ended up in VA-55, well we ended up in the RAG [replacement air group] together, VA-125, when we left there. Then we ended up in the same squadron, VA-55, when we finished the RAG training. We made the two cruises together. Jack and I stayed in touch over the years. He went east and I went west, but we ended up in Washington in time. Jack was selected for rear admiral and did a couple of tours, and then retired as a two-star.

EM: He wasn't by any chance related to the Callahan Medal of Honor recipient from World War II?

SA: No, he was a Calhoun, Jack Calhoun.

EM: Oh, okay.

SA: Yeah. Jack's family I think was from Missouri, farmers from Missouri. He was a great guy.

EM: How did you like jets?

SA: I loved it.

EM: What training aircraft did you fly in?

SA: We had the F9F Cougar, it was a swept wing Cougar. It was a really a fun time. I was more than convinced that I had done the right thing.

EM: More power, more maneuverability?

SA: Yeah, you had a lot more maneuverability and power. I did find that the engine response wasn't exactly what I was used to with the props and we talked about that earlier, the instant response that you have from props and you don't with the jets. The F9 [F9F Cougar] was not a very responsive engine, but it was a nice airplane to fly. When I'd gone through training initially, that version of the Cougar had just started being used in the training command. The Blue Angels had just transitioned to it. You know it was a nice flashback to my earlier days of saying boy, that's a nice airplane, and now I got to fly it. [Laughter]

EM: They didn't have any A-4s there?

SA: No, the A-4 hadn't entered the training command yet, it was still only a fleet squadron airplane.

EM: Okay, anything more about Kingsville?

SA: No, it was just a good experience. It was fun to watch the young guys that hadn't gotten their wings yet.

EM: You're an old guy now, right about 30. [Laughter]

SA: Their outlook and stuff, it was sort of fun to come back and see the difference that a few years out in the fleet had made compared to when you were a student.

EM: Right. So you're off to Attack Squadron 125 at Lemoore?

SA: Yes.

EM: And you got your Skyhawks there, right?

SA: Yes. VA-125 was the A-4 [Skyhawk] RAG, the replacement air group training squadron. It had about 100 airplanes out there on the ramp, and by that time there were several variants of the A-4, so they had A-4Bs, A-4Cs, A-4Es. The fleet was still A-4C or A-4E, and the difference was the engines. One had one type of engine, and the other had a more modern engine. They tried to keep you flying the same platform so that you didn't have to mentally try to make the difference between "am I in a B, or a C, or am I in an E," and sort of point you toward when you went to the fleet what kind of squadron you were going to go to. But the fleet was moving so fast, and airplanes, that finally at the end, we were sort of jumping between the airplanes. You know, one time you'd be flying a C and the next time you'd be flying an E.*

EM: Were they that different?

SA: No, they really weren't. The E had some more modern systems, the trim was a little bit different, the engine start procedure was a little bit different, but not much. The difference was in the power. The E was much more powerful, and so you could drive it a little bit harder through maneuvering and all.

The training itself was good. Since I was old, I ended up being sort of the section lead of a group of young men—they put us in sections, and you went through the program until you finally got to the weapons stage, and then you went to Yuma, Arizona [Marine Corps Air Station] or Fallon, Nevada [Naval Air Station], for weapons stage. It made it nice, as you stayed together as a group. You knew everybody, because when you went to the detachment you knew everybody in that particular det, you had flown with them before and you were all pretty much on the same page. That was my introduction to the weapons work down at Yuma—really, really fun.

EM: What weapons did you employ?

SA: We fired rockets and dropped—

EM: Zunis [air-to-surface rockets]?

SA: Yes, and 2.75s [air-to-surface rockets], and we had the little practice bombs that we used for the bombing targets, and then we had—I'm not sure whether we did, I think we did one or two flights with live ordnance down there. I can't recall.

* The workhorse of the air war in North Vietnam and Laos was the Douglas A-4 Skyhawk. A-4s carried bombs, rockets, and anti-radar missiles. Skyhawks carried out more attack sorties than any other Navy aircraft during the war, but the enemy took a toll on them, downing 195 planes.

EM: Mark 82s [general purpose or conventional "iron" bombs]?

SA: Mark 82s. We did the nuclear, checked out in the nuclear delivery.

EM: So that was still a part of the training, even though the war was going on?

SA: Yeah, we still had to do that. So that was the fun, and there were always some crazy incidents that happened there. I was going to lead a flight of—I think there were three ensigns who were in the class, and one of the guys was really colorful. It sort of goes back to the story I was telling you about when I was flying the SNJ and flying formation—the same thing happened here. We were going to go on a night bombing run and the target was just east of Yuma. You came off the runway, and basically you did a right-hand turn, a slow climbing right-hand turn. You were on run-in line for the bull's-eye over on that. So it was a night hop.

While we're briefing, the guy that was going to be number 4 says, "I don't want to be number 4. This guy number 3, I've flown on his wing before and he's hard to fly on. I don't want to fly."

I said, "Look [laughter], this is the order that we are flying in."

He's grumbling about this. We take off and we are in this turn and of course everybody starts rendezvousing on the turn, and just as we roll out on the run-in line to come over to the target I hear him say, "God damn it, turn your lights back on." [Laughter]

EM: Your lights were supposed to be on?

SA: Oh yeah. I'm saying what, who, and so I look over my shoulder to see what's going on, and here comes an injection seat out of number 3. He had flamed out. He had not checked his fuel load.

EM: Was this the squirrelly guy?

SA: Yes, this was the guy that the number 4 didn't want to fly on—so number 4 was a guy, he just said turn your lights back on, so I knew that there was something going on back there, and then I look, here's number 3, he ejects from the airplane.

EM: Did he punch the wrong button?

SA: No, what happened was he didn't check his fuel. The plane didn't get fueled and he ran out of gas. The lights went out because the engine shut down. [Laughter]

EM: Isn't that like checklist number 1?

SA: Yeah, number 1, that's the first item on your checklist is fuel.

EM: Did he go far in the Navy?

SA: No, that was the last flight. He was a nice, nice kid, but he had a little bit of a drift factor there. It was so funny to me, that here I was, all these years later, with a number 4 and number 3 who were still arguing over who they wanted to fly wing on. As you're finishing up, especially for us guys who were more senior there, they sort of put you in, identified as you possible "must pump" guys for combat losses.

EM: If somebody goes down, they send you out right away?

SA: They send you out right away. We were told, "We don't know exactly where your final assignment or final squadron is going to be, but you're in line that if somebody—they have a loss out there that needs to be replaced." Jack Calhoun, for example, myself, and a couple of other guys were in waiting. We didn't have to go, so Jack and I both got orders to VA-55. It was still on cruise when we found out that we were going to 55. We joined the squadron just as they got back, and they had left a couple of airplanes on the carrier that weren't ready for the fly-off. And so the first flights we—Jack and I—had in

55, we got a ride down to San Diego and picked up those two airplanes and flew them back to Lemoore.

EM: Going back to the training on the A-4, you flew other aircraft later—attack aircraft. What did you think of the A-4 in comparison with the others?

SA: The A-4, of course, will always be my favorite airplane. You sort of strapped the airplane on. It was small, I had to squeeze pretty hard to get into the A-4; in fact, my squadron gave me a great big shoehorn they made for me. It was a very straightforward airplane to fly, and it had nice manners. It didn't have any real surprises. And you fit in it. I mean it was like driving a sports car. You got in, you were there, and everything was right where you wanted it to be.

EM: Some people called it a scooter?

SA: Scooter, Tinkertoy, but it was, I think, one of the best airplanes that has ever been built. It was originally built for just a one-way trip, you know, to carry one nuke and go as far as you could and drop it, and maybe you could make it home and maybe you couldn't. So it didn't have a whole lot of exotic backup systems, but as time went on they kept getting added on. But it was great around the ship, beautiful carrier airplane. I liked it a lot. When we got to 55—

EM: Admiral, before you go there, just wanted a couple more things about Lemoore. Did you see a lot of impact from the Vietnam War in the training, like, "Here's what's going to happen to you shortly"?

SA: Oh, yeah. By that time, a lot of the instructors has already flown a couple of combat tours. They were pretty pointed in telling you, this is the type of thing that you've got to be watching for and you've got to be aware of. The community of Lemoore, the town of Lemoore was very small. The next town over, Hanford, was not that big either, so the

Navy had put in a lot of housing. There was a great percentage of the people who lived right on base.

When we went up there, I had a brand new Plymouth, '66 Plymouth. Mine was dark, midnight blue with a white top. Well, the official cars up there were Plymouths. They were black, with either a white or black top, and so my wife was told very, very pointedly, don't drive up to our house at night in your car without calling us ahead of time, because they were worried about the many calls. You know, on a very routine basis there was a combat loss, and the whole community knew it. It was really from day one, once you got into the circuit there at Lemoore, it was a wartime footing. I mean, it was always there.

EM: What was the scuttlebutt about the operational side of things?

SA: That the tempo was really hard. That you know there were lots of losses. Some not only from the combat losses, but operational [non-combat] losses. You had quite a few operational losses out there as well. And there were peaks and valleys, there was no constant or continuum—it was hard to figure out what was going on. People weren't too sure what was really working and what really wasn't working.

EM: No shock and awe approach?

SA: No. It was just plugging ahead.

EM: All right, and you're off to *Hancock* with VA-55.

SA: Yeah, VA-55. Our skipper was Bob [Robert E.] Kirksey, really a fine gentlemen, later Vice Admiral Kirksey. He has since passed away here a few years back. Jeff [John F.] Wellings was the XO. Great XO. Jack Calhoun, myself, Frank [C.] Green and Dale H. Osborne were all lieutenant commanders who joined the squadron at the same time as 55 came back off of their *Constellation* [CVA-64] cruise. Their leadership team from the department head structure left, and we came in and we are all transition guys. Jack had been a seaplane pilot, he had flown seaplanes. Frank Green was a seaplane guy. Dale Osborne was an S-2 guy like myself, and so the young guys in the squadron are looking at all of these lieutenant commanders who came in and wondering what in the world is going on. Who are these guys?

EM: Seaplanes were taken out of operation in '67 or so?

SA: Yes. A lot of the transition guys came out of seaplanes when they were looking for—so the four of us were the new guys. Bob Kirksey was, like I said, just a great leader, and Jeff Wellings as well. They were not worried about us, I mean, Bob, when we finally deployed and went to combat, he took time to fly with each one of us on our first combat hop.

EM: You were his wingman?

SA: Yeah, yeah. And he laid out the rules of the game as he wanted them played. He was very precise in everything he did, he was just a very detailed guy, but very personable as well. He had made up his mind that he wasn't going to lose anybody to dumb stuff. He went over and over and over dumb things that you don't do. That you don't make multiple runs. You know if you have to, if you get to a point where you have to do a second run for something, you don't come out of the same place, you know. Basically he just laid down the rules as he wanted them followed. He treated us just like we had been flying the airplane forever.

For the most part, we did okay. We went on the first weapons det to start the buildup to Fallon, and Frank Green had a slight midair [collision] with one of our new nuggets [new guys], Clark [A.] Wilson. They bent up some wingtips and stuff, but they both got back on the ground okay. He let us get right into the middle of it and watched us pretty close, but also never expressed that he thought we were second-class citizens at all. Whatever we could do, he would endorse it.

EM: So you had a good chance to work up with the squadron before you actually went back?

SA: Yeah. Let's see, we got there in December, I think, and that next deployment was—

EM: July?

SA: I think July.

EM: July '68?

SA: Yeah.

EM: Just a couple of things. The ship, *Hancock,* you were also onboard *Kearsarge* and *Bennington.* Anything that stands out as differences between the ships?

SA: Well, they were all *Essex* class carriers, but of course, *Hancock* was a more modern version as they called them in those days, 27 Charlies. So there were lots of similarities, but there were some differences. I felt very comfortable because the differences were in systems more than anything else. The two other carriers, *Bennington* and *Kearsarge,* both had hydraulic cats. *Hancock* had steam catapults, much nicer ride than a hydraulic catapult.

EM: The hydraulic had more of a jolt?

SA: The hydraulic was sort of like being shot out of a gun. You had one great big hammer and then you just sort of dribbled off of the end, and the steam was constant pressure on you all the way down the cat track, so it was a much smoother ride. Gave you a lot better end speed and was very good. We still had some wooden decks. We had those issues of the teak to cherry rating as time went on, and the pans that set in had leaked into the ready rooms and stuff. *Hancock* was really tired, she was really a workhorse. Of course, I kept thinking, we kept saying this was *Hancock*'s last cruise. Well, I ended up doing four cruises on her and never did make the last one. [Laughter]

EM: An old war horse.

SA: She was a great ship. Fond memories of *Hancock.*

EM: Do you remember who the COs were?

SA: Let's see, Howie [Howard E.] Greer was, I think, the first CO when I got there. Jack [Albert Jackson] Monger.

EM: Actually, I've got Howard Greer and Harold Terry in your squadron.

SA: Terry was the CAG. Howie Greer, of course he went on to—let's see, he was a three-star eventually, and he was a good CO. He had the combat time, and he ran a really good ship. This wasn't the first cruise on *Hancock,* but it was one of the second or third cruises.

Hancock had always had problems with leaks in the fuel tanks, like all the carriers do. There was also a little interface leak between fuel tanks and fresh water tanks and stuff like this. It wasn't uncommon to sit down at the evening meal and pick up your glass of water and it tasted like jet fuel.

One of the fighter squadron's COs—and we were still doing things like the old formal Navy—like we would have the second sitting in the ward room at night for the more senior officers. It was a sit-down. The XO had his table. He had his eight o'clock reports done right before the sit-down, and he and the department heads would come in and sit down at the head table, and then a blessing, and the evening meal would come. So we still had sort of the formal Navy structure of mealtime. While the XO and the department heads are out doing the eight o'clock reports, the CO of the fighter squadron, who had a seat at the table, took lighter fluid and put it on top of the water. The XO and chief engineer and everybody comes in and sits down, and the fighter guy takes his cigarette lighter out and goes—[flicking sound]—and a puff of flame—and he says, "XO, there's just too much jet fuel in this water tonight." [Laughter] The engineer I thought was going to kill him.

EM: He didn't appreciate that? [Laughter]

SA: But it was a good laugh.

EM: You were the OP's training officer on this first cruise?

SA: Right.

EM: For the squadron, what did you have to do in that regard?

SA: Well, the training officer's job basically was, as you can imagine, to conduct admin and operational inspections. You've got to show a track record of the qualifications that the people have got. So you've got qualifications for swimming, ejection seat, high altitude chamber. Then you go into your specific navigation, your bomb records, your rockets accuracy, your gun accuracy, your NATOPS [naval air training and operating procedures standardization] manual exams, your instrument training exams and instrument check ride, all those things. You have a long list of things that everybody has to complete and sometimes, in those turnaround times, the time got very compressed. You ended up with not as many airplanes as you would like to have.

Basically you had to manage with the ops officer and the maintenance officer to make sure that we didn't waste any flights. In the limited time, in the limited number of airplanes, you sometimes didn't really have enough flights to accomplish everything that everybody needed. You really had to make the call sometimes when like the skipper would want to go somewhere and pick three guys to go with him for a—you know a practice bombing run, derby or something like this. You'd have to say, "Skipper, you know, we still got three guys who need that same qualification"—and so it was just managing the limited flight time and availability hours to get everybody their qualifications.

EM: Could they satisfy their training requirements during combat operations? Or were these separate?

SA: They were separate, yeah. In other words, before you got there, everything had to be done. Like we talked about the SIOP yesterday. We still had that whole process to go through, so we had all the training for the nav, the debriefing and briefings that each pilot had to give on what their assignments were. Then they had to have the practice runs and the delivery accuracy that was required.

EM: Where did you do your training when you were actually deployed in the Gulf of Tonkin?

SA: The training back home was easy. We either went to Fallon or Yuma. Overseas, there were some ranges. You had ranges in Okinawa and you had some ranges in the Philippines, so there was a chance for some pilots to complete their training there. If you had a port visit in the Philippines or you were going up to Japan, that you could get some practice, particularly for a new guy who joined the squadron. He could get in his qualifications there.

EM: So as training officer, you weren't necessarily training them for the combat operations? Sounds like it was more administrative.

SA: Yeah, it was more administrative than anything. We had a syllabus, and of course the training officer was responsible to make sure everybody got through it. And the syllabus had the flights in there that we needed, like for the ECM [electronic countermeasures] runs, Shrikes [air-to-surface missile], so we would have to go over to China Lake and run their EW [electronic warfare] range and also be able to get live fire.

EM: You practiced those strikes when you were out at Fallon?

SA: Yes.

EM: What did you think of the weapon then?

SA: It was good. It brought a new capability to the thing. Before the suppression technique was to run a bunch of guys in there with some cluster bombs and try to knock down some of the triple-A [antiaircraft artillery] as the bombers were getting in. With Shrike, you had the ability to reach out there and grab as soon as the radar came up, you could focus in.

EM: Shred the radar.

SA: When we got to 55 they had been flying A-4Cs. We were one of the first squadrons to get the new A-4F, which had a lot more capability in the EW world than the older models of the A-4. We became the Iron Hand [anti-surface-to-air missile operation] squadron, the squadron predominately using the Shrikes.*

EM: VA-55 became this, an Iron Hand squadron?

SA: Yes, we were one of the Iron Hands.

EM: On your carrier? Hazardous duty.

SA: Yes, yeah.

* The Navy and Air Force took direct action beginning in August 1965 to deal with the threat from SAM batteries—in Operation Iron Hand. Navy attack aircraft employed AGM-45 Shrike and other air-to-surface missiles to destroy the radars that guided the North Vietnamese Soviet-made SA-2 Guideline missiles. Attacking these sites, often defended by numerous antiaircraft guns of many calibers, was essential but inherently dangerous. In April 1967, for instance, Lieutenant Commander Michael J. Estocin earned the Medal of Honor (awarded posthumously) for his attack on an enemy SAM sight during which he exhibited great courage and professional skill.

EM: All right, describe your daily routine during workups for combat.
SA: Okay.

EM: Deploying over there.

SA: When we were back on home base in Lemoore, the routine was pretty much—we'd write our flight schedule for the day of things that we could do around Lemoore that we needed. In other words, you'd be low level nav flight for your nuclear qualifications, so you needed so many of those that you could do. The guys had to plan this out and make their strip charts, and then one guy would act as the instructor and one guy would act as the guy who needed the qualification.

We'd roam all around the West Coast, up and down the coast, back over in Nevada and Arizona and places like that on the low-level routes. We would have a fairly modest flight schedule, one that didn't tax the maintenance too much. We would have a morning launch, an afternoon launch and then if we were doing some stuff at night, then we would have a nighttime launch. If you weren't flying you were doing your admin work.

EM: It doesn't sound like a nine to five thing. Every day was different?

SA: Every day was different.

EM: Did you like that or did you prefer more routine?

SA: No, it was good. Sometimes you'd get compressed on your paperwork. Everyone thinks that the pilots just went out and flew airplanes, well, now we all had other jobs. The surface community always believed that all we ever did was go out and fly airplanes. But we had the same paperwork demon as everybody else. Sometimes you'd almost get to the point where you'd beg to get off the flight schedule just so you could get some paperwork done before it crushed you. When it got that bad, it was really bad to give up a flight.

There was always something in the future. We were turning around so fast in those days from cruise to cruise, sometimes as short as five months between being back home and going on the next cruise. You knew the next week you were going to Fallon and you were going to be gone for two weeks up there and there you just flew just—

EM: Nonstop?

SA: Nonstop, and then you knew you were going out to the boat. If you were going to go out to the boat, you had to get the FCLPs. You had to have so many field carrier landing practice flights until the LSOs were satisfied that you were ready to go to the boat. And then you'd go out for a week or two for carrier qualification and workups. There was always something just next week or two weeks away that you were aiming for. It was sort of a steady beat and one that you just didn't—there weren't a whole lot of periods of down time. That's why when we came back off cruise they tried to give you 30 days of pretty much everything—

EM: Decompression?

SA: Decompression, yeah.

EM: The first combat deployment seems like the *Hancock* got involved in was in July of '68. Is that accurate? Well, as I'm sure you know, by July of '68, Lyndon Johnson had called for a bombing halt. He did so at the end of March. You were allowed to bomb below the 19th parallel, but not above it.*

SA: Not above it.

---

* On 31 March 1968, President Johnson announced the end of bombing in most of North Vietnam, called on the leaders in Hanoi to join the United States in peace talks, and expressed his intention not to run for reelection in November. On 31 October, he ended the Rolling Thunder bombing campaign begun in 1965.

EM: So your operations were fairly constrained in terms of geography. What was the view in the squadron, in the wing, in the ship about the bombing halts and the fact that you were not allowed to go into central North Vietnam?

SA: It was really a downer because you couldn't get to where the big staging areas were. Now you were trying to dig them out from all the mini paths that they had found down through Laos, down the trail [Ho Chi Minh Trail] and everything to bring stuff in. You didn't have the alpha strikes.* You were all on sort of road recce [reconnaissance] flights or assigned targets by FACs [forward air controllers], and so you were working ops with the FACs. Sometimes you had some decent targets, small ammo dumps and things like this. Those were worthwhile, but a lot of times you'd end up going over and the weather would be such that you couldn't get into the target, so you would end up going up and flying on the wing of an Air Force F-4 [Phantom].

EM: You would operate together?

SA: Yeah. And then we would do the radar bombing, where they had a beacon and their guy from the ground would vector them in, and then you would just be in level flight and then it would be drop, drop, drop, and you would just pickle off your bombs.

EM: That's unusual to me. Was that common, to fly with the Air Force?

SA: We did quite a bit of it.

EM: No kidding?

SA: Yeah, yeah. We didn't like it. [Laughter]

---

* These missions, on occasion involving 100 aircraft, employed attack aircraft, fighters, Iron Hand anti-radar bombers, and various electronic warfare and other support aircraft.

EM: Why?

SA: Sitting on the wing of an airplane and just being told when to pickle the bomb was not—[Laughter]

EM: Because of his advanced radars he had the—

SA: Well, they were equipped with a special beacon, and the ground radar could track them.

EM: I think that was called Sky Spot [radar-directed air strike]?

SA: Yeah, Sky Spot. We had lots of other names for it, but we did enjoy that, especially at night. At night, you know rendezvousing with this guy at night and just going over—

EM: Were there procedures such that you were at times a little nervous that you guys weren't on the same page?

SA: No. There were briefings leading up to it, and we knew what to expect. We knew how to do the rendezvous, and amazingly it came together pretty well. I mean for guys who didn't operate together very much, we did it.

EM: Where was the coordination between your squadron or your CAG?

SA: Basically, what happens is that you get your target assignments into the ship. You would get the schedule put together of what kind of ordnance you were going to carry, and then when you got to the briefing room, who you were going to check in with. You always checked in with an Air Force—they had the command set up over the beach. You'd check in with their Airborne Command Post and verify who you were, what your sortie number was. They had the script. It was sort of a precursor to the ATO [air tasking order] concept.

Then they would hand you off to a more local guy. Either a local FAC or, in the case of where the weather had gotten out and you were going to do one of these radar bombing things, they would hand you off to the controller for that flight. He would vector you to the rendezvous to join up with the lead. You would rendezvous on the lead and basically they just talked to you all the way.

EM: The schedule was developed with the fleet and the Air Force together? Or did the Air Force just come out with the schedule and the Navy fit into it?

SA: No. The schedule was sent to you via the Air Force, from a joint base out of Saigon. You basically had a planning cell there that took whatever they were getting from Washington and then it came out to you and—

EM: There was a Seventh Fleet liaison there?

SA: Yes, right. It would come in and you'd say, "Okay we got the following strikes for tomorrow," and then all of us ops officers would gather with strike ops and we would horse-trade for, well, I want this sortie, and I want that one. They evened it out so that was an even load on the squadrons and if, for example, one squadron was having some maintenance difficulties and needed a breather for a day or so, then we had the opportunity to—okay, give them some room and we will pick up an extra sortie here on this event, or an extra sortie on that event, to give them a little break.

EM: This is off the beaten track, but your view of the antiwar movement and the media at that point in the war. We are talking '69 from a squadron aviator's standpoint.

SA: It was very disappointing, of course. All of us—I probably shouldn't say all, but I think the great percentage of us believed that we were doing something right. That it was an important thing to be doing and so we were a little shocked at how people back home were reacting to it. You know, that they didn't seem to see the same things that we saw. But on the other hand, we didn't let it bother us too much. It was one of those things we would talk about and grumble about, but it wasn't going to change what we were about. As the war went on, for example, Jane Fonda, who later on, when I was CO of VA-164 in the '72 timeframe, under our flight suits our colors were black and orange. Under our flight suits we wore these—one of our young lieutenants had found a place to do some t-shirts for us. We had our A-4 with bombs coming off of it and it had Fonda's friends on it. [Laughter]

EM: She had a rep even then?

SA: Oh, yes.

EM: Well, what was the perception in '68? There's a turning point in the war here. Things are changing because of the Tet Offensive. I'm sure you knew that had happened. Did you think, "Well, it's just another blip in the whole thing," or did you think it was a significant change in the war?*

SA: No, we thought it was a significant change. We were really wondering where we were going to go from here. We were pretty sure things were going to heat up again and we would probably halt the bombing below the 19th parallel and stuff would change. We were all looking to say okay, are we going to end up going back up north on the alpha strike business up in the real Indian country, or will we still stay sort of south? There was a lot of concern about that, because quite a bit of difference in your tactics depending on where you were flying. When you were down south, after you made sure they hadn't brought any new big guns—like the big 85s [North Vietnamese antiaircraft artillery] and stuff they had up north, you could work a little bit better. As Bob Kirksey said, his rule to us was you're never below 3,000 feet ever, ever, ever, ever. But there were times down south where there were opportunities to go lower.

---

* The enemy's Tet Offensive, begun in late January 1968 and lasting for months, resulted in the death of tens of thousands of Communist troops but proved to be a strategic victory for North Vietnam. The ferocity of the fighting, which also caused high U.S. casualties, convinced many Americans, including President Johnson, that the war was unwinnable.

EM: The antiaircraft threat wasn't as serious?

SA: No, it wasn't. Not again until the '72 timeframe, when they got the shoulder-fired stuff.

EM: Strela [Soviet-made SA-7 shoulder-fired surface-to-air missile]?

SA: Yeah.

EM: In '68, while all of this was happening—I mean before that you were part of an operation, the bombing was increasing in '66, '67, and then all of a sudden Johnson said no, we're going to stop here at the 19th. Was there a morale problem in your squadron, guys saying, "What the hell are we doing this for? Victory doesn't seem to be an option right now."

SA: Yeah. The guys who had made the previous cruise, it influenced them more than those of us who this was our first cruise, our first combat. They reacted more toward that than we did, because we just didn't have the same visceral feeling that some of these guys did who had been in some pretty heavy stuff. Then all of a sudden, their hands were tied behind their backs. Those of us who were just starting to learn combat—well, we've got some combat, we're learning. But the guys who had been in more heavy stuff earlier were saying, "Hey, we were starting to make some progress, and if we have to go back, it's going to be worse than it was when we left. If we are going to go back, then we need to get back there now, before they bring everything to bear on us the first time we have to go in." So there was that friction and dialogue and, like I said, the guys who were on their second cruise were more animated about it than us new guys.

EM: Did anyone in the wing say, "Well that's it. I don't want to fly anymore, this doesn't make any sense to me"?

SA: No, we didn't have anybody. I think we had a couple of guys close, but we didn't have anybody who said that's it.

EM: Anybody relieved, sent back because they were dodging responsibilities?

SA: No, not with us. I'm aware that there were some instances of that, but not in our air wing.

EM: In this first deployment, any memorable missions, things that really stick out as out of the ordinary, if you can call it ordinary?

SA: Well, like I say there were four of us transition guys who went in the VA-55, and we lost one of us fairly early on. Dale Osborne was shot down in September. Dale was flying with the XO. He and the XO were one section, and myself—I'm not sure who was with me. I had the other section. We were on a road recce flight [aircraft operations along road segments to locate and destroy trucks or other targets of opportunity] south of Vinh, as I recall, and I guess we were maybe two miles, three miles apart. My wingman and I had just come off of our bombing run, and the XO and Dale—I think the XO thought he had a truck down there, so he rolled in while Dale stayed high. In those cases we'd leave one guy high so he could watch and tell the other guy, who was rolling in, if he was taking any flak. The XO pulled off, Dale rolled in, and then Dale got hit and went in. We didn't have radio comms with him right away, as I recall now.

EM: He bailed out?

SA: Yes, he ejected just coming off the target. We had a good shoot. We thought that he had gotten to the ground okay, but as I recall we didn't hear from him right away. We heard later from him and then he was gone. We were never able to pinpoint to get in to get him. The rule then was, you had to have good voice contact with the guy on the ground and verify where he was because you didn't want to get a helo in there starting without a specific spot to get to.

EM: Was he taken prisoner?

SA: Yes. He returned with the returnees at the end of the thing. That was the first thing that we had in our squadron, but by that time we'd started picking up some losses. We had one of our F-8s [Crusaders], the F-8s had jumped or a couple of MiGs [Russian-designed fighter aircraft] had jumped them over Vinh. I guess this must have been after there was some relief. This was in the September timeframe, so we were back up in the southern part in Route Package 1 or 2; we were still back up in there.* They got engaged with the MiGs and the young F-8 guy ran out of fuel. He ejected out over the water and we got him back. Then we had D. J. Wright at the end of September, he got hit over land and made it back out over the water and we picked him up, and then a sister squadron, VA-164, lost their XO in October by 85-millimeter up by Vinh.

EM: Quite a few squadron commanders and XOs were killed.

SA: Yeah, we started losing some guys. The lieutenant JG, the F-8 guy who had to eject and get picked out of the water—we now go on into February, and he had an operational loss. He hit the ramp coming back aboard and was killed. He tried to eject but had an operational problem. The ops officer of 164, we lost him when the nose gear collapsed on a cat shot, so again the operational losses, they just continued to go on.

EM: Dangerous business. Personally, do you recall any scrapes with antiaircraft, close calls with antiaircraft fire?

SA: Let's see, on that cruise, no real close calls. I saw a lot of it, but no hits on me.

---

* To improve the efficiency of the bombing campaign, Admiral Ulysses S. G. Sharp, CinCPac, established separate Air Force and Navy operating areas or "route packages" in North Vietnam. Air Force units carried out strikes in Route Packages 5 and 6A, the areas closest to their bases in Thailand, while the Navy executed its attacks in Route Packages 2, 3, 4, and 6B close to Gulf of Tonkin.

EM: None of your planes with bullet holes in them?

SA: Yes, we had planes with bullet holes, but not mine. I was pretty lucky. One of the fun events was when Jack Calhoun and I, we went out one night together. Pretty rare for two senior guys to be flying together, but for some reason we ended up on the night section to go road recce up by a Vinh, which was always an exciting time. You fly with your lights out once you get over the target area, the area where you're supposed to be looking for somebody.

We were running with our lights out and he said, "Well I'm north of Vinh," and I said, "I'm south." We were going to rendezvous; we were going to get back together because we had separated to go after—thought we were going to go after something and nothing was there, so we were orbiting the area.

We wanted to get back together, and I said, "I'm south," and he said, "I'm north," so I said, "Okay, I'm going to give you two blinks of my lights so you can let me know if you see me." I did that, and as soon as I did the whole sky started lighting up. [Laughter] We said, "No, let's do this a different way." [Laughter]

EM: How about SAMs, did you have to dodge any?

SA: On that cruise, I don't think we saw—I didn't see a SAM on that cruise.

EM: But you were below the 19th parallel?

SA: Yeah.

EM: How about MiGs?

SA: No. The closest we came to the MiGs was that episode with the F-8s coming out of Vinh. The great thing about being on the small deck carrier was we had the F-8s and by that time, the F-8s had made their mark as being the MiG killers. The MiGs stayed away from us when they knew we had F-8s with us. [F-8s accounted for 18 MiGs during the war.]

EM: Being the Iron Hand squadron, you were going after radar sites and fire control radars?

SA: Yeah, were going after the fire control radars and the SAM radars.

EM: How successful were you doing that?

SA: We thought that we were pretty successful. We didn't always know whether we destroyed the van or hit the van or not, but we always knew if we shut it [radar] down. We always knew. Sometimes they could see us. They knew when we launched, and sometimes they would turn it off. You still accomplished the mission, but sometimes you don't know—you really want to do two things—you wanted get it shut down, and you'd like to get it shut down permanently, but sometimes the second part is harder. They were pretty clever about coming up and shutting down.

EM: All right, at some point did you have SERE [survival, evasion, resistance, and escape] training?

SA: Oh yes, yes [laughter], we had SERE training. There was SERE training back in ConUS [continental United States]. Everybody had to get their SERE qualification at some point in time in their life back in California. We all had SERE training out in the Philippines. The school was run by some of the natives, which was always an exciting time, because they would drag you out through the jungles, and you'd try to eat iguana and all that kind of stuff.

EM: How did you fare in that training?

SA: It was okay. I wasn't looking forward to it, but I survived it and we had a good time. For some reason VA-55 had, from the previous cruise, invented a fictitious person, and the job of the junior officers was to get this fictitious person every qualification that he could possibly get. I still don't know how they did it, because you had to check in and go through lots of ground school before you went on the survival trip with these guys. They were able to get a certificate of completion for this guy. [Laughter] I'm still not sure how they accomplished that.

EM: Were you locked in a cage and beaten, waterboarded?

SA: No. You were told the techniques. The survival school back in ConUS that was run out of San Diego, I had done that earlier, and that qualification lasted for a long time. By the time Vietnam rolled around, it had gotten a little more aggressive, I think, than when I had been there. But when I was there, they would lock you up in the boxes, beat the boxes, interrogate you. One guy hit me and knocked me clear across the room in an interrogation room. Of course, they knew how to do it so that you didn't do any permanent harm, but it was sort of a shock to me.

The funny thing was, in my group, when I went through there, we had one of the Korean War POWs, and they made him go through SERE school. Well, when it got to the interrogation phase, he had a flashback. He thought he was back in Korea. He bounced back out of it, but for a period of time he was in a different world and just scared to death.

EM: That was the real thing in Korea?

SA: Yeah. The natives there in the Philippines, most of that was teaching you the techniques on how to find food, how to find safe water, how to evade, really more survival techniques.

EM: Psychologically I'm sure you thought about, "Well, I could get shot down." What did you do to prepare yourself for that?

SA: You gave a lot of thought, you sort of had to go through, "Am I going to try to shoot my way out with my little trusty .38 pistol?" You spent most of your time thinking about, "How do I evade. How do I keep hidden." For me, I didn't think I had much of a chance at staying hidden very long.

EM: So was word getting back to your squadron that the POWs were being tortured?

SA: Yes.

EM: That was a common occurrence?

SA: Yes. It was. They kept us pretty much up to date as best they could, without divulging too much about what was happening to the POWs. We knew there was a lot of torture going on. When Dale got shot down, our nickname for him was the Wizard. Dale, he was an intellectual, and he thought of things in different way than a lot of us. He was a really fun guy to be around. We all had a great time with him. But he'd have these visions of things, the way he thought the world was going, and they didn't quite match what the rest of us thought was happening. We were all convinced that if they had really captured Dale and he was in good shape, they'd probably turn him back to us in a couple of months, because they wouldn't be able to figure out [laughter]—he'd be asking them questions. We were all convinced that Dale was going to come back right away, because they would be tired of trying to figure him out.

EM: Did you talk to him after he was released? After the war?

SA: I had one opportunity to talk to Dale, and he was still at the hospital there in Balboa [Naval Hospital, San Diego, California] at the time. He was not too communicative. His wife had left him during the time he was gone. I tried to reach out several times over the years since, through the friends in the POW network. They said they would see him every now and then, but he hasn't reached back. I'm sure it's a—

EM: Rough experience?

SA: Yeah.

EM: All right, so on the 31 of October, Rolling Thunder is over. There's a bombing halt, at least in North Vietnam. What did your squadron do at that point?

SA: This is where we spent a lot of our time with Hillsdale and Moonbeam, the Airborne Air Force Controllers, where we would report into them with our bomb loads and time on station and they would assign us—

EM: Strikes in Laos?

SA: Yes.

EM: In South Vietnam?

SA: Right. We didn't have as much flexibility like we did when we were up north, where you would be sent on a road recce and basically you were given an area where you could work in, and you were free to do whatever, as long as the targets were there.

EM: In Laos you're working against the trails, you're looking for trucks?*

SA: Trucks, yeah, and dumped ammo, you know—ammo dumps.

EM: How effective do you think those strikes were?

* From 1964 to 1972, Navy and Air Force aircraft bombed the Ho Chi Minh Trail to interdict the flow of enemy supplies from North Vietnam to South Vietnam via Laos.

SA: Sometimes you'd have this big surprise. The FAC would say, "We got a truck down here," and trying to see a truck in the jungle—anyhow, they would mark them with a smoke rocket, and you'd come in and they would say, "Okay, from my smoke 100 yards to the left," or north or south, east, west, or whatever, and you'd try to hit it, and then he'd tell you whether you did or you didn't, as far whether you damaged anything. Every now and then you'd have either a bad run and not hit where you thought you were going to hit, or you'd have a bomb hanging up and it would come off late and wander off into someplace, and then the whole jungle would explode. [Laughter]

EM: So didn't hurt to miss once in a while?

SA: Every now and then a miss was pretty good. We had several of those surprises.

EM: But it seems ludicrous that multi-million-dollar aircraft would be looking for a truck.

SA: Yeah.

EM: Now, about your first combat deployment. Were you satisfied with you your A-4 and the weapons performance?

SA: Yes. Like I said, we had the brand new A-4F and it was really a difference. The other two squadrons both still had A-4Es, basically the same engine, but we had better EW gear, and that's probably the reason we were Iron Handers.

EM: After you were below the 19th parallel, was EW as much a concern as up north?

SA: No.

EM: So you did fewer Iron Hand missions and more just regular strikes?

SA: Yeah, yeah.

EM: Did you do close air support in South Vietnam to help troops in battle?

SA: We did. We had some of those where the troops were engaged. We had both ground FACs and air FACs [one-man, fixed-wing planes that identified enemy targets and helped attack aircraft carry out strikes], depending on the situation for those. We didn't do a whole lot of that, but every now and then there would be a call, and we'd be available to go in and do some work for them.

EM: Did you ever get a FAC calling back and saying you did a great job, you're on target?

SA: Yeah, they always gave you some feedback on every flight. Every time you worked with a FAC, they almost always gave you a good debrief on the way back.

EM: Okay, anything additional about this first deployment? Something you want to bring up?

SA: No, I don't think so. It was a good experience. We had several different spectrums of warfare to work with, basically, but we had no alpha strikes on that cruise.

EM: Were the alpha strikes out as a way to do things? Or did they come back later, during '72?

SA: No, they came back later in gangbuster fashion in '72. With the bombing halts and all, we just weren't going—once we pulled back down, then there was no target that was of sufficient definition that it would warrant something like an alpha strike.

EM: Okay, so you're back to the States—you went on leave? How did you feel after that first foray into combat?

SA: I was glad to be home. I knew we were going to go back pretty quick, and I felt comfortable that I had learned a lot, and learned again from the leadership of Bob Kirksey and Jeff Wellings, that this was a game where you had to set a set of rules, because if you got careless, if you just didn't pay attention, things could go wrong. Not only from the warfighting perspective—in other words, you could put yourself into a situation where you got shot down, but if you didn't pay attention to the normal stuff that you've got to pay attention to, you're also liable for an operational loss. A lot of people got caught up in the moment and lost some of the perspective on what it takes to stay alive just flying an airplane. Getting enough rest, making sure that mentally you're ready.

EM: Make sure you had fuel?

SA: Yeah, make sure you had fuel.

EM: How did the weather affect your operations?

SA: A lot. Depending on the time of year you were there, you had a lot of rain or a lot of times a lot of fog. You didn't want to be in a threat zone over an overcast, so if you were where the weather—whether it was you had low-hanging clouds, yeah, you might be able to get to the target, but then you might not be able to see the target, but you also couldn't see SAMs until it was too late. In the recoveries back at the ship—we had some really bad recoveries aboard ship due to weather from low visibility and pushing the limits getting back.*

EM: Guys that had to eject?

SA: We had the ramp strikes at the ship. We had more of those on the second one when we tried to double-cycle the F-8s, put them up on the bar CAP [barrier combat air patrol]

* During the northeast monsoon (November–March), huge clouds, sudden storms, and dense fog significantly reduced visibility over North Vietnam and the Gulf of Tonkin. The southwest monsoon (May–September) brought similar weather to Laos and South Vietnam.

station for two cycles instead of one. So the guys were in the cockpit for three and a half, four hours, and they'd come back and were really fatigued. We started having uptick in the ramp strikes.

EM: Those are almost always fatal, aren't they?

SA: Yeah. Not always; some walked away, but not often.

EM: What was your squadron or your air wing's tactics about SAMs, how did you deal with those?

SA: You always wanted to see the SAM. So the first thing was if anyone saw the SAM, make sure that you got it broadcast out, so that the people knew where the right place to look was. Because the sooner you could see it, the more opportunity you had to evade it. Basically what you tried to do was, get the SAM nose-on if you could, because if you lost trail of it behind you, you didn't know when to make a hard right.

EM: So you would head right toward the SAM?

SA: You would try to get it up into your front quarter view so you had it on either side of your nose. Basically you would watch it until you thought you had to make a move. Then you had to make a dramatic move—and usually it was a down, rather than an up and away—it was a down and under the SAM, so that you would try to get it to go past you. If you went up, you'd let off too much airspeed, so you needed as much energy, because they could pull a lot—the SAMs could pull more Gs than you could. But they were going a lot faster, and so even though they could out-G you, they couldn't always out-turn you.

EM: Did you have to make that maneuver?

SA: Not on that cruise. I did later on. Again, later on the SAM shooters changed tactics as time went on. What they did was, they ended up firing two. They fired one low with a normal low profile, and then coming up at you and hopefully you're paying attention to it, they would launch a second one up and then bring it down on you. So while you're watching and trying to do this—you get it to a point where you can maneuver against the one coming up at you—there's one above you coming down at you.

EM: Like a scissors tactic?

SA: So they got sort of clever about that.

EM: For your bombing operations, what were the altitude restrictions?

SA: We tried to always pickle at 5,000 feet and make sure that we were nose up and had bottomed out at no less than 3,000 feet above the ground. We in VA-55, on all the cruises that I was with in the A-4, my A-4s had the CP-741 bombing computer, which was a device that basically calculated—did the bomb equation for you, so that you didn't always have to be as precise on altitude and airspeed as you had to be if you were doing a manual bomb. In other words, with a manual bomb your mill setting on the gunsight was set for a release at a specific altitude and specific airspeed. The CP-741 let you get the target in your pipper, designate the target, and the computer would say okay, at this airspeed and at this altitude, I will release the bomb at this point.

EM: You would release the bomb, or the system would?

SA: The system would release it. So basically it gave you a little more flexibility than if you were in the bombing run and you had wings level, you were tracking the target, you thought you had a really good solution with the CP-741, rather than wait until the next 500 feet where you were originally going to pickle at. You could pickle right then and just start to pull, and the bomb would come off. We spent a lot of time calibrating them, it was great. So it gave you a little more flexibility to not be in the bombing run as long as you normally would be with a manual run. Basically, most of the time the accuracy was

EM: All right, I guess your next deployment was August of '69?

SA: Yes.

EM: President Nixon was now the commander in chief. Did you think he had a different approach to the war?

SA: We thought that he was in it to win it. We thought okay, we're no longer going to fool around with this thing and we were going to get it done. So we were encouraged with Nixon's arrival.

EM: All right, so in August of '69, this was before Linebacker [bombing campaign], so you're still doing operations in Laos and South Vietnam?

SA: Yes.

EM: Any difference this time from the last time? Still *Hancock,* right?

SA: Yes, still *Hancock.* Pretty much the same. It was pretty much the same where we still had the same people to talk with over the beach. We still had Moonbeam and Hillsdale, and we had some new FACs; we knew names for some of them. Some of them were still the same outfits that had been there before. We had our new skipper, at this time it was Fred Lawler. Fred had served some time with the Marines, knew a lot of the Marines.

EM: He was the CO of the ship?

SA: CO of the squadron.

EM: Lawler?

SA: Yes, Fred [Frederick W.] Lawler. Jack Calhoun and Frank Green and I are still there. Now, we had the young guys from the cruise before, but we'd lost the younger guys who had been there when we first joined 55. We had below us pretty much a new structure for the young guys.

EM: Sounds like there was a lot of continuity.

SA: There was a lot of continuity. It was a very comfortable arrangement, and things were pretty much on automatic, as far as the procedures. We were all very comfortable with the ship. The ship and air wing really worked well together.

EM: Did you notice any difference in the enemy's air defenses on this go-around?

SA: Not a lot. We started getting suspicious about their use of optics. We thought a lot of times we weren't—no matter where we were, whether we were over in Laos or down South Vietnam or in the bottom of North Vietnam, we were getting shot at with—when we never saw or heard anything with our electronics. So we were thinking that they had developed a new optical tracking system that—we weren't sure, but we had the suspicion that they had some better optics than they had before, because we didn't think they were using their radars as much as they had. Intel said no, they're just not using it, because you've got the Shrikes and stuff and they're just trying to protect their radars.

EM: Do you think they got more skilled at the use of their radars?

SA: I think they did, yeah. I do, I think they had a lot of help.

EM: From whom?

SA: The Russians. [Laughter] Yeah.

EM: When you were bombing in South Vietnam, was it mostly up toward the northern part of South Vietnam, or did you get down in the Saigon region? You mentioned the other time that you were around Nui Ba Den, the Black Virgin Mountain?

SA: Yeah. When we were south, very seldom did we go down into anything south of Saigon. We were almost always north of Saigon. The thing that I mentioned the other day about down there in the plantations, that was one of the rare times that we were down there doing some real heavy bombing at that time. A lot stuff off against the western border, over against Laos and Cambodia and over that area. I don't think we were in Cambodia on that second cruise, whether we got there. I know later on I ended up over in Cambodia, but I think that was in '72.

EM: Your cruise here is August '69 to April '70, and we went into Cambodia April or May, or May or June, actually, of '70. You might have already left.*

SA: Okay, so probably did. Maybe had a couple over there. Close to there, there was a lot of traffic on those rivers, over against the river that comes down sort of alongside Cambodia. I'm trying to think exactly where that was.

EM: The Mekong?

SA: What did we call the place over there that was a reference point? Something peak.

EM: Parrot's Beak.

SA: Beak, yeah.

EM: Because it looked like a parrot's beak?

---

* The Nixon Administration inaugurated bombing in Cambodia in 1969, and in the late spring of 1970, U.S. and South Vietnamese troops launched an offensive into Cambodia with the object of destroying or seizing enemy supplies and aiding the pro-American Cambodian government of Lon Nol.

SA: Yeah, yeah.

EM: Well that's all river around there, I mean, it's marsh and whatnot.

SA: Yeah. We did some stuff in through there.

EM: Because that Parrot's Beak pointed right toward Saigon.

SA: Yes.

EM: And they used to come through there and make their attacks.

SA: We spent some time over there. That was good. They couldn't hide quite as well over there.

EM: Do you remember any missions where the word came back that you guys missed, hit some friendlies, or you missed the target altogether?

SA: No. I can't remember the details. I remember we had one scare in the air wing. It wasn't with us, but I think one of the other squadrons had an episode with one of the FACs where they got a wrong vector and there was some concern that they got into some friendlies. I don't think it was ever confirmed one way or the other whether they did or they didn't. I can't remember the details on that. I just remember that we had that one sort of scare where there was some concern that they had got into some friendlies. Fortunately, most of the time with the FACs, I really have to say that the FACs did a terrific job. I really don't know how those guys did it.

EM: They did a terrific job in what regard?

SA: Well, I mean the guy was a command center all unto himself, yet he still had to fly and keep himself alive and shoot rockets to mark the target. They always seemed to know what was going on. I still don't understand why they all weren't shot down, but I guess they had an understanding that if you shoot me down, then everybody knows where you are. I don't know.

EM: Because they were unarmed aircraft, right?

SA: Yeah, they had just had those smoke rockets—

EM: O-1 Bird-Dogs.

SA: And the guys, I think they used to carry a M1 in the cockpit with them. [Laughter] They really exercised great control over us. If they didn't like what you were doing, they'd call you off if they thought you were a bunch of wild Indians up there and not going to pay attention to them. They didn't hesitate to say, you know—

EM: They had the conn [control] in those operations, right?

SA: When you checked in with them, they were the lead. When we were south, we were almost always with FACs.

EM: How did you like Commander Lawler?

SA: Oh good, he was a fun guy. We lived sort of across the parking lot in Lemoore and so got to know his family really well. And Terry, his wife, went with Jennie Lou to the hospital to have Brad, our youngest, when he came along. We still stay in touch with them. They live up in Virginia on a farm. He's a New York boy, I'm not sure how he retired on a farm, but he's up there on a farm.

EM: How did he compare with Kirksey?

SA: Very different personalities. Bob was a very smooth guy. I mean he's very unruffled, very precise, very warm and friendly. Fred was more of the hail-fellow-well-met—little more boisterous, more outgoing, more tolerant of being a little off the edge a bit, but still very, very strict on air discipline. I was really fortunate to be with the guys in those early days, really strict air discipline guys. They really believed in it.

EM: Which of your superiors had the most impact on your development in this Vietnam period, early Vietnam period?

SA: I think probably Bob Kirksey from the standpoint he was the first, and he was also very committed to making sure that you knew what he expected. You just knew that he had a way of doing business that you either accepted or you didn't. If you didn't, then you'd better find some other place. From the standpoint of, I think of being the first, he taught some real good skill sets. Jeff Wellings and then Fred Lawler both followed that trend. They believed in it. As I say, Fred had a little more flair to him. He had a little more outgoing to him.

EM: Who was Jeff Wells?

SA: Jeff Wellings, middle guy. Jeff was a funny guy.

EM: Squadron commander as well?

SA: Yeah. He was the XO and then CO after Kirksey. Jeff had a little bit of comic relief to him. He had a real dry wit and had some good jokes. Again, he was a good, good keeper of the flame. In other words, he had bought into Bob Kirksey's way of doing business and he kept it going. The transition to him from Kirksey was—there was just really no transition. So he really did a great job. I think he lives out in Phoenix now. He made captain, then retired up in the Jacksonville area, then I think moved to Phoenix. Really a good man, a lot of fun, really a lot of fun. He had good wit. Like I say, Fred was—he took on and had a little more color. He was more outgoing, and a little more amenable to having wild discussions about almost anything in the world.

EM: Sounds like you had a pretty close squadron.

SA: We were very close-knit. I feel very fortunate to have been with that group, because I watched other squadrons at that time. None of them were as close as we were. We were really a close-knit squadron. Part of the reason was when VA-55 came back off of the *Connie* [*Constellation*] and I joined them and then we attached to *Hancock*—well the squadrons, the other two A-4 squadrons, were VA-164 and VA-163, who had come off the *Oriskany* and had come off of that really bad cruise where they had lost so many people.* We had those two squadrons who were very close to each other, and we were the odd man out. We were the odd man out because we had bright new shiny airplanes and they had these war-battered E's [the A-4E Skyhawk variants had seen much action]. There was a little bit of showmanship on our part, where we showed off our new shiny airplanes. The tactics—they had grown up in a much more different tactical environment than VA-55 had, although they had both been out on the line at the same time. Kirksey and whoever, whatever the other A-4 squadrons were with them on *Connie,* had a very different approach to warfighting than the 163 Air Wing 16 guys did.

EM: Can you specify? What was the difference?

SA: A lot of the difference was attitude. As far as when we first went to Fallon together for our workups and ordnance delivery and stuff and how to plan strikes and all, the 163/164 guys were still in the mode of—sometimes we want to come in low and then climb in on the target. Kirksey said we're never coming in low. They also had a roll-in

* On 26 October 1966, a sailor mishandled a parachute flare that ignited a blaze on the hangar deck of the *Oriskany,* operating in the Gulf of Tonkin. Firefighters fought the flames while other crew members pushed 343 bombs over the side. The ship was saved but suffered serious damage. The tragic event took the lives of 44 officers and enlisted men, many from the ship's air wing. After repairs in the United States, the carrier returned to battle off Vietnam in June 1967.

where they sort of rolled in line abreast, which we would never do, because that just sort of put everybody in the same—

EM: Sight line of the enemy?

SA: Sight line, so we always fanned, we always did a fan roll-in. They were not as much against multiple runs as Kirksey was. There had to be a special occasion before you'd make a multiple run. For them sometimes multiple runs was just, "Okay, we'll do one run and then we'll come back again."

EM: Do you think that Kirksey was benefiting from their experience?

SA: I think so, yeah, I'm sure he had learned a lot. We had quite a bit of friction right there at the start as we melded the squadrons together. The CAG sort of had to step in, and basically you ended up doing business more Kirksey's way than we did.

EM: The other squadrons, they really didn't want to change, or were they reluctant to change?

SA: They were a little reluctant to change, because that was what they had gone with. Quite honestly, many, many of the folks in the squadrons attributed their high loss rate on the previous crews to bad luck, not bad tactics. A lot of other people thought it was bad tactics, not so much bad luck. A lot of great people, a lot of great people in both squadrons. Some of them I've stayed close with over the years. We did have that friction right at the start, and I think that's why—I think it probably worked in their squadrons as well, but I think it really particularly worked in our squadron to make us so close. We really sort of wrapped around the—

EM: The "band of brothers."

SA: Yes.

EM: Okay, anything you want to say about that second deployment that we haven't touched on?

SA: No, it was more of the same as the first deployment. Fits and starts, with bombing halts and stuff like that.

EM: How many missions had you completed at that point?

SA: I think at the end of that second cruise, I was right at about 250 missions.

EM: A lot of action?

SA: Yeah.

EM: Was that comparable to others in your squadron?

SA: Yes. I think Jack Calhoun and I both were somewhere around 240, 250 somewhere like that. All the guys who had made the two cruises were in the 200s.

EM: All right, we will move onto your time at Lemoore with VA-122, which was May of '70 to July of '71.

SA: Okay, I got orders ashore to VA-122. VA-122 at the time was flying the brand new A-7E. The A-7As and Bs had replaced the A-4s for the most part. We were sort of the last of the A-4 guys as I came ashore. A-7E was a whole new machine. Lots of bells and whistles on the airplane, and everyone was trying to develop how we were really going to use this airplane. Can it be more than just a daytime bomber? Can it be an all-weather, you know, mini A-6 type thing, because we had a very good navigation system, radar, everything in the airplane. We were trying to stretch the boundaries of the airplane. When I had checked into the squadron, the CO was a little put off by me showing up and the fact that there were plenty of A-7A/B guys around now coming ashore, and here was an A-4 guy who not only had to learn the new airplane, but then also had to get up to speed to be an instructor right away.

EM: Who was the CO?

SA: Commander [W.] Scott Gray [III]. I jumped into the fray and got checked out in the airplane, and it was really an amazing airplane. I mean the cockpit was huge.

EM: Did you need a shoehorn to get into that one?

SA: I didn't need a shoehorn. I could get in there, and it was nice and big and comfortable. The airplane had a lot more speed and power to it than the A-4. It wasn't quite as nimble, but it had more power, could carry more ordnance. It had a great weapon system on the thing. A little complicated. I mean now everything was computer controlled and you had to learn the computer and how to type stuff in. That was a little bit different. I was there for just a short time, and then the skipper was relieved by Nick [John L.] Nicholson, who I knew, somewhat, from the Lemoore area. He and I had crossed paths before, and I knew Nick. He was a real outgoing guy. Real people person. He was just like a magnet. I mean he just attracted people to himself. He took over the squadron and sort of changed the nature of the squadron—as far as I think, he started making it more like a fleet squadron rather than just a training squadron. Everybody got back together a little closer. I was still sort of in the weapons business. I was the VA-122 rep to the weapons training school there at Lemoore. I was responsible for the SIOP training and stuff that we did for the A-7E. Skip [R. L.] Leuschner [Jr.] was my counterpart over in VA-125, and he was the weapons guy for the A-7A/Bs. We both worked out of the weapons training center.

EM: Any new weapons coming into the inventory?

SA: We had some new weapons coming in; the Walleye and Bullpup were around still. Shrike coming, anticipating coming to HARM [high-speed anti-radiation missile] later on, but still Shrike. Walleye was there. We'd had Walleye already, in fact we carried some in VA-55.

EM: What was your impression of the Walleye?

SA: Walleye was an interesting weapon. It was a TV-guided weapon. And the big problem with Walleye was one, you needed the light just right for contrast on the target. It tended to walk on horizontal or vertical lines, so if your target was too well defined with a very hard horizontal line, a lot of times it would lock on one end of it and it would just sort of walk down the line. You wanted the sun angle to be about a certain point, which sort of softened those things. It took you a while to get into the scope and get the crosshairs on your aim point. So you had to do that good. You needed to be wings-level and pretty steady flight, and depending on how dexterous you were in getting those crosshairs lined up, you might be in steady flight a little bit longer than you would want to in gun range.

EM: Was it fire and forget?

SA: It was fire and forget, which was the good part. You had to have fairly good luck. It didn't have enough penetrating power for a lot of the things that we wanted to use it for. It never lived up to the capability, but it was certainly a step in the right direction for precision weapons. Something other than a manual delivery. Something that you could lock it up on a window and it would fly through the window. It set the stage for where we went later on in life. I guess that was probably all the new stuff that we were working at that time.

EM: No LGBs? I guess the Air Force had them?

SA: Not yet. No, that came later. In the middle, somewhere along the way, the new commodore there at the base decided that the weapons training school ought to be part of his empire, not a shared responsibility between the squadrons. He commissioned it as ComFAir [Commander Fleet Air Wing] Lemoore Weapons Training School, and I ended up, just because of seniority by a couple of numbers over Skip Leuschner, I became the first OIC [officer in charge] of the Weapons Training Center.

EM: What did you have to do to get that up and running?

SA: It was pretty much in place. All we really had to do was marry up the difference in instructions between the VA-125 and the VA-122 and make it a one-instruction, ComFAir Lemoore instruction, so we had to so some paperwork cleanup. Clean up some of the details about who reports to who and when, and between the squadrons and the now the commodore. It wasn't really hard work, it was just a lot of administrative details to make it into its own entity. One of the criteria was that when the commodore was going to appoint me to be the OIC, he had gone to the skipper and said, "Now, I'm going to do this. You've got to agree that we're going to leave him over here for at least a year. I don't want this to be a revolving door thing, you know, every six months somebody else comes in and takes over."

Skipper said, "Oh yeah, he's your guy for the year." [Laughter] Well, things weren't going as well as the skipper wanted over in VA-122, with the airplane availability and maintenance.

EM: What caused that?

SA: The new airplane, the A-7E, was a bear. I mean it was not very well supported initially. There were lots of problems with parts, spare parts, and there were also lots of problems with some of the pieces not working exactly right. Skipper wanted to work with full systems airplanes. This was a systems airplane. This was an airplane that had all the fancy stuff, and he didn't want to go back to the days of an A-4, where if you had an engine and wings and a pickle button, you could fly.

Things weren't going well. The commodore was all over him about, you know, "You're not getting the people through training on time, and you've got all these derelicts sitting out there on the ramp. Brand new airplanes that can't fly," et cetera, et cetera, et cetera.

So the skipper said, "Well I need help, Commodore, and I'm going to have to renege on my agreement with you. I've got to pull Arthur back here, and I want him to be my maintenance officer. Commodore had a fit."

EM: Who was the commodore?

SA: I'll get all of these down someplace. He finally agreed, okay, you got him. I came back over to be the maintenance officer, which was okay with me because I loved maintenance anyhow, much like I did the weapons business. I love playing with those things. It was okay, and I had a good team, and we were able to get things off and running. This was almost about halfway through my tour there at VA-122.

EM: What did you do to set things right?

SA: Well, the problem was we were close to 100 airplanes—between 80 and 100 airplanes, I guess, and they were all virtually new. We had two versions—again, the Navy procurement system had brought the airplane into being with the wrong engine. They brought it into being with the same engine as the A/B knowing full well that they were going to—they really wanted a fan-jet engine, the TF41, instead of the 30. The TF41 engine finally came alive, so now we had new birds coming with the TF41 engine. Completely different engine in all respects than the TF30. We had that problem of supporting two engines. Some of the first of the electronics in the new airplane weren't as reliable as they should have been. We just had all of these airplanes sitting out there on the ramp, but we could only get maybe two or three airplanes up a day to fly. You just couldn't accomplish the mission at that rate.

The first thing we did was try to take it in small bites. We would say, okay, how do we get one more airplane on the ramp tomorrow than we had today? Look out there on the ramp and see what the easiest airplane for us to fix is. To get it ready to fly tomorrow. That's the way we started. We started this—I just want one more. Find the one that we can—whatever it takes, we are going to, we are going to cannibalize airplanes. I mean, there is just no way around it, we are going to cannibalize the airplanes, but we are going to cannibalize them smart.

Let's not do something really dumb and wreck an airplane and have an airplane that we could never put back together. If we had one of the airplanes out there that had the 30 engine in it and it had enough systems discrepancies that we weren't ever going to get that thing up in a day or two, but we had another 30 airplane that was due for an engine change, there was the mate. We had an engine: extra work, extra time, a real pain in the neck, but there's a 30 engine that we could move into that airplane. It was that sort of thing, and once that got started, it sort of blossomed.

EM: Was the problem that the A-7 was a new aircraft, or there was too much additional material?

SA: It was both. A lot of people thought they had lived through all the growing problems with the A-7, with the A and B. Sadly, that was not so. We still had some of the basic A/B problems with the new airframe with the A-7E that were still maintenance issues. We had a whole bunch of avionics that were completely different, and it was—and they were a stretch. Trying to keep up—like the projected map display, we had a projected map display in there that sort of tracked right along with the radar. In other words, like our cars today with the navigation system. You turned it on, and you didn't need to carry any charts with you or anything else. Sometimes it just didn't work. [Laughter] The skipper's drive to have full systems airplanes, which was absolutely the right thing to do—I mean if the Navy had spent this much money to build a systems airplane, then we were not doing it any favors to start flying it without the systems.

EM: Did that improve over time?

SA: Yes. The lag time was way too long.

EM: One of your responsibilities was to be an A-7 training officer, right?

SA: Yes.

EM: I don't know what you call it, test pilot?

SA: Instructor pilot.

EM: Did you have time for that with your maintenance duty?

SA: I didn't get to fly as much as I would've liked to, but I still flew. You had to learn the plane first. It was complicated enough that you had to really stay on top of it. We'd get new tapes for the computer and stuff like this, and you really had to stay engaged, because the computer was—you could go off and get lost in a hurry [laughter] if you didn't know what you were doing.

We were also, at the time—from the weapons standpoint, the skipper thought we could really turn the A-7E into an all-weather bomber, because it actually had a better navigation system by then than the A-6s did. It was the next generation beyond where the A-6 was. He was saying, "Can we have a single-pilot, all-weather airplane bomber? Can we really do it?" We were working with the INS [inertial navigation system] system, alignments on the INS system, inertial nav system, to see how tight we could make the box that thing stayed in. We were experimenting with taking guys off and dropping the hood on the one A-7 and the guy with the chase, and see if he could get all the way up to Fallon and get on the nuclear bomb track and deliver a bomb with any kind of accuracy without ever sticking his head out of the cockpit.* We were getting pretty good at it.

* The pilot of one plane would be made to rely on instrumentation only as he flew up to Fallon, where he would rely on other instruments to drop his dummy bomb. He would be accompanied by another plane whose pilot would be able to see outside his cockpit.

A lot of people were not in favor of doing that, because they thought it was going to impact the acquisitions system on developing a follow-on to the A-6. In other words, if you could do this with the A-7, why would you want to follow on to the A-6?

EM: Politics rears its ugly head.

SA: Politics started jumping up.

EM: Did the A-7 conduct any night operations or foul weather operations in Vietnam?

SA: We flew them at night just like we flew the A-4 at night, but they did not do the A-6 type missions. We never did in Vietnam, as far as I know, because instead of going back into combat the next time in the A-7E, I went back in the A-4 again. [Laughter]

EM: Your students, were they mostly veterans at this point or transitioning to the new plane, or were they young guys?

SA: It was a half and half mix. You had the returning guys coming back to the fleet who had been A-7A/B guys or A-4 guys, and you had the brand new guys out of the training command just showing up. We had the standard mix of people. One of my students that I took out to the boat for CQ was Bob [Robert L.] Rasmussen. Bob was coming back to be CAG, and so we were checking him out in the A-7, and of course Bob was an F-8 driver, and the A-7 had a lot of similarities to the F-8. Looks similar, but it was sort of pudgy and not as graceful and pretty as the Crusader was. Bob's one of the guys headed for air wing command, and I'm the OIC of the carrier qualification ship. He's out to qualify in the A-7 prior to heading out to his CAG tour, and so he's out in the pattern and he's doing—of course Bob could fly anything, and he's doing great.

While he is out there, I get a call from the beach that says terminate Rasmussen, send him back home. They need him right away. Get him back here as fast as you can. Ras walks into the ready room and I say, "Gee, you really looked good out there Bob, but afraid I'm going to have to send you home." [Laughter]

He said, "What do you mean?" [Laughter]

I said, "Well, apparently they need you sooner than later, so you're on your way."

EM: Where did he go?

SA: You know, I can't remember exactly which wing he was going to, but they needed him right away. I pretended like I was sending him home because he had flunked the pattern. [Laughter]

EM: Pull his leg a little bit?

SA: Yeah.

EM: During this tour, in terms of your family situation—time back in California, and I guess you wife was there? Were all of your children born at that point?

SA: All of them were born. Brad had been born on that previous cruise in '55. The four of them, they were all there. I guess we only moved twice while we were in Lemoore. Still in government housing there on the base.

EM: Did they like the place?

SA: They liked the place. The kids really thrived there. The school was good, the base school was good. A little confining; it's not a big community, and so there wasn't a whole lot for them to do. Although the kids, when you talk to them today, they have fond memories of Lemoore: the swimming pool, the bowling alley, the gym, and their schoolmates. They all had friends whose fathers didn't come back, so they remember all that. But all in all it was very nice living there. I mean when you're only three or four miles from the ops site of the base to where the housing was, once you got home, you were home. It was a quick commute. It turned out well.

EM: I think before we get into your time over there in Vietnam, again with VA-164, the Linebacker strikes and all that, we will have to hold off until next time.

SA: Okay.

[End of 24 June 2015 interview]

Interview Number 3 with Adm. Stanley R. Arthur, USN (Ret.)
Date: 19 August 2015

EM: Okay, we're on the air. This interview is being conducted on the 19th of August 2015 with Admiral Stanley R. Arthur. The interviewer is Edward J. Marolda of the U.S. Naval Institute. Good morning, Admiral.

SA: Good morning, Ed.

EM: Good to see you. We have a day and a half of material to go through.

SA: Okay.

EM: You have a long, distinguished career, so it's a lot of things to talk about. We had discussed the Vietnam War last time, most of the Vietnam War experience, but I really wanted to get to the point where you were in the thick of it toward the end, with Attack Squadron 164, and I guess that was from July of '71 to June of '73?

SA: Right.

EM: So you were selected first as XO. Was this like a culmination of all you'd been working toward, and what was your reaction to that?

SA: Well it was—I think as we ended the last session I was over in VA-122, which was an A-7E RAG [replacement air group], as an instructor, and of course the A-7E was a brand new airplane at that time, the follow-on to the A-7As and Bs. It had all the fancy gear, and I was blissfully playing maintenance officer over there at the time, when I was selected for command. I was an alternate. The detailer called me up one day and said, "It's going to be a while before we can slate you into an A-7 squadron." I'd only been ashore a year or so; it was okay. But he said, "Looks like we're going to have to turn around the A-4 squadrons again. How would you like to go back to A-4s?" I said sign me up, and off I went.

EM: So it didn't concern you that you were going back to an aircraft that you had already flown?

SA: No. The A-4 was such a wonderful airplane to fly, and I knew that I could jump right back in the game with it, and I probably knew—I felt that I probably knew, still, a lot of people over there in the community. It was the same air wing that I had left. I left from VA-55. So going back to 164 was sort of a familiar thing.

What I hadn't really appreciated at the time was the squadron had just been back a very short period of time from the cruise on *Hancock* that I had missed after I left 55 and came over to the shore duty in 122. But they had been basically decommissioned. They were told, "You're going away and you're not"—so all the airplanes were gone except four. The XO from the previous cruise had already left the squadron to be XO in another squadron and fleet up to be CO. They brought him back to be CO. That was Ramsay Lawson, so Ramsay came back to be the CO, and I came over to be XO. Ramsay and I had never interfaced for some reason, and he was a light-attack guy, but we had never had any interface. So that was interesting, getting caught up with him and looking around to see where all the airplanes were. And there were none, except for those four. They had been painted out. All of the squadron markings had been removed and they were ready to be transferred. They put a freeze on it and said, nope, we're going to stand up the air wing again.

EM: What caused this turnaround?

SA: Well, Nixon was president. Nixon, of course, had been very active in trying to push to get peace talks going, and they had sort of stumbled, and his reaction to that was, if you're not going to move forward with the peace talks, we're going to come back and we're going to come back big. Intel had said that the next Tet [enemy offensive], the next year's Tet, the '72 Tet would be bigger than anything that had ever been seen. So everyone was wringing their hands, saying we've really got to be prepared because, as you recall, from a high in '68 or '69, I think, of almost a half a million troops in Vietnam, by '71, late '71, '72, we were down to 125,000 troops. So if anything was going to be done, it was going to be done from the air, not from the ground. I guess the powers-that-be said, boy we'll need more airplanes out there. So they decided not to stand down Air Wing 21 and *Hancock.* We got busy from the time I got there in July. We were able to go through all of the wickets, get the planes back, get everybody—we filled ourselves up with lots of people coming from lots of different places. We deployed in January, so it was a real fast buildup for us, and off we went.

EM: Just in time.

SA: Yes.

EM: For the show.

SA: It was—they were right that things were going to get busy. We left in January, and before we left Hawaii we had already got our first extension. The cruise was supposed to end in July, and we were told by the time we got to Hawaii or just as we were leaving Hawaii it would be August, late August, before we would come back. That was our first clue that things were going to be a little bit different this cruise. The other clue was we traditionally went from Hawaii to Yokosuka for the formal turnover to pick up our EW [electronics warfare] gear and stuff like that up in Yokosuka. But we went straight to Subic Bay. It was a direct nonstop cruise over to Subic Bay. We were in port there for a short period of time, and then we were off over to the Tonkin Gulf. We got to the Tonkin Gulf—we started combat ops on the 8th of February of '72.

EM: All right. So that wouldn't have been against North Vietnam. This was into Laos?

SA: This was mostly in Laos and Cambodia at that time. But there were some up north. In looking at some of the history, I ran across the statistic that said 8 to 13 February was the highest raid counts up in Vietnam up to that period of time. Things really escalated, and they escalated fast. We were on the line for 30 days. We were there all the way up to the 8th of March.

EM: Did you notice anything about the operational environment from your previous times when you were operating Cambodia and Laos? Was there more triple A [AAA, antiaircraft artillery] and enemy defenses?

SA: Yeah. There were a lot more defenses no matter where you went to. This whole cruise was completely different from the two cruises I'd made with 55.

EM: In that regard?

SA: In regard to the amount of AAA and SAMs and stuff that we had seen. We hadn't been over in Cambodia during the other cruises. We still didn't know whether we—we couldn't talk about it. We didn't know if anybody else knew about it. It was a very long haul for us to get over there and back to the ship. When you were going over to Cambodia, you had to get up high and go fast to get over there, and usually had just a couple of passes that you could make and then head back.

EM: These were on the border, targets along the border?

SA: Yes. One of my lieutenants and I, we were over there one day and working with the Air Force FAC. Of course when you're working with the jungle canopy and all, they can see a lot of stuff that we couldn't see, so they would mark the target with a smoke rocket and then we would come in, and they would vector us off of that smoke rocket, you know, 100 yards to the left of it or east of it, or whatever. We were picking up our BDA [bomb damage assessment] as we were coming off target: there are so many killed and destroyed, two heavy movers. What are heavy movers?

EM: Not bulldozers, right?

SA: No. We got back and we were scratching our heads, and one of the guys said, "That's what they are saying when you kill the elephants." My airplane ended up with two little elephants on the canopy. [Laughter] I think it was from that line period. Then we went up to Yoko and had a short in-port period up in Yokosuka, and we came back to be back on the line. Let's see, we came back on the 21st of March.

EM: So that was just before the big Easter Offensive,* which was the end of March or early April.

SA: Yes. The one thing that sort of ended that first line period was Lieutenant Jim [James M.] Loughmiller, who was with me when we were over in Cambodia. We were now up just at the bottom of Route Package 1, sort of over by Khe Sanh [U.S. Marine combat base near the demilitarized zone between North and South Vietnam], and we had been on a strike over there. And coming off of that target I smelled oil, some oil vapors, seeping around through the oxygen mask. We always did a check of the airplanes after we came off of target and are safe and out of triple-A range. One guy would go underneath the other guy and go all around the airplane to see if there was anything. I told Jim I was smelling oil, and I didn't think I'd been hit but he had better take a look.

So he came up and said, "Yeah, I don't see any big holes, I don't think you got hit, but you are venting oil, I can see oil smoke coming out."

So, I said, "Okay let's head for Danang and we will see how this goes as we go along. Well, as we went, as we were headed down toward Danang, my oil pressure started dropping so I said now it's going to be a matter of time whether I make it in there or not, with the oil pressure going down. We were getting pretty close to Danang, and I called and said I needed—we were over an overcast, so we couldn't see the ground, and I knew where Monkey Mountain was, which was north of the field, and we were coming from that way. They gave me a radar vector and then wanted to put me on the GCA [ground-controlled approach]. It's a radar-controlled approach, precision approach.

* On 30 March 1972, North Vietnam launched a massive invasion of South Vietnam, prompting President Nixon to initiate the Linebacker bombing and mining campaign in support of America's ally.

We were going downwind, but they would clear me for the downwind landing and wanted me to come down, to get down on the prescribed glide slope. I said, "I'm staying up here above the overcast right now until I know I've got the field and I know that I'm passed by the mountain, because I'll need some altitude if this thing quits on me."

Finally I'm crossing over Monkey Mountain so I said okay, I'll come down through the overcast now and try to catch up with the glide slope. About that time I get a couple of really bad bumps out of the engine, and Jim Loughmiller says, "You're on fire. There is a lot of fire coming out of the tailpipe. There's all sorts of stuff coming out." About that time I break out of the clouds and see the runway ahead, and I'm over the bay, and the bay is full of ammo ships. After he made that call, I said okay, I'm shutting the engine down, and I think I can make the field. I shut the engine down, and sure enough I was able to make the field.

EM: On no power?

SA: No power. Of course, luckily I had held on to that airspeed and altitude, and so I was able to drop the gear and get in and take a midfield arrestment on the plane. Got off and out of the airplane. They towed it away and I hopped on the—

EM: Had the plane taken fire?

SA: No, the engine had come apart and was throwing stuff out the tailpipe. Got a quick ride back to the ship on a COD and flew another hop that afternoon or that evening.

EM: That was your first real mishap in the air, wasn't it?

SA: Yes.

EM: What was your reaction?

SA: It was one of those things—you're not supposed to do that. You're supposed to jump out of the thing, especially if you're under 10,000 feet. I could see the field. I knew I had plenty of airspeed. I thought for sure I could make it. My only worry was if the airplane was really on fire, so when I finally got in the arresting gear on the runway, I jumped out of the airplane. And I've often wondered if I could ever find the guys that were in the C-130; they were getting ready to take off, coming at me, and they pulled them off the runway. So they were sitting off to the side of the runway, and the next thing they see is this A-4 coming in, catching the arresting gear, and smoke is rolling off of the thing. I crawled out of it. There was smoke all over the place, and started running across the field to get away from it, and then turned around and looked at it and often wondered what they thought, seeing this guy jump out.

EM: The plane wasn't consumed by flames?

SA: No, apparently it was just some residual from the oil and all that had got into the top part of the engine. Lots of smoke, but no fire.

EM: Your wing man, I guess that's standard practice, that you have to stay together whatever happens. He didn't go back to the ship separately. He stayed with you.

SA: He stayed with me, and now I'm trying to remember—yeah, I think he landed with me after I landed. Then he flew back to the ship, and I went back on the COD. We always stick together. Always one guy stays with you. That was sort of exciting.

EM: That was the opening gun to the Easter Offensive for you.

SA: That was the bell ringer. But then lots of things went on after that.

EM: Well, now we are into the Linebacker campaign and countering the Easter Offensive that the enemy launched. What was different this time with Linebacker, compared to your Rolling Thunder experience, in terms of rules of engagement and tactics?

SA: The big thing that was going on at that time was General [John D.] Lavelle, Air Force, 7th Air Force guy, who had instituted what he called protective-reaction strikes.* Basically, prior to that directive your hands were tied relative to what you could strike on an opportune basis. In other words, before that you could fly by a SAM site and see SAMs sitting on the rails, and you couldn't do anything unless it shot at you. You could go by a big flak site and if they didn't shoot at you, you weren't supposed to do anything. Under the guise of these protective-reaction strikes, the authority was if you think you've been shot at, you can react.

EM: A lot of guys thought they had been shot at.

SA: Of course, everyone thought they were shot at. If they saw a SAM site or they saw an AAA site or anything else that looked like it could be a threat, you know—

EM: Was there any baiting, I mean flying really close to the enemy defensive weapons and hoping to draw a reaction?

SA: What we did, one of the things that we did in our air wing was the photo recce flights. Normally the photo recce flights were escorted by another F-8. We had only F-8s on the *Hancock.* The photo F-8 would be escorted by one of the fighter squadrons with their F-8s. We started escorting on some of the routes that looked lucrative for targets. We started escorting them with A-4s with a lighter bomb load, so they could keep up with the photo bird and so we could react; if we saw something on the route or he saw something on the route that was of significance, we could strike it right then, which was completely a departure from what we had done up to that point in time. So there was a lot of relaxation in some of the rules of engagement. Not as much as we wanted because during the bad weather days, a lot of times we would get over our primary target and couldn't get into the primary, and we always had secondary targets.

---

* Lavelle was removed from his position as 7th Air Force commander in 1972 and forced to retire, due to alleged misconduct over bombing missions in Vietnam.

I was leading a strike one day that got close to the Thanh Hoa Bridge, and west of Thanh Hoa was a beautiful airfield. I guess it must have been 20 miles or so east of Thanh Hoa. Our target was weathered over when we got there, and we couldn't get into it. And they're shooting SAMs and we couldn't—you don't ever want to be above the cloud base with SAMs underneath you. I wanted to take the group farther east, and we'd make our alternate target instead of the one that they had given us: this airfield, which was sitting out there and looked like a ray of sun was just shining right on it.

EM: To crater the runway, or were there aircraft there?

SA: I couldn't see whether there were airplanes there, but I assume there were. We would go in to try to crater the runways. We couldn't do it, so we had to find something to just sort of dump on, something that was safe so we didn't have to fly back over the overcast when there were a lot of ordnance and SAMs chasing us.

EM: Going back to the protective-reaction strikes, did you ever feel that you had been shot at and then called in a strike?

SA: I don't think I ever called back for a—we did strike, we did strike some things that we saw on the way. I don't think we called for anybody to come back and go after it anymore. There may have been one. We got engaged one time, when all of a sudden out of the clear blue sky, we found a daggum train when we thought there were no more trains left. We had an "ollie, ollie, all home free" on that and sent back word to the ship. The next strike, we changed their targets and came in after the train.

EM: So you knocked out the train, pretty much?

SA: Yeah.

EM: Where was that? Do you recall?

SA: That was up north. It was south of Haiphong, maybe 10 miles or something like that. We came back; the second line period was like you say, that was the big offensive, everybody kept waiting for it and waiting for it and waiting for it, and then finally it came and it was Linebacker. We lost men in the air wing there; we had some problems with the other squadrons. Hank [Henry H.] Strong [Jr.] was skipper of 212. He was leading an alpha strike into Vinh, and he had the division ahead of me. He was the lead division, and I'm number two, so I was coming down the same pipe as he was, and I had two other divisions from our squadron; it was a pretty big strike.

They were over to the right of us. Hank got hit on his bomb run and went straight in. So I'm coming off target and he's down. Out of the corner of my eye I see an airplane, I felt must have been one of ours, and it was Lieutenant Rob [Robert] Barber. He had taken a hit in his centerline drop tank, and he looked like a flaming rocket. The gas had ignited. We were pointed toward the beach so we kept going, and his wingman was with him, and so he headed out over the beach.

EM: Punched out?

SA: No, he stayed with the airplane, and as soon as all of the fuel came out of the tank the flames went away, and so the guy said stay with the airplane, and he stayed with it and brought it back. Skipper of 55, George J. Fenzl, was hit but took a hole, I'm not sure where, I think it was in his wing. Vinh was always—during the whole war, if you ever wanted to see somebody shoot at you, all you had to do was go to Vinh. And they just shot the living daylights out of you.

EM: Quite a few squadron COs and XOs ended up POWs or killed. Why was that? Because they fly right up front, leading their group?*

---

* Killed in the line of duty during the war were 67 wing and squadron commanders and squadron executive officers.

SA: Yeah. They are up front and they are usually flying a little bit more than the other guys, especially on the big strike days, you know, the lead is always the senior guy.

EM: When you're the lead, you actually are in the front of the formation?

SA: Yes. We changed that some later in the war, to give our fighter guys more opportunities when all the MiGs were going for them, get them some recognition, and although they didn't lead from the front, they put the strike package together and briefed the formations and how we were going to get in and out of the target area. But they were not actually in the front of the group.

EM: You hear a lot about by the time of Linebacker, the A-4s weren't operating as much up north because of the more lethal environment, but obviously your squadron was up there. What's the story there? Were you one of only a few A-4 squadrons?

SA: Well, yeah. We were the three only A-4 squadrons left out there at that time.

EM: These were advanced A-4s?

SA: Yes. All of us at this time had A-4Fs, so we had the latest airplanes. They liked to keep the big deck carriers up north more than they did the *Hancock.* We did our time in the barrel up there quite a bit. We flew lots of—lots of strikes, a lot of alpha strikes, three a day, so we were up there. The interesting thing is, we never saw a MiG. Of course our F-8 guys said, well, you know, the MiGs will never come after you guys because they learned their lesson back in '67, '68, and they didn't want to fight the F-8s. They would fight the F-4s but they wouldn't fight the F-8s.

That's a little prejudice of the F-8 community. There was some truth to it. I mean, we would be up there flying alpha strikes right along with the big decks [carriers], and we would go in and we would never see a MiG. They would come, and they would see MiGs. In fact, the MiG activity was a lot greater than we recognized, because we just didn't see them that often. But the fact is during that whole war, we didn't have our first ace in the war until Linebacker.

EM: [Randall H.] Cunningham and [William P.] Driscoll.

SA: Cunningham. And then that was followed in August by Steve [Richard Stephen] Ritchie, an Air Force guy; he and his RIO [radar intercept officer, Charles B. DeBellevue] got five. Actually RIO was really some sort of guy, I guess. His weapons officer had also flown with another guy who got a MiG, so his RIO or weapons guy had six MiGs and Steve had five. [Laughter] But you know, both aces from then, only two aces or four aces from Vietnam from the U.S. side, all got their MiGs during this period, during this cruise.

EM: Do you recall the period of the Christmas bombing, when our B-52s were hitting Hanoi?

SA: Yeah, but we were gone, you mean in '72? We were gone by that time.

EM: This was in late '72, November, December.

SA: Yeah, we left, we got there in February of '72 and we left in September of '72.

EM: Okay, well that was a peace-is-at-hand kind of pause there, before the second strike period.*

SA: Yes, I recall the B-52s.

EM: How was the morale in your squadron during this Linebacker phase?

---

* On 26 October 1972, Henry Kissinger, the national security adviser and primary negotiator with the North Vietnamese, announced at a White House press conference that negotiations were going so well that "We believe that peace is at hand." Nixon then ordered a bombing halt.

SA: It was really good. We had a lot of young guys, and we had some good steady lieutenant commanders in the department head slots. It was even more so—as I have gotten older and think back on what was going on at the time, these young guys were just extraordinarily good. They stayed focused, they were easy to lead. They bought into our program. The squadron Ramsay and I had, both of us were sticklers for flight discipline, that there was not going to be any, no cowboy acts; this was serious business. As I used to reflect to them, I said, it's like playing poker. You never throw away an ace.

I said, if you are over the beach and you either bust your altitude limits or you're making multiple runs in an area where you shouldn't be making them, you're just throwing away an ace. You need every one of them in your hand and not in the other guy's hand. So you've got to pay attention and believe us when we say this is the way it's got to be done. And they bought into that. They were really good.

EM: How was their training? I mean, Strike U [Strike Warfare Center, Fallon, Nevada] had been established. Did that make a difference in their preparation?

SA: These guys had not—when we formed up, when the squadron got ready to go back on this hurry-up, we've got to turn this around and got to stand the squadrons back up, these guys all came from various and sundry places. Some of them were straight out of flight training. Some had been instructors. We had one guy who was a transition guy from the old HAL-3 [Helicopter Attack Light Squadron 3] community, helo gunships down in the [Mekong] Delta. They were sort of a pickup game, and most of them were rushed through the RAGs to get finished up.

EM: So they had to learn on the job.

SA: A lot of them learned on the job. We had a cadre of folks from the earlier cruise that had not left yet. It was interesting to me at how well they came together and how well the training went. We had only a short turnaround time in the States for training. But the weapons detachment that was sent up to Fallon was really one of the best I've been on. It was really a pleasure.

EM: You came in. You were XO and then you became CO. At what point was that?

SA: Let's see, on the first of July of '72. I became the CO. Ramsay left, and we did a change of command in our flight suits in the wardroom, and Ramsay got on the COD to leave the ship, and I jumped in an airplane to continue the strike for the day. It was a quick turnover. And Fred [Frederick J.] Gosebrink came in to be my XO. He had just been promoted to commander. Fred and I had not crossed paths anywhere before. He had also been a VS guy in his early life and transitioned to jets, so we had that in common.

In fact I just wrote him an email the other day, because I had been thinking about all of this stuff that we are going to talk about today. I said, you know, thinking about the environment that you had to jump into as XO, if you remember flying, everybody flew two hops a day.

We had a rule that we had three man-ups. You could man a spare only if you hadn't already flown your two flights. So we always tried to get the guys, because we were short of pilots and we were flying at such a high rate that we had to get three man-ups out of almost every one of us. So you'd man a spare in the first flight of the day, and then you would—if you didn't get launched as a spare, then the next, you'd get two more flights.

EM: What time were the launches?

SA: We ran a 12-hour cycle and then we rotated the cycle around. You'd have the day cycle, six to—well they changed the times, but usually an early morning start would go through to about the first night recovery, and then that 12-hour period, we used to call them midnight to noon, noon to midnight. They did rotate at that time. You would be on the day cycle or the night cycle.

EM: You launched at what, 9:00 in the morning first strike, and it took you what, two or three hours to carry that out?

SA: What you do is, if you launch, let's say you went on the 0900 launch, that recovery time was 10:30, an hour and a half after you launched. You would man-up the airplane at 8:30, launch at 9:00, recover somewhere between 10:30 and 10 minutes to 11:00 or so, and get that recovery period in.

EM: Grab lunch?

SA: Come down, debrief, grab lunch. Start another brief. You started your brief an hour and a half before your launch. If you did a spare and two man-ups, you almost made the whole day. [Laughter]

EM: That's a full day.

SA: Yeah, a full day. We limited the guys because we did full line periods. I don't think we ever had a line period less than 30 days. Guys were pretty well beat after that.

EM: All right. What was your reaction and that of your squadron mates to the Paris Peace Agreement? The war is over? What kind of a feeling?

SA: For us, we left in September. I think we got back in October when we finally landed at Lemoore.

EM: You would have been gone at that point. I'm just thinking your reaction to the news.

SA: We were—I don't know. It wasn't the same elation. I can't—there was just some, there was a little bit of—well, it's over, but what did we get done? A lot of reflection about things in the ROE [rules of engagement] that bothered us all the time we were out there no matter which regime we were under. Of course, at the end it was Nixon and there was a lot of relaxation. But there was still so many dumb things that we did. I remember when I flew my 500th strike, which was just a couple of weeks before we left the line for the last time, it was in September of '72. Where we struck that day was within visual range of the first place that I had ever dropped a bomb in Vietnam. It was down south of Vinh and I thought, well, 500, that's a lot of missions.

EM: Still hitting the same place? [Laughter]

SA: Still hitting the same place. [Laughter] Certainly we were glad that it was over. It had gone on too long. It wasn't like VJ Day [Victory over Japan], or we'd remember from our youth. There was none of that excitement about it at all.

EM: In hindsight, here we are 50 years later, what's your view of the Vietnam War? Did we go in there for the right reasons or not? Was it worth fighting?

SA: We talked a lot about that right after, because right after my change of command, the next July, I came back and went to ICAF [Industrial College of the Armed Forces], and while I was in ICAF we talked a lot. We had the POWs back in our class with us. We were all talking about stuff. I was always a great defender of the fact that we went out and did what we were supposed to do. That the domino effect was truly—there was some validity to that. It had to be stopped someplace, and this was probably as good a place as any to see if we could make that happen.* It would have probably been better if it had been some other country, something more on the national scene other than that place out there in the jungle. But I thought it was important that we made that stand.

As the years have gone by and you take a look at the cost, and you take a look at the government that we were trying to hold together down in South Vietnam, it just didn't all fit together. It just didn't fit together.

---

* President Dwight D. Eisenhower first referred to the domino effect in 1954. He suggested that if one country (such as Vietnam) fell to communist aggression, its neighbors would fall one after another like stacked dominos.

EM: On the operational level, can you talk a little bit about jointness? How did it work, or did it work? Did you operate with the Air Force to any extent?

SA: We did a lot of work with the Air Force, and one of the things that I really—we will go back and cover, is our airplanes were unique. We had the only Navy A-4 squadron that ever went out to Vietnam that had a laser capability. We had a laser spot tracker. The Marines had been working it. They hadn't taken it to war yet, but we got it and took it for the first time. It almost looked like a Sidewinder [air-to-air missile] nose in the nose of our airplane, and it could see laser energy. If somebody had a laser designator and said okay, I've got my designator on the target, you could pitch over and get it in the field of view of your nose, and it would reflect up into your gunsight, so you would have the target right in your gunsight.

Then you had laser-guided bombs, so as long as you were in the right envelope for the bomb, which was easy to do, you could release and it would hit that spot. We didn't get those until July. The trouble was we had no designation capability. We didn't have any laser guns. China Lake had been working on a hand-held laser designator. They sent a tech rep out with two boxes, and they were boxes like big brownie camera boxes, and we got two combat-configured TA-4s, the two-seater A-4s from the Marines, and brought them aboard ship. The training command had been bringing TA-4s back and forth on ships to CQ the students with. But the fleet hadn't used them on board the carriers for anything. We had those, but we didn't have enough pilots to spare to have a guy in the backseat of the TA-4s. We went around the air wing and said who wants to fly over the beach and get some extra strike flight air medals? We had E-2 guys and the photo recce guys from the F-8 squadron, and volunteers from the fighter squadron themselves.

EM: You had no trouble getting volunteers?

SA: No, no, they said they would love to come. We taught them about the gun and then, it was sort of heavy, so the guy who was flying, you'd have to time it so that he would be up over the target in a zero-G maneuver, so that the guy could hold the box steady [laughter], and he would illuminate the target, and by that time we were at our roll-in point, and then all we did was we rolled in, we would pick up the laser spot on the gunsight.

EM: Was that effective?

SA: Very, very. The Air Force loved us. The Air Force had been doing laser bombing for quite some time, but by this time, I found out after the war, they had only had six designators in-theater, and two of them had been shot down. So they were running out of designators. If we could show up with a designator, that was good for them. For us, if we could show up where we could see their designators—and we were the only guys—I think the F-4s started to get some capability a little bit later. They were happy to see us.

A lot of times we would just go over and if we were in a safe area where there wasn't too much AAA and we had somebody who could look out for the guy who was in the run, we would go single bomb at a time, which is something we didn't like to do at all. But because we could pick off the little "wiblicks" [water-borne logistics craft]—those guys, we could pick them off one at a time. It was just like shooting—

EM: They were still running those boats down the coast?

SA: Yeah, yeah.

EM: That's amazing.

SA: And the little rivers. One thing that we couldn't do was—well, we didn't have the designator capability—we got into this thing with the Chinese merchant ships offloading, so we were out there at night trying to bomb those daggum guys shoveling the stuff to the beach. We couldn't hit them while they were too close to the merchant ship, because if you did any damage to the merchant ship that was going to be an international incident.

So we had to wait until they got far enough away that we could hit them. You had to bomb underneath the flare, which is just crazy kind of stuff. It was not a very effective program. We spent a lot of time being around these daggum merchant ships.

EM: Was it hard to target the shuttling craft?

SA: It was. Of course they knew our cycles, so they could wait us out. They would keep the guys alongside. As soon as we left station, well then they would race those guys in before the next guys could be on station to get them. So it was a cat and mouse game. It was not much fun. We didn't like those flights at all.

EM: Well, they had to do that was because we had mined their ports, which made a big difference. Did you notice that? I mean, what was your reaction to the mining operation?*

SA: Luckily we didn't participate in that. Those mines are not much fun. Like when we rushed out there, the thing that we had to check off in the boxes as we came into Subic Bay after the race across from Hawaii, we had to do a MINEX [mine warfare exercise] exercise out there in Subic Bay to show that we had the wherewithal to lay a minefield. That was something that none of us were excited to do.

EM: Why is that? You're not hitting, seeing something explode, or—

SA: You don't see anything explode.

EM: Right, just plops in the water.

---

* On 8 May 1972, Task Force 77 aircraft mined North Vietnam's primary port, Haiphong, and in following weeks the other ports through which North Vietnam imported 85 percent of the munitions the country needed to fight the war. For the remainder of the conflict, no sizable merchant ships steamed into or out of North Vietnam.

SA: Plops in the water. You're flying this nice straight line and dumping stuff along the way.

EM: Were you ever involved in the strikes on the Thanh Hoa Bridge?

SA: Oh yes, yeah. We designed our own special—and I'm still convinced that we, although an A-7 squadron now takes credit for dropping that.

EM: Well, the Air Force takes credit too.

SA: Yeah, they take credit too. What we did was once, we got comfortable with the TA-4 and the operators in the back being able to hold a steady spot, we designed a special mission for the Thanh Hoa Bridge. What we did—our normal configuration for our flights was a centerline drop tank, and then two 500-pound bombs on a multiple ejection rack on the inboard wing stations and one 500-pound bomb each on the outboard wing stations. So basically you had six 500-pound bombs and the centerline drop tank. So for this special mission on the Thanh Hoa Bridge, we took off our drop tanks. We put a 2,000-pound bomb on the centerline, a 1,000-pound bomb on each of the inboard wing stations, and a 500-pound bomb on the outboard wing stations.*

EM: That's a big load for an A-4, isn't it?

SA: Yes. So we had a 5,000-pound bomb load on the A-4. We had the laser designator guy who could be higher, so his major threat was a SAM or 85-millimeters, and there were plenty of 85s around Thanh Hoa. We had it all timed out, and all these A-4s with this—what we called the slick wing load. We just pounded that thing. It was a thing of beauty.

---

* Air Force and Navy planes launched hundreds of attacks against the Thanh Hoa Bridge (called by the Vietnamese the Dragon's Jaw) during the Vietnam War but failed to permanently disable it. Finally in 1972, during the Linebacker Campaign, precision guided weapons and advanced conventional bombs enabled the two services to put it out of commission.

EM: Did the BDA show that you had knocked down some spans?

SA: Yeah. We had the BDA that said we had knocked down some spans. I don't know what happened but we all enjoyed that flight. The trouble was, with Thanh Hoa you just never knew, it was sort of like Vinh—anytime you were around the Thanh Hoa Bridge, somebody was going to be shooting at you with some big stuff. We were all in and out, no problem drops. It was one of the fun flights of the war, was that slick wing, because then when you came off target, I mean, you were as light as could be. You thought you were in a whole different kind of airplane. It was so fast and slick.

EM: Did you personally run up against SAMs?

SA: Yes.

EM: How did you deal with them? What was your tactic for evading?

SA: The big thing with SAMs was you had to see them. You had to translate the warnings to seeing them, and if you could see them, then it became an energy drill. You maintained as much energy as you could until you had to make a maneuver against the SAM. If you made the maneuver too fast, the SAMs could react to it. It was sort of a timing game. You didn't want to get your nose up in the air, because if you starting giving away energy you were going to get hammered with these things. Normally what you would do is you would work it, keep it as best you could, keep it on 30 degrees either side of the nose of the airplane, and then as it got closer, then you would take a hard break down, which was usually the best way to do it. It couldn't follow you. If you did it too early, you—

EM: What about your onboard ECM?

SA: That gave you the warning of the search and the launch so you'd know, you'd end up knowing where to be looking, but you really—you really couldn't evade it with the maneuver unless you had it in visual range.

EM: You didn't have any countermeasures like chaff [aluminum strips to confuse radar]?

SA: We had the chaff and we did have the countermeasures, and of course you would, any of your maneuvers, you would use the countermeasures. The one thing that we all became concerned with as time went on, especially toward the end of our period out there, was we were seeing the SAMs before we got any warning, and we had good EW gear. We religiously never went over—if the gear didn't check out, we didn't go. Some squadrons didn't rely that much, didn't make it rigid. We did. We absolutely—it had to pass a go-no-go check or you couldn't go. You'd stay above the ship.

We became convinced that they were optically guiding some of them. That they had a scheme where they were getting enough tracking information that they could launch on a visual and guide. I still don't know whether anyone has ever dragged that down to know for sure. The other thing that they did late in the game was, they started to counteract the ability of watching it visually and making that last-minute break on it as they started launching two. They would launch one low and then immediately follow it by one high.

You could almost always pick up the low missile, because it came off low and you would usually see the plume from the launch pad, and so you would be focused in on it, and in the meantime the second one they had launched up and were bringing it down on top of you. If you were maneuvering just against the low guy, you might be dead meat for the high guy that was coming down at you that you hadn't paid any attention to. We had to make sure that we didn't let the strike group focus in on: hey, it's just the low guys that you're looking out. There's guys up above you that are coming back down that you need to be watching for.

EM: Did your squadron lose any to SAMs?

SA: We didn't. The air wing did. The air wing lost some to SAMs. Our squadron was very fortunate. Ours was the seventh combat cruise for the Ghost Riders [squadron nickname], and it was the only combat cruise that we brought back every single airplane and every single pilot. The only thing that we didn't bring back was one of our linemen who apparently fell overboard our last night in port in Yokosuka. We didn't discover him missing until the muster after we got under way. People had seen him and thought that maybe he was still out in town and hadn't got back to the ship in time.

EM: Was his body recovered?

SA: Yes. I guess maybe a week or so after we left Yokosuka, they found his body underneath the pier. He apparently had got aboard ship and for some reason—normally at night, to board ship in port you didn't let people out on the catwalks after dark. Apparently, from where he was going back up to the line shack, he got out on the catwalk and fell over.

EM: What was your observation about your time as CO of the squadron? Were you happy with your performance?

SA: Yeah. It was one of the great adventures of a lifetime, really. We were very active. We had a lot of stuff happen. We won the CNO [Chief of Naval Operations] safety award. We won the Battle E [efficiency award], we did all the things that you want to do if you can do it. Plus we were able to acquit ourselves very well in combat, and being able to come home with all the airplanes and all the pilots was good. The rest of the air wing was not that fortunate.

I mentioned when I was in 55, the cadre of lieutenant commanders that had joined there who were all transition guys, and we lost Dale Osborne there on the first cruise. He became a POW. Then when Hank Strong, the skipper of 212 was lost, his replacement was Frank Green. He came in to be XO behind Duke [Robert W.] Peacher, who moved up from XO to CO of 212. Frank Green had been with me in 55. He was one of the transition lieutenant commanders when we all joined together.

So Frank came out to take the XO spot, and he'd been out on a flight, or I'd been out on a flight and came back in, debriefed, came over to the "dirty shirt locker" there in the wardroom to get a bite, because I was going back out on another hop, a night hop. Frank was also going out. He'd been out there about two weeks at that time, so we hadn't had a whole lot of time to talk.

We are sitting there. He's getting ready to go to brief, and I'm just getting caught up so I can go brief later on. We're talking, and I say, "Well, Frank, it's a little bit different war than we were in with 55."

He says, "Yeah, it sure is. It's busy."

I say, "Yeah, it is. How are you, are you doing okay? Is everything—"

"Yeah," he says. "Things are good. Good squadron. Good people. I'm a little—I'm not as comfortable at night as I'd like to be."

I said, "Well, it takes a little time."

"Yeah, I know. Take it easy."

He went out and that was it. He ended up—he was seventh spot on the mission.

EM: He was killed?

SA: Yeah. That was a real downer for me. He was a lot of fun, and it was too bad, well, it's always too bad, but I was interested to see the matchup between him and Duke Peacher. They were sort of like two peas in a pod, and they were both very trim, smaller stature guys. Both dark hair. Both outgoing personalities, strong personalities, and I thought this is really—they'll have a real good time. But nobody really got a chance to know him. He was just there that short period of time.

EM: How many other naval aviators got over 500 missions?

SA: I think there's probably—there's Denis R. Weichman, who had 700 and some missions. He had some missions with—I'm not sure, I think he flew with Air America or one of those in Laos at one period in time. And then there's Dick [Richard L.] Kiehl. So I think in the Navy, I think there are probably three of us over 500.

EM: That's quite a record.

SA: The Marines have a lot of guys up there in the 600s, 700s with the helos. A lot of Marines did several tours as attack guys and then came back as helo guys, or were helo guys and came back as attack guys. Of course, the helo guys—

EM: Shorter hops too, right off of Danang.

SA: They racked up a lot of stuff.

EM: Well, is there anything else you want to talk about of your Vietnam experience?

SA: There is something here that I want to—

EM: How was the intelligence? Were you happy with your pre-strike information?

SA: Yeah. The intel was good. For the most part we had good stuff. Our photo recce guys gave us good stuff on the pre-target. We usually had good photos of things like that. I was pretty happy with it. One of the other things, when An Loc and when that big offensive [North Vietnamese Easter Offensive] started, that's when I first saw the shoulder-fired missile.

EM: The Strela SA-7.

SA: Yeah. The SA-7s. That brought a new emphasis to the game. And you saw those a lot down south. You didn't see them so much up north, but you did see them, they brought a lot down with them when they came with that offensive back over in Laos and the [Ho Chi Minh] trail and stuff like this. You saw a lot of them especially in An Loc and Loc Ninh. I think the first day that we were told they were seeing them in theater we were at Loc Ninh. Sure enough, they were there.

EM: They were firing up against you?

SA: Yeah. Basically what we did then, what their technique was, they had to see your tailpipe if they were going to hit you with those things. They couldn't shoot at you as you were coming down the chute [following the flight path to conduct a strike]. They had to shoot at you as you were going off. Once we had seen them, then what we did was we made sure that there was nobody coming off target without somebody watching. We always had somebody designated to watch that section come off target, so that we could see if there was anything launched after them. If you could see it coming they could pop their flares, and we had countermeasures and they could also break a different way. They were wicked. They caught a lot, they didn't catch any of us, but they got a lot of planes there for a while with those things. That's something that we are still living with now.

EM: It's a lethal environment.

SA: It sure is. Going into places you don't really know.

EM: Yeah. We talked about the various aircraft. You never did fly an A-7 in combat, right?

SA: Never flew an A-7 in combat.

EM: Or an A-6?

SA: Or an A-6. The big deck guys—we were always away from the big deck guys. Let's see. We saw a lot of SAMs. We saw lots of alpha strikes. We saw just about everything on that cruise. We came back and then we went on another cruise while I was still CO. I left it as we got to Subic Bay. Those guys went on to—the peace accords were made, and then they had the Saigon extraction and stuff like that. They got involved with that.

The next-to-the-last flight that I had was—we still had our T-birds, our combat-configured T-birds, and so the day—I was going to fly the next day, which was going to be my last flight before the change of command. I called up the skipper of the ship, Jack Monger, who was an old A-4 driver. Jack was really one of the great guys. He just passed away last month. I say, "Skipper, how would you like to go flying today?"

He says, "What do you mean?"

I say, "We've got these TA-4s down here, and maybe you would like to take a flight off of your ship and fly around in a TA-4."

He says, "Yeah, I think I'd like to do that." We made sure he had all of the gear and got him down and got him briefed. We go up to the airplane, and we are walking around pre-flighting it.

I said, "Go ahead and crawl up and get in the front seat."

He said, "Oh no, no, I'm not going, what do you mean get in the front seat?"

I said, "Yeah. You can be up in the front seat. If you're not comfortable when we come back to the ship, I can land it from back here. You can be up there and take a catapult."

"Oh, okay." So he gets up, and we go flying, and he gets up and does some slow flight up there and gets back the feel of the airplane and the landing configuration.

We come back to the ship and I say, "Well, looks to me like you're ready to land this thing, so the first pass is yours."

He landed that thing. I tell you, he had a smile so big on his face. It was one of the real joys of life because he had been—he was just a magnificent skipper, and he's out there with all the big decks that had admirals on board. He didn't, so he was our guy. It was really good.

EM: You made his day.

SA: Yeah, yeah. And the squadron loved it, being able to say the skipper flew one of their airplanes.

EM: All right, so you're back to the States. The war is over. Did you take some leave and see the family?

SA: I left the squadron and came back, and we're headed for ICAF. Now, the interesting part of that was I get this call—I think we were in Subic when I got the call. The first stop in, just as we got to Subic out there. They said, "You're going to the Naval War College."

I said, "Oh, okay." I called my wife and said we're going to the Naval War College. She said, where is the Naval War College? I said it was up in Newport [Rhode Island].

She said, "Newport? I think I know somebody up there." So she called around and they said, oh you're coming to Newport. She said yeah, that's what he says. Well, housing is really tough up here. You've got to get on the list in a hurry, and most of the places are taken by now. She said, "Give me some hints on who I should talk to." So she scurries around and she finds a place to rent. She has a deposit on it and everything.

Then I'm talking to my detailer, he calls again and says "You're not going to the National War College."

I said, "National War College? I thought I was going to the Naval War College."

He said, "Oh no, no, you're going to the National War College."

I said, "Uh oh."

"But you're not going there now. We've rearranged the slate because we've got the POWs coming back, and most of them—we are going to try to get them into the service schools. There are some of them that we would prefer to go to National, and since you're a weapons system acquisition manager [WASM] subspecialty, we want you to go to ICAF. That's more in line with what your subspecialty is instead of political science. You're a WASM." In fact I was a WSAM 2, which was the next level up. "You're going to go to NavAir when you finish that to pay back your education over at NavAir.

I said, "That sounds fine, because I've done a lot of work with in my VX squadron days, with Pax River and Naval Air Systems Command." That was fine with me.

EM: Were you happy to come off the line? You had done a lot of sea duty, combat duty. Were you ready for a change?

SA: I was ready for some shore duty, yeah. I was ready, because through the '60s almost—I started the '60s on cruise. Finished the cruise in '62. Went to PG School for two years. Went to VX-1 for a year and a half, and I was gone almost all the time I was in VX-1 on weapons projects either in France or White Sands or Point Mugu or Pax River. Then I had ship's company on *Bennington,* so a cruise on *Bennington* and two cruises on *Hancock* in 55, and then right there in the early '70s, I'm back on the combat cruise. I started the cruise there in '72 and come ashore there in the summer of '73.

I was ready for a break and some time with the family. I called Jennie Lou and said, "We're not going to Newport." [Laughter]

She says, "I've got a house rented up there."

I say, "Oh, get out of it because we are not going there. We are going ICAF."

"ICAF? What's ICAF?"

"Industrial College of the Armed Forces. It's at Fort McNair in Washington, D.C."

"Oh, so we're going to Washington."

"Yes."

"Okay. Well, I think I know somebody there."

So while I'm doing the change of command and stuff out there and flying back to Lemoore, she finds this house out in Fairfax and we are buying a house. [Laughter]

EM: Sounds like she knew people all over the Navy.

SA: Yes, she did. She says, "What do you think? Should we go back and look at it first?"

I say, "I don't think we've got the time. We've got to get packed and get moving."

She says, "I really trust the taste of this lady. If she says she would live in it, I think I would live in."

I say, "That's good enough for me." We bought it sight unseen.

EM: Did you like the house?

SA: Yes, we did. You know where George Mason is, George Mason University?

EM: My youngest son went there. Got his degree there.

SA: On Braddock Road there. We were just right across on Braddock Road. The elementary school was just down the next block. The high school was two blocks away. It was perfect. It was just absolutely perfect. Off we went to Washington, D.C., and I'm off to ICAF. It became a very interesting—

EM: How was the curriculum there at ICAF? Was it challenging or did you find it was too easy? What was your reaction?

SA: This is going to be a long, shaggy-dog story.

EM: We've got time. [Laughter]

SA: First thing I decided to do was, since I was going to be in ICAF, and since they had an arrangement with George Washington University for a master's degree, I said I might as well pick up a master's while I'm here. All of my education has been technical. They didn't have an MBA, but they had a master of science in business administration. I decided I would get a master's degree in administration. So that complicated things a little bit. It took away the playtime.

EM: Pretty full slate.

SA: That sort of made it a little more—not as much fun as it could have been. But it had a great curriculum. They had these great speakers that would come, I mean all the government, high government officials would come, and they would have these auditorium speeches. Lots of reading, I mean tons of reading. One of the first things I did was take a speed reading course when I got there. They had some arrangements for some of the people to come in and give you a speed reading course. I loved to read, so that really was one of the smartest things I've ever done was to take that.

Very poor writer, and they had lots of stuff to write, so that was good. It helped me in that regard. But I ran afoul of the system. First of all I was a little slow in getting a thesis paper that they wanted a subject for. They said, "We sent you all these books and stuff out there."

I said, "Look, I never got them. I was on cruise. I left the ship in Subic Bay, and plus I was a late addition to you guys. I was supposed to be going over to National and they sent me over here instead. So I haven't had time to even think about it, but I guarantee you within the next month I will have decided on a subject I want to write about, so don't get too excited about this."

Next thing was we had these little critique sheets so that anytime we had a presentation in the auditorium, we had these little green critique sheets that we had to fill out.

EM: Of your fellow students. You critique their talks?

SA: You would critique the speaker, and whether it was a subject matter that was applicable or of interest. Well, what we had was, when these speakers would come in, I mean maybe we would have the Secretary of State, or maybe the head of the CIA or one of the unions, or some CEO of one of the major corporations. People that all had name recognition. They would have one of the faculty do the introduction of the speaker. Well obviously, from our standpoint, from a student standpoint, it was an opportunity for the faculty member to get some recognition from the higher-ups, and so the introductions would be these flowery, long-legged, you know, they would say, here's the Secretary of State.

EM: They had to showcase themselves?

SA: So I finally had it. I said, "Look, you've got this very important person. You've given him an hour to make the presentation. The first 10 minutes of the presentation are taken up by the introduction by somebody that you've already given us a bio on and had us do some reading on—things that they had written or speeches that they have given. So we already know who's going to be there. We already know where they came from. We already know—we know more about them than we know about ourselves almost, and you have somebody get up there and give a 10 minute introduction. Let's take full advantage of his time."

Then I get called aside and they say, "We noticed you haven't—we always have a question and answer period afterwards. You haven't asked a question of any of our speakers here so far. We want you to ask the speaker who is coming this Wednesday the following questions."

I say, "No, no, no, you know, we've got a lot of spring butts in our class."

EM: [Laughter] Spring butts? Yeah, they jump right up?

SA: They did that. You know these guys.

EM: Oh, yeah.

SA: I said, "But rest assured, if I really have a question, no doubt about it, I will stand up and ask a question. But as far as me asking a planted question, you're talking to the wrong guy."

We had a couple of iterations of this, and I think my little green sticker on the introductions was the final blow. So I get called down to the deputy commandant's office and he says, "Commander," he says, "It's obvious that you're not too happy with being here at ICAF. Perhaps it's time for you to get orders out of here."

I said, "Wait a minute, you know I'm frustrated. My fifth-grader has more academic freedom that I've got." I said, "We're supposed to be here expanding our knowledge, and you guys want to just railroad us down a narrow path. No, I want to be here. I love to learn. I've never been—I'm always happy when my student days are over, but I really love to learn stuff, and I'm here to learn something, despite your best efforts to make it not so." [Laughter] I said, "If you're asking me to go talk to my detailer and get orders out of here, no, I'm going to finish this program. You want to throw me out, well then that's another matter." I walk out, and I'm wondering if this is my last day at school or not. It all just sort of drifted away.

EM: It worked out?

SA: It worked out. [Lieutenant] General [Walter J.] Woolwine was a three-star Army guy. I saw him at an airport several years later. He still remembered me. [Laughter]

EM: Oh, is that right? [Laughter] You made an impression. Is that the Army's way to have more structure?

SA: It was the Army way. It was really funny that it got to that point. And here I was going to get into an environment that—from my previous school days, which was mostly math and chemistry and physics, here I'm going to learn something about the world. And they're grading me and spying on me to see how many questions I asked. It just rubbed me the wrong way.

Two things that I did there had a major impact on the rest of my career. The first was they had two major trips that the student body took in the academic year; one was a city trip, and you sort of went on a bidding scale of where you wanted to go—LA, New York—most guys picked the place where they could take their wives and see shows. Enjoy the social life.

I had a couple of questions in my mind about things I had seen while I was in the Navy. One of the things that I'd seen in my squadron and in VA-122 when I was a maintenance officer was, if you got a sailor who came out of Detroit, you had a problem child. The minute they walked in the door. It didn't take me very long to recognize that, so anytime I saw a sailor come into my organization—whether it was 122 in the maintenance gang, or it was in 164 when I was the XO or CO—and they came from Detroit, I had a one-on-one conversation with them.

Very hard core antiestablishment, most of them. Very much me-against-the-world type mentality. So I would sit down with them and say, "Look, I know where you come from. I know things must not be very good there because I can see some problems from people who come out of there. So I want to tell you something: the world is not against you. This place here, you do your job and you're going to get all the pats on the back that you ever thought you could get. We're going to take care of you. But I'm going to tell you one thing for certain: no mischief. There is no toleration for mischief in this organization." And I said, "Mischief can take many forms. Anytime you want to stray out of what is the straight line, you're going to get noticed and you're going to be called. We are not going to let it slide."

It sort of worked. So I wanted to go to Detroit to see why in the world this attitude came out of that place. They had all sorts of different opportunities for us there. We visited all of the major industries—of course I wanted to go see cars made. Did that, but spent a lot of time with the union guys. Not my cup of tea, but I wanted to find out, really see what they were all about. And then I wanted to ride with the police. They said, "Oh yeah, we can let you ride with the police." So, man alive, was that an eye-opener. I went out one night with the police. I mean, we went from guns being drawn in a domestic dispute—

EM: You stayed in the car?

SA: No, I was with them.

EM: Really?

SA: And the first thing that happened was, the two guys that I'm with are both ex–Air Force, MPs [military police] or air police, I guess they call them. And these guys, I mean, it was just they had the wrong attitude from the days of their—well they were going on, "I don't know what fuzzy head we're going to have to bang tonight," you know. I thought, oh man, this is not good.

That surprised me even more because they had the military background. So the first place we go is a Laundromat, and there's some disturbance there, and we go to see it. They see this car parked outside the Laundromat, and the guy turns to the other guy and says, "You know whose car that is, don't you?"

"Yeah, I know. I wonder what he's doing around here. I wonder if he's in there."

"Let's go see. I wonder why his car is here."

"We'll come back later and see." So we buzz around a little bit and come back. Now the car is up on cinderblocks and all the wheels are gone.

"Well, somebody got him this time." [Laughter] I thought, boy this is really something. Then we get into this domestic dispute thing, and they are dragging people out—but it was just one constant us-against-the-blacks. And I'm saying, boy, I just, you know—I mean they were looking for trouble. They were just actually stirring the pot. They had their suspect list and they were going to find some way to get some of them.

Then I had a better understanding of why people coming out of Detroit had a real problem with authority. I kept doing that as I went on up in other jobs, when I had the ships and stuff like that.

EM: You say there was a second thing at ICAF that really made an impression.

SA: The next was the overseas trip. These were the plums, the two-week overseas trips.

EM: They all picked Paris? [Laughter]

SA: Paris and London. There was a group of us that picked Iran, Kuwait, and Turkey. Of course, Jennie Lou wasn't happy because wives couldn't go on this trip. My rationale for that was these are places that a Navy guy is never going to see. [Laughter]

EM: Famous last words. [Laughter]

SA: Of course, Iran was—for me, the ability to go and see part of the old Persian Empire, that was really intriguing to me. Turkey, with all of the old stuff there, and Kuwait as well. What an interesting trip. I know why I chose it, and then for it to play back to me in the years later—

EM: At that time, what was your reaction to Iran? I mean, the Shah was still in power.

SA: The Shah was embarked on a campaign for a satellite network for school systems across the country that could be generated from a central place, and they have schools lashed up with satellites. Really impressive with where they were trying to get it to. The people that we met were just delightful. Absolutely delightful. Many of them educated in the U.S. Very, very polite. Very thoughtful. Very educated. I thought Iran was going in the right direction. The place was full of U.S. contractors billeted all over the place. The helicopters were all Bell Helicopters; they had a huge organization over there. They had F-4s. My Marine buddies who were with me on this thing had one of the guys who'd been a POW, by the name of John [A.] Dramesi. John ran two escapes.*

EM: He was lucky he survived. The other guy, [Edwin L.] Atterbury, didn't fare so well.

SA: A lot of the POWs thought John was off his rocker. John was just a different kind of guy. He was a really tough guy. His thing was a couple of drinks, and the next thing you know John's standing on his head and walking around on his hands. [Laughter] He'd get his feet up in the air and they'd say there's John.

But I was going to tell you about with all the contractors. You know, the F-14s were headed that way, and the F-4s. We were going down to one of the airfields one day, and all of the F-4s are sitting in revetments with their noses open, and I tapped my Marine buddy and I said, "See, there is radar in there. That's where you guys always put your ditty bags for cross-country trips. These guys got real radars in theirs." And they did. Their airplanes were absolutely perfection. They were all up, but they were all mostly contractor-maintained in the US. They had plenty of spare parts.

---

* Air Force captains John Dramesi and Edwin Atterberry escaped from a prison camp in Hanoi in May 1969 but were captured the next day. Both men and many other prisoners suspected of complicity in the attempt were severely tortured. Atterberry died as a result.

EM: Were there any demonstrations or religious opposition while you were there?

SA: No. The only thing that we had to watch out for was drinking. They had plenty of nightclubs, but they were contained. And you didn't wander out from one of those and go wandering the streets.

EM: No bar-hopping.

SA: No bar-hopping, none of that. It was all very well maintained. We really had a good trip. We traveled. We went up to one of the nuclear plants that the Russians were running for them.

EM: Was that Bushehr?

SA: Yeah, and the Russians wanted to take us out drinking that night. We went to their club for a little while and decided that we didn't want to stay there. We had a big embassy party there and this Russian attaché, he was Army, very well-decorated, you know. What's that big medal that they all wear?

EM: Order of Lenin?

SA: Yeah. He had two of them. He sidles up to me and says, "You're Navy?" I say yes. He says, "You should be Army. You should be Russian Army. You look more like Russian Army than they did." [Laughter]

I say, "No, no, I'm Navy.

He says, "Oh, that's a shame." He was knocking down the vodka pretty hard.

No, Iran was really, really—I was surprised with how there was not more pushback on the mullahs when they came in power, really clamped down, because from what I could see, the middle class in Iran was very well educated and had a very tempered view of the rest of the world. [Break]

EM: All right, we are back on the air. Anything more you want to say about your student trip to the Middle East?

SA: Well, let's see. Turkey was interesting. Again, the radicalism that we see today throughout the Middle East was really not that apparent at that time. This was '74, I guess. If it was there, it was under the surface. We had dinner at one of the—it had a church theme—favorite restaurants, and all the ladies there were passing the food up and down the tables. We were treated very nicely and warmly there. Kuwait was a different thing altogether. I think one of the first visits we had there was with the ambassador, and we went to the oil ministry and sat down with the old minister, and, of course, they were all in their traditional garb. Varies. As always in the Middle East, when you meet them, they are overly friendly. Very, very friendly, very welcoming. Nice briefs about what's going on with the oil and the refineries, et cetera. Lots of evidence that the Kuwaitis don't do any of their own work. It's all Pakistanis or Filipinos or whoever they can recruit to come and do anything. They didn't lift a finger to do anything.

We get all the facts and figures in the oil industry and how it works and what their role in the world is and this type of thing. Then we end up at, I think it was the Hilton Hotel, not very far from the embassy. As we were walking in or walking out, I forget the sequence of events now, but three or four of the very heavyweights that were at the meeting, at the oil ministry, are coming out. They are now in European garb. Suit and tie. Very boisterous. Very loud. Obviously they had been drinking. Overheard as they walked by is sort of a slam about the Americans around here. I'm thinking, hmm—

EM: An undercurrent.

SA: Yes, a very strong undercurrent. They seemed so nice earlier. [Laughter]

EM: So there are many faces?

SA: Many faces. Out of those three places, that was the only place that I heard anything.

EM: Just going back to Iran, when the overthrow happened in '79, were you surprised? I mean, you had been in '74, only a few years before.

SA: Yes, I was really surprised. I didn't think it would go anywhere. I thought it would fizzle out and order would come back. Although I'm sure that, you know, the Shah's regime had lots of problems. The things that we were seeing, the things we were seeing that they were trying to do for the future of the education, the satellite net to bring education throughout the whole—you know, Iran is a huge, huge place with very few real centers of population. They were trying to take the good stuff out to all the people, which I thought was a very noble thing. I was very surprised, and I have been surprised since that some sort of semblance of order has not been restored. I think it's there, if it could be tapped.

EM: Well, there is a sizable middle class in the cities, but there are a lot of Iranians out there in the countryside.

SA: And they are very rural. They are really, really out there.

EM: Anything more you want to say about your ICAF experience?

SA: No, I think as it turned out it was a little shaky start, but the end result was good and I learned some good things.

EM: What about your education at GW [George Washington University]? Was that helpful, that master's degree?

SA: It was. They made it reasonably easy to get. We spent most of our time—the instructors would come to ICAF, to Fort McNair, late in the afternoon, and we would do the courses there. The interesting part of that was we had to take a final exam, and it was like any other institute. There were no gimmes. [Laughter] A couple of our guys flunked, and one of them was a Navy intel guy who I thought was absolutely the sharpest tack in the whole world.

EM: Didn't do his homework.

SA: Something happened—I don't know whether he had a bad day or what. When I found that out I was surprised that I had passed. We lost some guys there.

EM: Was it pass-fail?

SA: Yeah.

EM: You had some POWs in the class, too?

SA: When the POWs came back they tried to get as many of them into a slot that would help bring them back into the mainstream of society. And so I guess all of the war colleges—I've learned since that the Air Force sent as many of theirs as they could to Maxwell [Air War College] and the Navy sent theirs up to Newport. They earmarked some of these guys for bigger and better things who came to National [War College] and ICAF. Like I said, John Dramesi was still carrying on. To the credit of all of the POWs, for all they went through, they did a wonderful job when they came back of keeping the interactions that they had experienced over there among themselves, out of play. You didn't hear any of them talk bad about the other guy. Now, over the years we start seeing some little—

EM: There were a couple.

SA: I really enjoyed getting to know John. He was his own guy. He was fun.

EM: All right. I wanted to go back to when you're still out there in the Gulf of Tonkin. In '72, as you know, the Navy was having—all of the services were having big problems—

with anti-establishment, anti-authority, racial troubles, drug troubles. What experience did you have on *Hancock?*

SA: We did not have any real serious incidents that I can recall on *Hancock.* Of all of the places, I would have thought we would have had more [trouble]. She was run down, she was really a tired old lady. The crew worked in just horrendous conditions. I mean, we had the crew berthing in the expansion joints, and anytime it rained, which it rained a lot out there, water was just streaming through the expansion joints. The kids were living in these berthing areas with water sloshing around—it was really, really tough. I can't recall one incident that we really had.

EM: You mentioned some of the guys from Detroit, but they got corrected pretty quickly?

SA: Yeah, they did. It was interesting. When I was the XO, Ramsay was the CO, he had a different view of non-judicial punishment. Ramsay didn't want to hold any NJP [non-judicial punishment]. He said, "Your job as XO is to make sure that these things don't have to come to me." I'm not sure where this came from. I've run into several people since that had sort of the same perspective. I'm sort of the other way. Not to rush to it, but to use it when it needs to be used, and use it quick and fast, is sort of where I come from. The job for me as the XO was to get the leading chief and myself on the same page and then make sure that the division officers were in step with us. It was to get things nipped in the bud very early, to look for where there could be trouble, and before there is trouble, take some action so it doesn't get to a point where we had to do something about it. We were so busy, and when we got in port, people were so tired. We certainly had some liberty incidents and stuff like this, but not of the racial variety.

EM: Did you have any racial polarization in the squadron or on the ship? You know, groups of blacks and groups of whites?

SA: No, it didn't seem to be. Now, by the time I got to *Coral Sea* a little bit later, I was—we will talk from that perspective. By that time, we had lots of problems.

EM: Looking at it from the larger picture, then, what were your views of [Elmo R.] Zumwalt [Jr.] as CNO? Dealing with that issue, but of course other things, too?

SA: I did mention that Zumwalt came aboard ship very early in our cruise while we were out there, by the way. Everybody was happy to see him. My first encounter with the Z-grams was before I got to 164 when I was still on 122, and the skipper brought me back over to be his maintenance officer. I walk in one morning into the maintenance office and the chiefs are there, they are looking at me with these big eyes, and they say, "Boss, we've got a problem. Have you seen the Z-whatever that number was?" No, I haven't seen it yet. He says, "Well, it says that we will have a locker for every one of our sailors so that he can go home in civilian clothes and come to work in civilian clothes and have his uniform and work uniform here in his locker."

I think my maintenance department had 1,000 men, those days in the RAG. I said, "Where are we going to get all of these lockers?" We don't know. "How are we going to manage this?" We don't know. "Well, who wrote this thing?" The man, he wrote it.*

EM: Came out of Washington.

SA: It came out of—it's a Z-gram, we've got to comply. And the skipper is asking me what we are going to do. Well, luckily we found out there was a Department of Defense activity where you turn in all of your—

EM: DRMO. [Defense Reutilization Marketing Office]

* Admiral Elmo R. Zumwalt Jr., Chief of Naval Operations 1970–74, instituted revolutionary changes to improve the status in the Navy of female, African American, and enlisted sailors. Many of the measures produced positive results, but others did not and generated heated opposition from traditionalists in the naval service.

SA: DRMO, yeah. There was one big huge one [storage site] up by Sacramento, so I call up there and they say, "Oh we've got tons of—" [Laughter]

EM: We've got lockers.

SA: We got trucks and then started doing that. The beards and the haircuts and all of this stuff.

EM: What was your reaction to all of that?

SA: I couldn't understand it at first. I said, "Why are we doing this? What's the big—why are we doing this?" If it's an important thing to do, why aren't we doing it within the normal way of doing business? You know, the chain of command is there, and people have already thought out the process and say okay, we're going to do this. Here's the instruction on how we are going to do it. But all we got was this little short Z-gram that says this is it. It was left to all of us to figure out how to interpret it. So I thought it was sort of a strange way of doing business. I was not happy, especially as maintenance officer. I think if I had not been maintenance officer I would have sort of brushed it off. But the fact that I had 1,000 guys that were now going to be wearing beards and long hair and civilian clothes to work was something that I had to figure out how to handle.

EM: Did you think it was unprofessional or a danger because of their hair getting caught in the machinery?

SA: Well, I just thought it was a trivial thing to be fussing with. That we had so much more important things to do, and here we are playing around with this stuff. And although he and I never talked directly on the subject, later in life I had a couple of opportunities to spend some time with Zumwalt and talk with him, and I sort of understood then where he came from. He rightfully saw that the way of doing business was not going to get to where you wanted to go in time. He thought there was a clock ticking and that he needed to get in front of this wave of the new generation. And he knew that the old guard would drag their feet all of the way, so he just went by them.

EM: He was kind of tough on a lot of naval aviators.

SA: Yeah.

EM: Was there resentment in your community that those guys were being singled out?

SA: Well, there was, but on the other hand, we knew things weren't really good within the community, that some people had sort of bypassed the system and were headed for bigger and better things, when they really hadn't earned—we would have said earned their stripes.

EM: Any individuals in particular come to mind?

SA: None that I worked for, but during the war there were people who did not perform at the level they should have, and I'll talk to one example that I know of, but I never worked for him directly. I've read about enough people saying things, and it was sort of the old Air Wing 16 *Oriskany* caper and their CAG, Burt [H.] Shepherd, who I think eventually got two stars, and then he retired out of the training command and went to be a minister. From all of the people that apparently were in the air wing with him, he was not a great leader. He didn't react to things that were going on around him in a way that people thought. And yet he was pretty much rewarded at least to the two-star level, and so you'd say if somebody were looking at Burt who hadn't served with him and seen this guy with all of the decorations and stuff, he obviously was right in the middle of the very heart of some of the toughest years—'67, '68 timeframe—still getting promoted, and then somebody comes in and sort of terminates him. If they didn't know the background they would say, well, why did they hammer this guy?

EM: Some of the CTF [Commander Task Force] 77s didn't fare too well.

SA: Yes, and I think Zumwalt—to me had an uncanny eye for people who could think forward and not backward. I think he went on a campaign to get rid of as many of the backward-looking folks as he could. Some of them had just—I mean on paper, had just splendid records.

EM: Was this altruistic, or did he have another agenda as well? Were there other motivations for what he did?

SA: No. I think Admiral Zumwalt understood way before his time that there was a transformation going on in society that was markedly different, and that if we didn't understand how to interface with it, we were going to be left behind. And it took me a long time to get to that rationalization. But as the years went by I looked back, and I know when I was Vice Chief I looked back several different times and said Admiral Zumwalt saw this coming. He really knew the business, the people. This was the generation that we were going to have to really do business different with.

EM: There were some reforms that he instituted that didn't work out as well. Do you recall anything that didn't outlast his term?

SA: Well, I know there were some, and I'm trying to think. I guess if I had kept all of those Z-grams—

EM: They are at the Navy Archives if you ever need to get them. In fact, they are online.

SA: I know there's some, but I can't recall right now.

EM: One thing that I think, he's held in great esteem by females in the Navy, and he made some strides there. What was your reaction to what he did to get women more into the service culture?

SA: I thought, again, that is one of the things that he absolutely recognized, and I will talk a little bit about this when we get to my BuPers [Bureau of Naval Personnel] days. I was there when we sent the first females to flight training. Of course the rationale was, as we had just learned in the Vietnam War, that we were in danger of running out of pilots at different periods of time. That's sort of where I got my chance to get in the attack community, when they starting having to come out into the other communities and say hey, anybody want to fly jets? He had seen that the demographics of the nation were changing. That if the manpower pool was going to shrink, which it was, then there were jobs that could be done by the nontraditional system. So why not take advantage of the female part of the demographics to do the jobs that don't require a male to do?

In those days, of course, nobody ever thought that females would go to combat, but why does JAG [Judge Advocate General's] Corps have to be all male? Why can't there be more females? Why doesn't the admin side of the house of the shore establishment have a lot more females in it? And for the aviation community, we had a whole community of aviators that were in what we called the support side of the house. Surely if we've got three combat aviators and we've got slots for three combat aviators, why are we having to take one of the combat aviators and put him over here in the support community flying transports or whatever? Why can't we train a female to do that job?

EM: So it was the needs of the service.

SA: The needs of the service were starting to dictate, because of demographics, that if we were going to have a viable force, and we could only use the males in combat, then there were lots of roles for the females. I thought he saw that right away and he reacted to it.

EM: Was there also a feeling for a need for greater equity between males and females?

SA: At that time there was not so much. It was—there are roles and jobs, but then rapidly it became—I can remember one of the early female aviators and her husband after she had completed training and gone on to her first fleet activity. They came into the bureau for some career counseling—you know, what's next? I had to look the guy in the eye and say, "Well, her prospects are better than yours." [Laughter]

EM: Why was that? More openings for her?

SA: There were more openings for her. The future for her and her community was going to be better than the future for him and his community.

EM: What community was he in?

SA: He was—

EM: Helicopter?

SA: No. He was, as I recall, an A-6 B/N [bombardier/navigator]. So I said, "When it comes time for the next detail, her needs will probably take precedence over yours. Just understand that. It's going to be one of the things that this marriage is going to have to work through."

EM: That's tough with two service members and they get posted at different places in the world. That can put some strain on the situation.

SA: Of course, all the nits and grits of how you manage those communities became more evident as time wore on, that if she was only going to be in, say, the VR [reserve squadron] squadrons or what translated later on into maybe VP squadrons or whatever, and the husband was in a fighter squadron, well, somewhere along the way, they were going to go that-a-way. Now, today, it could be different.

EM: That actually is a nice segue into your next assignment, which is the head of Aviation Assignment Branch in BuPers.

SA: Oh, yes.

EM: '74 to '76. Do you want to describe your duties?

SA: Well, I'll start by saying how I got there. I'm in ICAF, finishing up, and of course, I'd been told all along that since you're a WASM 2 you're going to NavAir. You're going to be a program manager over at NavAir doing something. I'm looking for my orders to NavAir, and my detailer, who was an old friend from the Lemoore days, calls me up and says, "Congratulations. I've got your next assignment. You're coming over here to BuPers."

I said, "Oh no, no, no, what am I doing in BuPers? I'm not a personnel guy. I'm a weapons systems acquisition manager type guy. I'm supposed to be going to NavAir."

"Oh, no, you're going to come over here and relieve John Disher as Pers-432. What's Pers-432? Well, Pers-432 is the outfit that details all aviators lieutenant commander and below to their assignments. So you're going to be the head detailer for all of the lieutenant commanders and below in the aviation community."

I said, "I don't know a thing about detailing people to orders. Ham, are you sure you're talking—are you sure you've got the right guy on the telephone?" His name was Ham [Hamilton] Bing. He'd been a neighbor of mine at Lemoore.

He said, "I know who I'm talking to. You're coming to BuPers." I said I couldn't do that. He said, "Tough luck, you're coming over to BuPers."

I finished up in August, I think, at ICAF. Those of us in the GW program had to stay a little longer to finish up our work. I go over to relieve John Disher, and John says, "Hurry up. You've got to get here and get me relieved so I can go on to my next job." I think he was headed out to be CAG. "The boss is not going to let me leave until your deputy shows up."

I said, "Who's my deputy going to be?"

He said, "It's a fighter guy out of Miramar by the name of Chuck [Charles] McGrail. Chuck and I are old friends. You'll like Chuck. He's supposed to be leaving Miramar in the next week and be here shortly so I can get on my way." John was relieved to get out of there. We were waiting for Chuck to show up, and it was one of these shaggy dog stories of Chuck and his British Rover.

EM: Breakdown?

SA: They break down. [Laughter] So it was one message after another. "I'm here in Yuma, Arizona, waiting for a part." "I'm here in San Antonio waiting for a part." Finally one day, here comes this guy down the passageway. He's got one of the old-style umbrellas, you know, with the cane, that we use to call bumper chutes, the British bumper chutes. He's strolling down the passageway toward us. In those days we all wore civilian clothes because in D.C. you couldn't wear a uniform, except one day a week. Here he comes striding in, and that was it. Chuck was there. John left and I had the job as 432.

EM: Who was the chief of BuPers at that point?

SA: The chief of BuPers at that time was, I'm not sure if it was Admiral [James D.] Watkins.

EM: [Charles K.] Duncan maybe?

SA: It probably was Duncan, and then Watkins relieved him. There went the two new guys sitting there trying to figure out how we are going to run this detailing shop. One of our lieutenants was Jay [L.] Johnson, future CNO, and a lot of the guys, they were all top-running guys. A lot of guys made flag and some of them didn't. It was really an interesting tour. Again, there were several moments there where things didn't go exactly as we wanted them to go.

EM: For instance?

SA: Well, let's see. I think the first thing that hit us—I may be out of sequence, but one of the first things that hit us was a lieutenant commander promotion list came out. Of course, neither one of us had been there to be on the board when it was in session. For the aviation community, a lieutenant commander promotion opportunity was dog meat.

EM: Because after the war the services contracted?

SA: We couldn't figure out what had happened. We couldn't figure out why the aviators promoted at such a poor rate compared to the surface and sub-surface community.

EM: Maybe related to the CNO?

SA: We didn't think so. But what it really meant was there was an extraordinary amount of Naval Academy graduates that did not promote. And that caught everyone's eye. Chuck and I formed up the office into teams, and we pulled together all of the service records—fitness report records of every aviator that had gone before that board, and we conducted our own mini-selection board. Sort of went through the records and said: should have promoted, might, and no way. Our results were completely different than what the board results had been.

Then we went first through the top group of absolute should-have-been selected, no question about it. What in their records did somebody see that would have given them an opportunity to be soft on them for promotion? We did that for the second group, but we wanted to look at that first group first, just to see if there were any dramatic things that popped out. And, of course, there were. The aviators had stacked up the training command to an overload situation, and it was people who would get their commission from the Naval Academy, and they wouldn't report to Pensacola for nine months, sometimes a year. So no valid fitness reports. No valid fitness reports while you're in training, because it's just training and duty under instruction, and we didn't have any leadership roles for student aviators or any of that sort of stuff. Now you have nine months to a year of dead time with no fitness reports. It was taking them two years to get through the training command, so now you've got possibly three years of no fitness reports.

EM: No way to evaluate their leadership.

SA: And so finally the guy gets to his first squadron, and he's making lieutenant within about a year of getting there, and he goes to one squadron tour and comes ashore, and now he's up for lieutenant commander. And he's competing against a surface guy that three weeks after he graduated from the Naval Academy is a commissioned officer on a ship.

EM: Why was there such a lag between graduation and that nine months to a year?

SA: Well, they had over-primed—for several year groups—they had over-primed the pipeline with aviators. The second story where I get in trouble, I will talk to this, that part of the equation. So what we found was that the aviators were presenting a very narrow career pattern for the board to look at, and nobody was explaining why it looked that way. We had to go back in and say that now the boards have to be aware that when an aviator comes up for lieutenant commander, he may have only had X amount of time in an observed job that was other than flying. You need to take a look at that. You need to not pay much attention to the front end, but you need to pay a lot of attention to the back end of this and what he's doing today. Is he in the right kind of job? Does it look like he is on track to go further and not penalize him for this dead time that is not any fault of his own?

One of the dumb things we did was we said, "Hey, guys, since you have all of this dead time, there's a master's degree program you can get into. You can get a master's in aero-engineering or something like this while you are waiting." Well, what that ended up turning into was, it just made the wait longer, because it didn't complete in time. So you had an extra six months or so involved in that program. So it was another delay in getting meaningful reports.

They didn't understand—they don't understand even today—the necessity for the aviation community to be a safe community; it's very dangerous work, and to be as safe as it can possibly be, you need to have people indoctrinated by people who know that. The fact that we have so many aviators that have to come back into the training command as instructor pilots is—yeah, you say, all he is is an instructor pilot. Well, he's an instructor pilot teaching a new guy how to be a good aviator, and without that we're going to have bad aviators on the other end. Don't think of this guy as a second-class citizen. He's a first-class citizen doing a real meaningful job.

EM: Did you sell that reappraisal?

SA: We sold it somewhat. We stemmed the tide. The next few boards looked good. However, if you paid attention in the last year, the whole thing happened again.

EM: Right now?

SA: Yes. I got in trouble on that one, too.

EM: How did you get in trouble?

SA: I wrote an email to six of my closest friends, one of whom—I don't care who—let it get out of our group, because I used some intemperate language. [Laughter] And let it get out into the general population, and, you know, I was very highly critical of the present aviation leadership for letting this happen again.

EM: Déjà vu all over again.

SA: Déjà vu all over again. Everybody had forgotten some of the same issues were there. And the sad thing is that some of the leadership in positions still don't recognize it. But that's another story. It was learning the job under fire. It was a great body of work that we accomplished in a very short period of time. I think we did all of that stuff and had all of our recommendations and everything inside of a month, and all of those records were stacked all over the place. It told Chuck and me that the management of the community was really—we were the heart of the management community. We needed to really pay attention and make sure that our detailers understood.

EM: Your aviation assignment branch, were you under the chief? Or were you under a subordinate commander?

SA: My job billet was not under BuPers. It was under the Navy Support Activity something-something. I forget.

EM: Naval Support Activity, Washington?

SA: Something like that. I'll remember the acronym sometime. But that was my first—I said, "Wait a minute, why doesn't my job say I'm part of BuPers? Why does it say that I'm under this Navy Support Activity?" Well, you remember that the Navy had the personnel drawdown here a few times, and we had to remove 15 percent from the staff assignments. What we did was, we made another activity and your billet went into this other activity. Doing the same job, it doesn't show that you're part of BuPers. If somebody wants to know how many people are working in BuPers, your number doesn't count. It counts over here in this other activity. Crazy things that we do. We had a 431 shop that did the commanders and above, a 432 shop that did the lieutenant commanders and below, and a 433 shop that was the placement office. They are the ones that create the billets and we fill them. We had the placement guys and the two detailing shops. We had some other things, but those were the three main pieces. We all worked for a captain, who at that time was Tom [Thomas J.] Kilcline [Jr.], Pers-43.

EM: He made flag.

SA: He made flag, and he was relieved by Gene [William G.] Sizemore [II], who also made flag. Then Pers-4 was on top of this organization. Pers-4 was Admiral [Carlisle A.

H.] Trost. Rear Admiral Trost. He reported directly to the chief. So we had the chief to 4, then to 43 to 432. That was the organization.

EM: In the aftermath of the Vietnam War, did you have any trouble with some officers not wanting to be assigned, or were they any problems in that regard? Moral or ethical problems stemming from the war?

SA: No. I'm sure we must have had a couple, but we didn't have many. The big thing that we had going on at the time, I had a RAD [released from active duty] officer. It was her job to call somebody up and say goodbye. That affected the reserves.

EM: I'm not sure I understand. What was she doing?

SA: That was when we had to draw down our numbers. We had released from active duty the shop that she ran.

EM: So she had to tell people they're out. Not a fun job.

SA: She was great. I mean, she was tough as nails.

EM: What was her name?

SA: Tory was her first name. I won't give you her last because she got married later. I'll tell you, she had the toughest job in the whole building, and she did it well. We'd get these mandatory orders, you've got to release X number of people, and basically she had a list of who was next.

EM: Who got the axe the most? Were they helicopter aviators?

SA: No. It sort of went by that community, but it went more by the reserves, who went first, and then if we had to go into the USN, it was performance. If you had somebody from one of the support communities that was marginal, they would go before a marginal guy in a, say, VF [fighter] squadron or a VA squadron. So there was some community-leveling there. It went mostly by demand. Where we had more people in excess than others. But it still had some very tragic results. There were a lot of people that went home who didn't need to. There were some people that went home voluntarily who wanted to come back, and of course, trying to come back in those days was very hard because we were pushing people out. Some people who I flew with in Vietnam had gotten out and wanted to come back. A particular case in mind was Eliot [F.] Tozer [III], who had been with me in VA-55.

EM: That name is familiar. Has he written?

SA: His name has been written. He's dead now, but his name is probably pretty well known around Annapolis, because I think that is where his wife may still be. Eliot was a young aviator that came to us in '55. Superb, superb aviator. Superb guy. Very energetic. Sort of a free spirit. Was not totally enamored with the chain of command. He liked to deviate out of the chain of command, but we could always drag him back in. He was just bright, you know? He was just one of these guys that not only could fly an airplane, but he asked the right questions and he pressed you. He made you smarter than you wanted to be. He got out and got into the reserves, and then he wanted to come back on active duty. He happened to call me one day and asked, "What can you do for me?"

I said, "T, we don't have any slots on active duty. We're releasing people today. I can't get you on active duty. But it just so happens that today the TAR [temporary active reserve] board—" Temporary, what is it? The training and the TARs, training active reserves—

EM: I don't remember what the acronym stood for.

SA: They were sort of the home community for the reserve organization, you know, people that worked on Monday, Tuesday, Wednesday, Thursday, and Friday, while the reserves worked Saturday and Sunday. I said, "But the TAR board is meeting as we speak. I've got a very close friend who is down there and is managing that board, and if you thought you would want to come in and take on that kind of role, you know what the TARs do. There's always some movement later on, if there's ever a thing that you could do."

He said, "Yes, yes, yes, put my name on the list."

So I walked down to Frank [X.] Azzarita and I said, "Frank, here's a real top-quality guy. If you want a top-quality guy in the TAR community, here's a guy that can really do the job." He took it and they selected him. T got back into the flying business. Flew with the reserve unit there in Anacostia. Flew the photo recce birds, the F-8s, which he just loved. And then he got killed in a fly-by high-speed pass, coming by the carrier and the wings came off.

EM: Really?

SA: Yeah.

EM: Well, that's what he wanted to do.

SA: What he wanted to do, but it was sad. It wasn't what his wife wanted him to do, I think.

EM: All right, well, during your time at BuPers you had a transition from Zumwalt to Admiral [James L.] Holloway [III], still with us. I know him very well. In fact, I interviewed him a couple of years ago. Great guy. He was an aviator, and some of the things that Zumwalt had instituted he walked back. What was your opinion of Holloway?

SA: Holloway was a great guy. Of course, he had an outstanding Vietnam record, very highly decorated. Very well thought of. I had only seen him once in my career up to the point when he became CNO, and that was when the *Enterprise* [CVAN-65] came into Pearl Harbor. I think I was on *Bennington* at that time. I remember him walking into the O Club with all those ribbons and that big shiny ship, the *Enterprise* out there. He has that jaunty stride. Boy, there's a colorful guy, really something. Then later, got to see him in action, there and over the years. I mean, he didn't let any grass grow under his feet. He was off and running.

EM: What was the reaction in the aviation community? Because Zumwalt, for various reasons, the aviators weren't totally happy with him, and here comes Holloway, who was one of your own. Was there a feeling, oh, now it's our time again.

SA: There was a feeling that it was time that there be an aviator in charge of the Navy, and here was somebody that I think that everybody felt they could rally around. He was really an aviator's aviator, and that he would do good things for aviation. I think it was sort of a sigh of relief that there was an aviator in charge.

EM: One of the things that I think he's taken credit for is changing the carrier configuration so that you wouldn't have CVSs and CVAs [attack aircraft carrier], you would just have a carrier that would have various functions. What was your opinion of that evolution?

SA: Having grown up in the early part of my career as a CVS guy, it was sort of hard to imagine blending the CVS and the CVA into one configuration. Now, in Vietnam, we had taken one of the CVSs and made it an all-attack carrier for a while. No fighters.

EM: Was that *Intrepid* [CVS-11]?

SA: *Intrepid.* Ineptitude as we used to say. [Laughter] I think there was a concern that could you really do both missions at the same time. As a CVS, our flight schedule was that we just flew 24 hours around the clock, but at a much slower pace than the CVAs, where you were bumping off 16, 17 airplanes every 90 minutes. We would launch four, five, or six and go to sleep for four hours, recover, and off we'd go again. So it was interesting to see how that would come, and there were several different experiments that they ran trying to find the right sort of thing. It turned out to be the right thing to do, because in the long run we weren't going to be able to have carriers with such a focus.

EM: Holloway was very big for the big deck carriers, nuclear powered carriers and it was kind of a reaction, I thought. Zumwalt's big thing with the "high-low mix," where he would have a bunch of small carriers and some big carriers—Holloway didn't seem to go for that. Which approach did you think was the most valid?

SA: It's interesting, because we're certainly going to come back to this argument again. I mean, we visited this argument several times over the years. But I think Holloway, having been leading the nuclear carrier with his CO of *Enterprise* and all, he sort of put the nuclear carrier on the map. I think he recognized that the fact was that nuclear was really a great strength, because the more time you could free up the carrier to not be alongside an oiler or something was of great benefit. But of more benefit was the fact that if there was an event somewhere far removed from the nearest carrier, the nuclear carrier could really make the difference in getting there and being ready to go when it got there, where the conventional guys could run fast to get there, but the tankers and all that were supporting them would take a while to get there.

I think he saw that the response of the nuclear carrier was a big plus that we ought to take advantage of. Is there a place for the small deck? That argument still goes. Again, I spent almost all of my career on the small decks, so I can see that the small decks can contribute a lot. But in an era of constrained resources, if you have the possibility to have two nuclear carriers instead of one nuclear carrier and two small carriers, you're probably going to opt for the two nuclear.

EM: Of course the big thing, and this is current events, is survivability. The thinking probably is if you've got more ships out there, more targets, it would be harder to take them out. In the big deck carrier, the nuclear powered carriers, we've only got 11 of them now, right? You lose one of those, that's a big increment. Are they at greater risk now?

SA: Well, I guess if you buy into the intel, what everybody's reacting to on the intelligence, they are at risk. I continue to believe that we have continually, over the span of my career, always given the enemy or potential enemy much more credit for their capabilities than they've actually ever had. So I'm still on the doubter side. Just take the current circumstance, that the Chinese have this wonderful, over-the-horizon missile that can destroy any carrier anytime, anywhere. We can't do it, so why do we think the Chinese can do it? Now, is there a possibility they can? Yes. Certainly. Are there some countermeasures? Yes, and we've got them. I sometimes think we give the other guy so much more capability.

EM: The enemy is 10 feet tall.

SA: Yeah. But it is a concern. If I were running the Navy today, I would be looking very hard at what we are doing with our large-deck amphibs. Now that the Marines have the F-35B, I wonder how willing they are to join the fray alongside a big deck nuke. Or do they still want to be tied to their air ground task force?

EM: They usually are. [Laughter]

SA: If we were going to use the large-deck amphibs as surrogates for small-deck carriers, there are a lot of things that we missed doing in the construction of them that we shouldn't have. That's where I think we sort of have lost a little bit of focus. We have a potential capability sitting in the wings. I know when I used them in Desert Storm, my concern was the ammo-handling. I was just scared to death that we were going to blow one of them out of the water.

EM: One of the amphibs?

SA: One of the amphibs just handling too much more ordnance than they were used to handling.

EM: Didn't a Styx missile impact close by an amphib in the harbor in Bahrain?

SA: No, it was up there in the port city over in Saudi Arabia across the way. One landed right in the water. I was holding my breath.

EM: Admiral, anything more you want to say about your BuPers time? Did you love Washington, being in D.C.?

SA: I'll tell you another flap that goes in with the one that we did on the lieutenant commanders, and it's still got the same issues that keep going on today. This is the one that got me thrown out of the Pentagon. As part of that study that I mentioned, the one way of getting people through the pipeline, we started taking a look at what the pipeline really should consist of. How many people needed to go through that pipeline. How fast it should go. My pipeline, when I went through flight training, was 13 months from the day I entered until the day I got my wings. That was sort of what was expected, and then we let it grow to 18, sometimes 24 months. Rudy [E. Rudolph] Kohn [Jr.] was in 55 right when I first reported, and left just as I got to 164, and Rudy and I served together many times over the intervening years. Rudy was over in the 431 shop in the job- placement side of the house, and he's going to his class reunion up at Penn State University for a football game. He calls me up and says, "Okay, I will be going over—I'm leaving on Thursday. You know, if any questions come up on the pilot-training rate, there's going to be a hearing next week. I know you and I have been working this thing close together, so I've left your name if somebody's got a question for you to answer."

I said, "Okay Rudy, got it. Have a good time." And sure enough, Friday afternoon Admiral Ferdinand [B.] Koch in OP-05 in the building [the Pentagon], I think he was the 51 training guy—

EM: DCNO [Deputy Chief of Naval Operations for Operations]?

SA: Vice Admiral [William D.] Houser was OP-05, and Ferdinand Koch was one of his two-star section assistants. Admiral Koch calls up and says that he's going to testify on the Hill [Capitol] on Monday or Tuesday, before Rudy gets back. On our pilot-training rate. He says he needs to know what to say, because he hears that there are some people on the Hill who are against the number that we have given them.

I say, "Well, Admiral, here's the story. Right now the planner record that you're going over there to defend has 1,000 people to put in the pipeline this year. The money is reserved for 1,000 people to enter the program. The Hill doesn't believe we need 1,000 aviators trained this year. As you recall from what we learned on the promotion list for lieutenant commanders—that we've had a big backlog in the training command, and that's what they are reacting to."

He says, "What are you telling me?"

I say, "Admiral, it has been one of the well-respected numbers among the aviation community for a number of years now that if you ever train less than 1,000 aviators a year, naval aviation is dead. That's not true. The real number that we ought to be training to this year is 600, but nobody can stand that, so when the Hill starts banging on you that surely you can train less than 1,000 and survive—when they get you down to 800, just throw yourself on the floor and say, "Oh my God, you just killed naval aviation." [Laughter]

He says, "What the hell are you telling me?"

I say, "I'm telling you that 600 is the real number. None of us can stand that number. If we go down from 1,000 to 600 in one year, we will never get ourselves back. So we can't go that low, although that's what it really calls for."

Shit, next thing he goes up to Admiral Houser. Houser calls Admiral Wes [Wesley L.] McDonald, who is the deputy to Watkins, and asks what in the hell is going on over there. McDonald says, "What do you mean?"

Houser says, "Koch calls this, whoever this Arthur guy is, and he gives him this bullshit number." I don't know if you've ever heard of Ferdinand Koch, but he was a little guy who would get all hyped up. Wes says he will check it out, and he had just come from the head of the training command, CNATra [Chief of Naval Aviation Training].

So he comes down, and he knows me real well, and I say, "Admiral, you know the problem. You know we've been stacked up, and you know what happened on the promotion list. We've been training too many aviators. We don't have enough room for them. Half of the class of 1971 through the Naval Academy never got a fleet seat. Never went to a fleet squadron, ever. Half of the aviators of your primary institution did not get a first shot."

He says, "Oh shit, it's really that bad?" I say yes it's that bad, but I'm just a stand-in for Rudy. He says, "I want you to give this brief to Admiral Houser and his staff." Because in those days they put the number together, not BuPers. We sort of assisted them, but they had the final say.

First of all I had to go through Admiral Trost and Admiral Watkins. I give the brief to Admiral Watkins. I say, "Admiral, when I came into the Navy, this was the number of people we were training, and there are just years of peaks and valleys, peaks and valleys, peaks and valleys. What it has cost us now is, you're aware of what happened on the promotion board. We've got aviators who are spending too much time waiting to get into a fleet seat, and therefore they are not competitive when it comes time for promotion to lieutenant commander."

He says, "I understand that. We need to fix it."

I say yes sir. So Wes now tells Admiral Houser to come on up, we will give you the brief. I go through the same brief, and they turn around and look at Wes McDonald and say, "You let that son of a bitch give this brief to a black shoe?" [Laughter] By this time Admiral Kilkline has left. Gene Sizemore was the Pers-43.

Gene calls me up the next day and says, "You're not to ever go down to the Pentagon again. [Laughter]

I say, "What?"

He says, "The OP-05 guys don't ever want to see you at the Pentagon."

I say, "Okay, I understand."

EM: So they wanted to stick to the 1,000 figure.

SA: Yeah. I never went to the Pentagon again until I got orders to be OP-50. [Laughter]

EM: Keep him out of here.

SA: I thought that was pretty good; not being allowed to go to the Pentagon was a mark of honor.

EM: Do you want to say a few more things about BuPers?

SA: One of the interesting parts of the job was what we called "cousins." Your constituents all had friendly cousins. You would get these calls from out of the blue from somebody like Admiral [Noel] Gayler out at CinCPaC [Commander in Chief, Pacific Command]. "Hey, Commander, I want to talk to you about my aide out here. What have you got in mind for his next assignment?" You tell him and: "Oh, I don't like the sound of that." [Laughter]

I would have to run down to Admiral Trost and say, "Admiral Gayler's going to give you a call here in a minute." Stansfield [M.] Turner was at the Naval War College. You will probably remember this. He was sent up there to try to turn it around.

EM: He did, he turned it around.

SA: He calls one day and he says, "I've got this young lieutenant commander here by the name of [Ronald J.] Zlatoper, 'Zap' Zlatoper. Do you know him?"

I say "No, he must be one of our guys. I can look him up in our books."

"Well, you know this guy is really special, he is really special. He needs to get back to the fleet soonest, because he's going to be a flag officer one of these days and we need to get him moving."

"Yes, sir. Happy to say now that I've had a chance to look at his jacket, that's exactly what we got in mind. As soon as he comes out there, we are going to send him off to the squadron and he's on his way."

"Boy, that's great, that's great." Then, you know, he [Turner] left there and went down to be CinCLant [Commander in Chief, Atlantic Command]. So the next thing, I get this call from Admiral Turner, and he says, "You remember this guy by the name of Zlatoper?"

I say, "Yes sir, we got him going just like you requested."

"Well, that's not what I want."

"That's what you told me you wanted."

He says, "Yes, but now that I'm down here in CinCLantFlt [Commander in Chief, Atlantic Fleet] or CinCLant—"

EM: Well, the two commands were together, weren't they?

SA: Yeah. He says, "I got a special job down here that I really need him for."

"Well, Admiral, you know he really needs to go to that assignment." [Laughter] With that name, I could not forget that name.

EM: Zlatoper?

SA: Yes. Years later I run into Ron. I say, "You know, I know you—I don't really know you, but I know a lot about you that you probably don't remember. I was in the detailing shop when your boss, Admiral Turner, had special desires on your—"

He says, "Oh yes, I remember."

EM: He was under you in the [Persian] Gulf?

SA: Yep, yep. He had one of the battle groups. Really good guy. We did a lot of things in Washington together. We still stay in touch. Good guy.

EM: All right, let's talk about your time back to sea. CO of *San Jose*.

SA: Yeah, *San Jose*. I wasn't supposed to go to *San Jose*. Toward the end of my tour there in BuPers, the captain's list came out and I ended up being selected a year early for captain. At the time I was slated to take over one of the air wings out at Miramar. My immediate boss comes over and says, "Well, I've got good news and bad news." I asked him what's the good news, and he says, "The good news is that you've selected early for captain."

EM: That's good news.

SA: Yeah. I say, "What's the bad news?"

He says, "The bad news is we don't send captains to be CAGs. Your air wing command has just gone."

I say, "Wait a minute. Can I not accept captain early and still go to my air wing?" No. The deal is captains go to captains' jobs. I say, "Oh, dear me, I really want to go to that air wing."

He says, "Well, you're not going to go to that air wing." I ask him where am I going to go. "The good news, the other piece of good news, is that you've screened for ship command, so you're going to go to *San Jose*."

EM: Is that good news?

SA: Well, it was good news that I had screened to go to a ship. I thought that was going to be good, but I wanted to go to an air wing much more than I wanted to go to the ship. I was sort of crushed that I wasn't going to go back to flying.

EM: Was it traditional for aviators to go to a stores ship, or would they go to destroyers, cruisers?

SA: At that time, aviators' first at-sea command was one of the service force ships. An ammo ship or a storage ship or the AOEs [fast combat support ship] or an amphib. *San Jose* was a great, great ship. The best ships, great ships to command. I went off to Alameda to join the *San Jose*. The guy I was to relieve there was James Austin. Jim was a fighter pilot, nuclear qualified. He was going to go to—I think he went to *Enterprise* to be CO.

EM: Well, how do you prepare for that? I mean you're an aviator, you're a flyer. What do you know about driving ships and—

SA: If you go back to my late lieutenant days and early lieutenant commander, I did a ship's company tour on *Bennington* as assistant navigator. While there, I qualified for fleet OOD underway, CDO underway, CDO import, so I had all of the shipboard quals that I needed, and I felt very comfortable—loved to drive ships. I mean, ship driving is a lot of fun, and I know why some of the surface guys don't enjoy driving the ship. They enjoy the command, and they enjoy the camaraderie and bringing the team together and all of that, but I found a lot of them didn't really enjoy the moving of the ship.

EM: So you liked getting out of port and coming into port?

SA: All of that.

EM: It's kind of hairy, isn't it?

SA: Yeah, anything about moving the ship around was. I guess that's the aviator instinct, you know, you like to see things moving. I got there and relieved Jim, and I think it was on a Friday. On Monday we got under way. I had come up to the bridge. Of course, I had been under way with [Eugene B.] Red McDaniel, one of the POWs.

EM: Oh yeah, I've met him.

SA: He had command—

EM: In fact I worked with his son.

SA: Oh, is that right?

EM: Yeah, Mike McDaniel worked under me, actually.

SA: Well, Red commanded one of the other AFSs [stores ship] out of Alameda, and so I went to sea with him for two weeks, or a week—something like that—before I came over to relieve Jim. I got to watch the routine on the AFS and get familiar with the class of ship. I decided that I might as well show the troops I'm not afraid of this thing and said to the OOD, "Okay, I'll get it under way, and as soon as the pilot and I get it free from the pier and we are headed fair out the channel, I'll turn it over to you." I recognized right away that there was a lot of hesitancy on that bridge team. They all sort of backed away, watching carefully. I finally figured it all out. I guess Jim had been a little volatile on the bridge. He was prone to—

EM: Outbursts?

SA: Little outburst every now and then, so they were all waiting for the first outburst. We got under way and headed fair, and I turned it over to the young OOD. That hadn't happened quite that way, I guess, before, so he was really happy to have the ship. We went off to sea. Wonderful ships. The mission was a good mission and the ships were, too.

EM: How involved did you get with the UnRep [underway replenishment] procedures?

SA: Very. The UnReps—being on the carrier side of things, it looks a lot different when you are over on the other side. When you are the UnRep ship, you are the guide. So you sort of dictate to the big guy, okay, this is the course and speed. And then you've got to have a really good bridge team and make sure they are well trained. I had brand new UnRep gear. During the time that Jim had been in port they had installed a whole new series, a new mod to the UnRep gear that had not been tested out in cold weather before. One of the first things I got to do—well, it wasn't the first thing, but it was one of the fun things—was they wanted me to take it up to Alaska to get UnReps in the cold weather. So they sent me up to Alaska in November. Well, Alaska in November is a little problem, you know, at least Anchorage. It's a little problematic about how bad the weather is going to be, whether it's really going to be cold or whether it's not. It turned out to be almost like fall.

We were up in Anchorage waiting for this freezing cold weather, and none was available until right at the end. I think the day we finally got under way to come back out of there was snow. We got to see some snow, then we got to see some pretty heavy weather coming back down to Vancouver for a port visit there. The thing about Anchorage, when you moor in Anchorage, your range of tide is 30 feet, so you've got to have line crews handling the lines almost the whole time you're alongside the pier, or else you're—

EM: Up and down?

SA: Yeah, 30 feet for the range of tide. They have a pretty swift current when the tide is coming in and going out, so you had to time your departures and arrivals pretty closely. Nobody's around to bother you. You certainly are your own boss. You check in by message to your guy at ServGru 1 [Service Group 1].

EM: You reported to ServGru 1?

SA: Yeah. Other than that, you are out there by your lonesome doing your thing until you finally get somebody that wants to UnRep. The size of the crew was good, 500 and some. You get to know everybody.

EM: Was it a fairly new ship?

SA: It was. *San Jose* was the last of the AFS class, and so she had all of the mods. There's a big passageway down below that traverses the ship, and you can bring some stuff from the store rooms up to this big alley, really, that the fork trucks could go back and forth on. The first AFSs, the alley would only take a forklift one at a time, so you had to go back, drop a load, and back, whereas AFS-7—I think 5, 6, and 7 are the same, I'm not sure, maybe it was 6, not 7—but we had the double-wide alleyway so you could run forklifts going both directions. We could move a lot more stores a lot faster. They were good. We had a helo det on board. I got to fly with those guys, fly the H-46 a little bit just for fun. I couldn't UnRep; that was beyond my skill set.

EM: How about VertRep [vertical replenishment]? They [helicopters] have got to hover just in a certain way?

SA: Oh, yeah. What they did was if we were really going to push a lot of carriers, they wanted to get done just as fast as they possibly could. With the H-46, the good drivers of those things could hover over the pallet, pick up the pallet, and then they would rotate around the aft rotor. They would just spin right around and do a 270 spin and be pointed right at the flight deck of the carrier, drop it, do another spin, and be right back on top of you. I mean, they could move those pallets. It was just amazing how they could do that. So I had a really good time with *San Jose* and got to deploy with it. We spent most of our time over in WestPac dodging typhoons that year. Ended up going into Taiwan right after one of the typhoons had gone through, and the berth that they assigned me to was in the channel, a really narrow channel where you come into Taipei, and the pier is in the channel, so you come in from one direction and go to leave in the other direction.

EM: Did you have a pilot?

SA: We had a pilot. As we were coming in, we're going by ships that are sitting on their sides.

EM: Didn't quite make it?

SA: It was sort of interesting. It was a great port visit. We had a good time.

EM: What challenges did you have on the ship?

SA: Well, we are in Alameda so we are part of the San Francisco scene, and we're talking about the years '76 to '77. UC [University of California] Berkeley is on the other side of us. So we were sort of caught up in the social scene around the Bay Area. We did have some issues with the crew.

EM: Drugs?

SA: Drugs and basically not wanting to be led; wanting to do their own thing. We tried to capture [engage] them and get them to do their own thing [within Navy structures]. For the most part, though, you would sort of think maybe the service-group ships would be second rate in the talent level. What we had were people, some of them are on their second sea-duty tour, and this might be their second time with this service group ship. So they were sort of tuned into the mission. The Supply Corps guys were first class. The Supply Corps and the supply ratings were really top notch. Engineers so-so, but the deck people were top notch. The guys with the UnRep gear, most of them had done multiple tours with UnRep gear so they knew it well.

It was just the young sailors coming aboard, and of course at that time we were into this—what was the word we called it in those days—we call it sensitivity training now. But we had these teams that came around and have to get you and your crew to form up teams, and you would go off-site for a couple of days. Race relations—that was it, race relations.

EM: So this is a follow-on from the Zumwalt years, and Admiral Holloway kept some of that in.

SA: Yeah. These sensitivity training sessions became sort of problematic in some aspects. Sometimes they raised issues that were not issues, so it was good that the officers went as well as the enlisted, because it was like—I was in one session, and the guy was saying that if you go someplace and somebody doesn't say hello to you, what does that mean? They were always trying to spin it that that person had some sort of issue with you. You need to understand that he may have some issue about who you are.

I said, "Wait a minute, that's not true. Part of my life I moved around, all the time when I was kid, so I kept going to different places. One place that we always came back to was this little town in southeastern Ohio. In that town, until somebody got to know you, you're an outsider. All they are doing is looking at you and waiting to see if you are one of the good guys or one of the bad guys. It was incumbent on you to show a side that shows that you're part of the team, and not waiting for somebody else to step up to shake your hand." I said, "Nobody there was going to shake your hand until they knew who you were. If they knew who you were, you know, you got a handshake. If they didn't know you, they wanted to know who you were. You had to sort of prove yourself, and it wasn't a distinction of whether you were white, black, or red. The distinction was that you didn't come from here."

I said, "You've got to have alertness to all of these signals, but this blanket saying that if someone's not really friendly to you the first time you see them that they've got something against you, that's just not my experience." Those programs were—I know what they were designed to do, but sometimes they would go awry.

EM: Since the ship was in Alameda, did you have trouble with demonstrators outside of your gates, or anti-war, anti-military?

SA: No. Most of the demonstrations and stuff took place over in San Francisco. Demonstrators were smart. They wanted to go someplace where there was some action, so they would be over in San Francisco or they would be up the hill at Berkeley, but very few outside of the gate at Alameda. Alameda was sort of a compound unto itself. Some of the kids went to school right outside the gate, and some went to school inside the gate.

EM: Did you have good officers?

SA: Yes. I had some great officers. My LDOs [limited duty officers] were super. I had an LDO first lieutenant. I had an LDO who was the mess-management specialist, the guy that took care of all of the messes on the ship. The commander was a Supply Corps guy. My SupO [supply officer] on an AFS was almost of the same breed as the SupO on the carrier. It was a very plum job for a Supply Corps commander. His second in command, a lieutenant commander, was a very bright guy. All in all, it was really a good experience. My XO, for most of the time that I was there—first XO was there just a short period of time with me and then second XO, Tad [Theodore] Derf, was an LDO. He had been a tugboat skipper and a salvage boat skipper. He was a real old salt. He made no bones about it: "An aviator captain, sheesh." [Laughter] "What have I done to deserve this?" [Laughter] He thought he had been shipped off to Hades.

We've stayed close over the years. What a delight. We had more fun, and he never fails to mention it anytime we get a chance to talk, what a fun trip he had with me. He'd never been across the line [equator]. He had never been at an across-the-line ceremony. Of course I had, several times, and that really burned his butt. So we come out of Singapore and we go down to meet the *Midway* [CVA-41] coming out of Australia, and we're going to cross the line, and he was just worried sick that he was going to be ignored in this whole thing. That he really wouldn't get to experience the full thrill of being a pollywog, becoming a shellback. Boy, I fixed it up for him.

I had a SEAL [naval commando] detachment on board at the time. A couple of the SEALs hadn't been across the line either. The master chief with the SEALs, none of his officers had been across, I don't think, but he had been. The master chief and I decided we were going to take charge of this and make sure that these guys who were so worried that they hadn't been across the line were going to get the full experience. We fixed them up good. We had them hung in a cargo net over the flight deck with fire hoses spraying them down, and we did everything we possibly could do, and when it was all over, we almost had to put the XO to bed, he was so worn out. He finally came up late at night and he says, "Skipper, that was the most wonderful day I've ever had in the Navy."

EM: Oh, that's good.

SA: He says, "I was so afraid I was going to miss out on all of this stuff, but you made sure I didn't."

EM: That's an age-old tradition.

SA: Of course today I would have probably been fired for some of the things that we did, but it was good. We opened the whole thing up with the SEALs duck-walking all around the ship. The whole ship got to see the SEALs going along, *quack quack quack* duck-walking, so that sort of opened the stage for where we were going from there. Then on the way back, we were released from our Seventh Fleet duties as soon as we got *Midway* UnRep'd, so what I did was, since we were down south coming back up, instead of coming up north of the Philippines the traditional way, I brought them back by way of the Java Sea through Leyte Gulf, and what I had done was get some charts made up of the history of the battles at Leyte Gulf and the battleships coming down and where all of the sites were that the battles took place. And then as we passed each one of them—

EM: Surigao Strait?

SA: Yep, Surigao Strait.*

EM: Admiral Holloway tells a great story; he was our after-dinner speaker once. He was a young JG up in the fire direction position by the radar, and he's in Surigao Strait with his destroyer division right next to the shore so the ship could be hidden. And here comes the Japanese fleet. He could see them—said he had binoculars, and the ships filled the lenses. The U.S. ships opened up, and he could see they were taking them [the Japanese ships] down. Quite a story.

SA: I'll tell you, the thing that amazed me was navigating through there and realizing the Japanese used to come racing down there in the full darkness. Those guys did some really fancy seamanship. It was really good for the crew. They enjoyed it and got a little bit of history, and then we got to Hawaii.

EM: Did you have one of your officers get up and read a little history lesson?

---

* During the World War II Battle of Leyte Gulf, a U.S. fleet of battleships and other combatants almost completely destroyed a mighty Japanese armada attempting to pass through Surigao Strait.

SA: As we passed each key point, we would point it out and say what the battle of the day was, and then we would always have a chart that they could go down to look at. Then we got to Hawaii and we had a Tiger Cruise, so all the guys could invite their kids to come out. And the new skipper who was going to relieve me came out so that he could ride the ship back to Alameda. We ran into my old friend from VA-122, my skipper when I left 122, Nick Nicholson, who by that time was the head of the JNROTC [junior naval reserve officers training corps] over in Sanger High School there around Fresno, California. Nick had sent me a message saying, "Could I send some of my JNROTC students over to come back with you on your Tiger Cruise?" I said sure. I said, "Now, realize that we are going to get into Alameda on Christmas Eve day, so anything can happen."

"Yeah, I'll tell all of the parents." So he sent out this group from Sanger as well. Two of my sons came out.

EM: You got back on Christmas Day as planned?

SA: On Christmas Eve day we came back, but I tell you, we had the roughest crossing that we had ever had. We were in terrible seas. Everyone was seasick. The kids were thrown out of the racks. The new skipper, Len [Leonard G.] Perry, was seasick the whole time.

EM: You never got seasick?

SA: No. It was just an awful crossing, and I was scared to death that we weren't going to make it in time. You know, the kids would miss Christmas and everything.

EM: Well, they got a taste of life at sea.

SA: Yeah, life at sea for sure. It was really good, the *San Jose,* and I really feel sorry for the aviators today because they don't get that experience now.

EM: They don't get to command a surface ship?

SA: Well, they do, but that will probably be their only command. They will go to an amphib, or they will go in the pipeline for the new carriers and go in as XO, and then they will fleet up to CO. But having that individual, out there by your lonesome, going wherever—

One of the stories on the trip up to Anchorage, Alaska: My navigator was a young lieutenant, and he came to me and said, "Skipper, I'm going to plan our transit up to Anchorage."

I said, "Okay, what's your thinking?"

He said, "I want to go through the inland waterway up to there." I said we'd take a look at it and see. So he mapped it all out, and I finally said no.

EM: Too dicey?

SA: "It's going to be a little too dicey for me. I think the work that you did is great." Because I loved my job as assistant navigator on the carriers, I said, "This is really good work, but it's just a little too tight for me."

EM: Too many shoals or narrow passageways?

SA: Narrow passageways and the winter weather. A different time of year, I might have tried it. I was pretty sure that ServPac [Service Force, Pacific Fleet] wouldn't have approved it. Years later, I'm ComCarGru [commander carrier group] out of San Diego, I'm on the carrier and the pilot comes on board, and I see the pilot, and I ask, "Who's the pilot on the bridge?" It's this young kid after he got out of the Navy.

EM: Is that right? He liked doing that then.

SA: He loved it. We talked about it then. It was good. The young kids on there, the surface guys, they weren't in the mainstream of the surface Navy. I had one young fellow—in fact more through his wife than my wife, they stayed in contact. He had made his decision to get out of the Navy before I got on board. We are coming into San Diego. I used to love to come into San Diego without tugs, which most guys didn't want to do. I loved to come in there with some of the destroyers that had the big bow-mounted sonar. They couldn't turn very well—that daggum bow-mounted sonar was like another rudder up there, so they couldn't bring their bow around like I could. I used to come in there, and for refresher training you would have to come back in port to pick up the training observers, and then you'd have to get back out. I would come in without a tug, and spin around and pick up the guys and go out, and the destroyers would come in and have to get a tug to push their bow around. I used to just give them grief. "Oh, you destroyer guys can't even get it turned around out there in the bay without a tug."

So I was letting this young kid bring us in one morning, and he wasn't correcting for the drift that was pushing us off to the right side of the channel. I'm coaching him a little bit, just saying, "Make sure you know what your drift is and where we are in the channel." I bring him out on the starboard wing, because then you could see the buoys as you line up going up the channel real well from the starboard wing. And he's still not getting the picture quite right, and he edges over a little bit, but he just can't get himself to bring the bow over to the left enough. I'm letting him try and try, and finally we are coming up on this buoy, and I say, "You've got the buoy?" Yes, got it. "Okay. Going to be a little close here." Okay. Finally it gets right down beside us, and we're going to hit it if we are not careful, so I say, "Okay, I've got the con."

I turned out of the channel just to swing the stern around. As soon as the stern cleared the thing, I hard-ruddered back into the channel. So when it was all over, he said, "I should have seen that a little bit sooner."

I said, "Yeah, but now you know how it looks when you get in close, how fast it narrows down."

He said, "Yeah. Thanks for letting me do that." The day he checked out, he said, "You know, Skipper, if somebody had done that for me when I first got here, I wouldn't be getting out."

EM: Really?

SA: I said, "Well, I'm sorry you are."

The commodore, at one point in time—we had a problem with one of the other ships—he called me over and wanted me to go over and investigate what was going on. I came back and told him what I thought. He says, "Okay, I've got to ask you something. Did you know 50 percent of my ships are commanded by surface guys and 50 percent are commanded by aviators? I'm finding that the aviators train their people better than surface guys." I said he was absolutely right.

EM: Training in what facet?

SA: On ship handling, and I said, "You're absolutely right. Do you want to know why?"

He said, "That's why I'm asking you the question."

I said, "Well, for aviators, we never stop training. We get our wings, we go to a squadron, and when we are in the squadron the old guys are training us, and we are training the next ones behind us. There isn't a day that goes by we aren't out flying airplanes or doing something, that we are not training somebody. Not just ourselves, but we've got somebody on our wing, or more or if we are in a multiengine plane, we've got a whole other crew that has to be trained in that airplane. We never spend a day in a mission just operating. You are always training. It's sort of second nature to us to see things that somebody hasn't seen before and to show them what it is we see. That's the difference. I don't know why the surface guys don't seem to have that same mental push. Maybe they do, but from my perspective, it's all from the way we were taught."

EM: Don't they go to surface warfare school?

SA: They go to all of the schools. It seems like that personal touch—just like letting that kid get into extremis. I found that the surface guys were very concerned about somebody getting dinged for being out of the channel or some breach of seamanship. Where the aviators were more in line to say, hey, we're going to let somebody take it as far as we think they can go, and if they can't, then we will take charge of it. I said, "A lot of people just can't stand to let it go that far."

EM: Let them make a mistake to learn from it.

SA: Let them make a mistake. I said, "The other thing about aviators is that we are taught that basically you can't live long enough to make every mistake, so if you don't take advantage of learning mistakes from the other guy, you're not going to live long enough to make them all. Let the other guys learn them for you."

I just had a great time on the *San Jose.* One of the best. To think about not flying an airplane was pretty brutal, but Tom Kilcline, my boss, told me I would like it. He was a great ice cream fan, so before I loaded up to go out to the Philippines—his job now was ComNavPhil [Commander Naval Forces, Philippines]—so when I got to *San Jose* all ready for deployment to WestPac, I got hold of the supply officer and I said, "Okay, enough of this commercial ice cream stuff. I want you to go over to Baskin Robbins, and I want you to buy a couple of pallets' worth of ice cream. Here are the flavors I want. We are not dispensing—that's for our ship's consumption, and for me to pay a couple of debts with."

He said, "Okay, Skipper, I got it. I'll go get you some Baskin Robbins ice cream."

We got to the Philippines, and I called Admiral Kilcline and said, "Here's this captain that you told had to go out and do a captain's job. We are going to be in port for a few days."

He said, "Why don't you come up to the house and have dinner?"

I said, "Okay, that will be fine. I've got something to give you anyhow, so that would be good." We got up there to the house, and I showed him what I brought. I had several gallons of ice cream.

EM: Different flavors?

SA: Yeah, different flavors. He said, "It's funny you should remember this. You know what I'm doing right now? I'm trying to get the Navy Exchange down here to offer a contract to open up a Baskin Robbins. [Laughter]

EM: Anything else you want to say about *San Jose?*

SA: Nah, just beautiful ship, beautiful people, great mission. A wonderful mission.

EM: You got promoted to captain in August of '76. Did you have, over the years—here you were are an 0–6—mentors, people that were ahead of you that helped you along?

SA: Yes. There was always somebody that took an interest in me almost every command that I was at. I finally inherited some mentors with the big M, meaning somebody high up in the chain that sometimes takes a look at you and pays attention to what is going on.

EM: Who were they?

SA: Wes McDonald was a mentor of mine from the standpoint of when I was at BuPers, and I really got to know Wes. I knew him from the Lemoore days, but we hadn't really interfaced since the Lemoore days. But when he was the Chief of Naval Air Training, I went down there every quarter to give a brief to his squadron COs all about what was going on in the bureau and what we were doing with the students. Where they were going and what it looked like for them, and also what their careers looked like, because for most of the training command Cos, that was sort of their last flying job, unless they became one of the wing commanders in the training command itself. They didn't usually come back to the fleet.

So every time I was down there, I would touch base with him and see what his concerns were about his people. Did he have any particular people that he wanted us to pay particular attention to or anything like that? So we got to be comfortable being able to talk together. Then when he came up to be the deputy to Admiral Watkins at BuPers, with some of those episodes that occurred, he and Admiral Trost were very supportive of me and sort of flew cover for me when things were difficult.

EM: And Trost was a submariner.

SA: He was one of the big M guys. I think I went to CinCPacFlt [Commander in Chief, Pacific Fleet] after my carrier command, because he was deputy CinCPacFlt at the time, and I'm pretty sure he steered those orders for me. He never said so, and he wasn't there when I got there. He had already detached. But I have a feeling that the relationship that we had in BuPers was one of the things—and then of course I worked for him in the building when he was CNO for a while. Wonderful guy. The story of when we were in the bureau—the days that I was in the bureau were when we had the big gas rationing in D.C.

EM: '76, wasn't it?

SA: Yeah, and he had a carpool, and at 1630 Admiral Trost would walk out of his office, walk down the steps, go out on—what's that pike that goes right by BuPers?

EM:110? By the annex up on the hill, you mean?

SA: Yes.

EM: Well, 95 or Washington Boulevard?

SA: No. It's Columbia Pike.

EM: Okay, yeah. Right.

SA: He would walk out, stand right there by the little bus stop, and his ride would come by and he'd hop in and go home. I mean, it was like clockwork. That was him. He was such a bright, bright guy. Gosh darn. He was one of my big M guys I think. But the mentors—

The first squadron I was in, or actually the second squadron I was in after we had split into VS-29, the maintenance officer was a lieutenant commander at the time. I think I told you about the first squadron, where I had the first class petty officer that always mentored me and gave me the squadron instructions to read in the mornings. Then Bob Colvin, the maintenance officer when we had a detachment to go over to an Air Force base while we were in Japan for one extended period on one of the deployments, he designated me as OIC. Now, I wasn't the senior lieutenant around, so he reached down a few levels to grab me, and I didn't work for him. At that time I was in the ops department, not the maintenance department, but he had convinced the ops officer that he wanted me to run the detachment over there for the maintenance part of it.

All along the way there was always somebody that gave me another little push. Bob Kirksey, when I got to 55, he took us transition guys, took time to understand each of us and what we needed, and spent some time with us and helped us grow into our role as a nugget jet pilot, but also as a mid-grade officer with other responsibilities. He took time to make sure that we felt comfortable doing that. Along the way there was these folks that pop up every now and then. Tom Kilcline from the days at BuPers, and then the relationship we developed when he was ComNavPhil and I was be-bopping in and out of port there.

EM: Were you too low-ranking for Admiral Holloway to take notice of?

SA: Yeah. I didn't really know him much until after retirement. We crossed paths several times then. In fact I convinced him to come back and help us at ANA [Association for Naval Aviation] one time when I ran into him at the airport out there at Washington National. He was sitting there and I said, "Admiral, I need your help. Can you come back and help us a little bit?"

He said, "Sure I can."

EM: Okay, you're off to Naval Reactors, Idaho Falls. Back to your science background, right?

SA: Remedial Reading 3, as I called it. This is the outgrowth of SOSMRC.

EM: What is SOSMRC?

SA: Senior Officers Material Readiness Course. At the nuclear training facility outside of Idaho Falls and, of course, this is where all of the "nukes" are trained. It was the nuke kingdom for [Hyman G.] Rickover. After we came out of Vietnam and the horrible conditions of our ships continued, we just couldn't recover from it. It just got worse and worse and worse. They had all of the ship fires and ship casualties and stuff, and so Rickover convinced the CNO that he could fix this. That the senior officers going to ships ought to be more schooled in the engineering aspects of keeping a ship alive.

EM: But that's nuclear engineering?

SA: Well, yeah, but he said the nuclear mindset is what you need. You need people that understand water chemistry—how and why it's so important to keep the boiler safe to steam and all of this other stuff. They need to understand that a steam leak is not an acceptable thing to have. You go aboard any surface ship and they've got boiler power and they have steam leaks all over the place. That shouldn't be. So, "I can teach this and we will improve the readiness of the fleet by order of magnitude."

EM: Rickover persuaded whom? Holloway?

SA: I think it was Holloway. So off we go. I'm not the first class. I think I'm the second class there. And of course I'd already had ship command, so I'm saying it's too late for me. I've already learned that. I'm already indoctrinated in the old way. Just let me go to my carrier. No, no, no. You've got to go to SOSMRC. I said, "Well, that's for the guys who haven't been to sea yet." No, we are sending you to SOSMRC.

I'm there with a great group of guys. Bob Kirksey is there. Rudy Kohn is there. Lew [W. Lewis] Chatham is there. Dick [Richard M.] Dunleavy is there. The surface guy that used to be BuPers, Bill [William R.] Smedberg [III]. Paul [F.] McCarthy, Hugh Webster, another surface guy. There is a whole group of us there at SOSMRC. It was sort of eye-opening to me, as we got into the academic part of the program, how many of our senior naval officers had no math background. I mean, they were all scared to death of algebra.

EM: I can sympathize.

SA: We had this old guy—I shouldn't say old, but he was probably in his early 60s, and he was going to teach us algebra. I thought, what the devil, you know, I've had enough algebra. I don't need any more algebra. Well, I learned something from him. But shoot, I had learned algebra all wrong. This guy had the right idea. His introduction to algebra was unlike our introduction had been in high school and college. I didn't get it in high school. I got it in college for the first time. Here's this thing with all these wild equations that you've got to solve. He walks into the classroom and says that algebra is something you use every day of your life. You just don't know it as algebra. And let me tell you what it is. You go in, and apples are sitting there on the counter. Three apples for 90 cents, five apples for a dollar. Which is the bargain? If you can answer which one is the bargain, you've just done algebra.

EM: Oh. [Laughter]

SA: So he taught it from the aspect that you use it all the time, but you don't do the formal part of it.

EM: Practical application?

SA: Yes. So he made the bridge between how you take the practical application to the more formal part of the equations.

EM: I wish I'd had him as an instructor in my days.

SA: This was great for my grandkids. They were in here moaning and groaning the other day about algebra, and I gave them the same sort of thing, and they said, "That's algebra?"

I said, "Yes, it's algebra. Now, what the teachers want you to do is be able to write that equation that answers that question from the equation standpoint. But you know what it is just from the standpoint of you being able—five for a dollar and three for 90 cents. Of course five for a dollar is the better deal. Now, there's an equation that tells you that."

It was funny watching everyone struggle. They divided the class in half. I think they did this deliberately. They put Bob Kirksey, who was a senior aviator there, as the head guy for half the class, and Bill Smedberg, who was a senior surface guy there, as the head guy for the other half of the class. The classes were divided: almost all surface guys went with Bill and all aviators went with Kirksey. We had quizzes and stuff. I mean, it was like Remedial Reading 3. It was like we were all in grade school. We had these little pop quizzes, and we had tests at the end of the week and all of this other stuff. Finally we are quite a long way into the program, and the instructors pulled Admiral Smedberg aside and said that his half of the class is not doing very well. This is all shipboard stuff. Those aviators are cleaning your clock over there. [Laughter] So then the competition. Smedberg is over there saying that okay, this week on Friday night we are not going down to the club.

EM: The club?

SA: We had to go off the site to get to anyplace. We were going to spend the weekend here studying and getting these grades up. It really got to be fun. What we did was, half the guys lived in town in Idaho Falls, so they got apartments and lived in town. The rest of us lived at the facility.

EM: Was this unaccompanied?

SA: Yes. And it was a long drive from there to out in the middle of nowhere where the nuclear site is. The other half of us lived on site in these old Quonset huts, and we weren't allowed to have TVs. So we bootlegged TVs. We hand-made antennas so that we could watch Jane Pauley in the morning while we were having breakfast. On Friday, usually early afternoon Friday, we were released for the weekend, and it was just like a covey of quail—we just zipped, we were out of there.

In Idaho Falls it was ski season and we were up at Blue Sky in Montana. We were over at the Grand Tetons. We were over on the west side by Squaw Valley or Snow Valley. Snow Valley, I guess it was. We were at all the big ski resorts, skiing on the weekends, hooting and hollering and having a good time. Wednesday night it was Howe, Idaho, a little town that was off from the site, still out on the valley floor, and it was about a 40 minute drive out to Howe—two turns. You come out straight for 20 miles, turn left, straight for 20 miles, and you're there. We used to have speed races and see who could get there the fastest, because there was nobody out on the highways. They always had fresh-baked pies. So Wednesday night was our night to go up there, have a few beers, have a big steak and fresh-baked apple pie or whatever, cherry pie—whatever your poison was—and then race back to the site and be back in. It was a great bonding experience because all of us that were there got to know each other really well.

EM: All captains?

SA: Yeah, all captains, except for the two admirals. Well, there was one other admiral, a submarine admiral there, too. We called him the ghost, because he would disappear.

EM: Did you have any nuclear-engineering training?

SA: We did get some nuclear. What we did was, we learned to do the water chemistry like the nukes did it. We got taught welding techniques that the nukes did. We got to do a little welding ourselves, play around with that. We learned the PMS [preventive maintenance system] schedule that the nukes used, so we learned a lot about how they approached the maintaining of their equipment, which for the aviators was second nature. For the surface guys, it was an eye-opener. To me, when I had *San Jose* it was an eye-opener, and when I was ship's company on *Bennington* and the other ships. I always used to grade exercises on any ship that I was on. If they needed an observer for an observed exercise, I would always volunteer to do that just because it was interesting to me. The difference between an aviator's perspective of maintaining airplanes and a ship driver's expectations on maintaining a ship were just at opposite ends.

EM: Do you think that is because when you're in the air, everything's got to work? There is no room for error?

SA: When you're in the air, if anything goes wrong, you're the guy that's going to pay the price. Whereas on the ship, yeah, maybe the ship will go dead in the water, but on the other hand, you're not the guy down there that's trying to fix the steam that's spraying clear across the fire room and stuff like this. You're not as personally involved with it as you always are in an airplane. If anything breaks while you are in the air, then you're the guy that has to sort of manage the consequences. You may or may not be the guy that has to manage the consequences when you're on a big ship.

EM: So that was a valuable experience for you?

SA: It was. The best memories of it are the fun times we all had jabbing each other and cavorting around on the ski slopes. Some of us had never skied in our lives, especially for a lot of the aviators. Myself, I had always wanted to ski, but I was afraid I would break a leg and I wouldn't be able to fly an airplane again. But I knew I wasn't going to fly an airplane again, so I could break a leg.

EM: You didn't break a leg, though, right?

SA: I didn't. A lot of the other guys did.

EM: Is that right?

SA: The funniest stories, we had. It was almost at the end of the season. It was probably March or April—we got out there in January, and I think we finished up the first week in May or the last week in April. So it was either the last of March or the middle of April, somewhere in that time span. We are over in the Tetons. The name of the town escapes my brain this second. We are skiing on this big slope in the Tetons, and all of the ski shops had just put all their gear on sale. The real hardcore ski guys have gone into town and bought themselves top-to-bottom—I mean, big fancy ski pants and jackets, boots and skis and bindings, and they bought the very best. Sam was a fighter pilot out of Miramar. He had really gone big time, and there is a group of us coming down this long slope together, and Sam just goes blowing by us.

Halfway down the slope is this little knoll off to the side that you could sort of come up and stop at the top of, and then roll down the other side and pick right back up going downhill. He's so far in front, and he's showing off. He does this big swooping, sliding turn as he comes up this little knoll to stop and look back at us coming down, and all of a sudden he sort of crumples. Well, what he had done, he had dug his ski in at the end of that slide and just popped his knee. He is lying on the ground in total agony. The first guy up to him looks down at him and says, "Can I have your boots?" [Laughter]

EM: Oh, jeez. It's called gallows humor.

SA: He didn't finish the course. We lost him from the course.

EM: Really?

SA: Bob Kirksey got hurt. He busted a rib. Hugh Webster and I had a major collision up at Yellowstone. Hugh was Naval Academy—played line for the Navy football team. He was really a big guy. And I wasn't that tiny then. I still had some of my heft left, some of the weight was still there. We were coming down this slope. Hugh is way off to the side of me, and I'm coming down the slope, and all of a sudden, out of the corner of my eye, I see this blur. Well, the blur is Hugh coming from about my four o'clock position, and he's going to hit me for sure, and I'm trying to ease a little farther away, and he's still coming. He ends up going across my skis, in the front. When he goes across the front of my skis, one of his skis lifts up and he ends up circling around in front of me, falling as he's going. So by the time he hits the snow, I run right across his back, over him, and on down the hill. We both recover from this and look at each other, and we take off our skis and say, "Our skiing days are—"

EM: Over? [Laughter]

SA: Over.

EM: Anything more you want to say about Idaho?

SA: At the end of the Idaho trip we take a tour. We go to different places and see some of the stuff that they want us to see. One of the places we are going to is Bremerton, Washington, to see the naval shipyard up there. And of course Bremerton has the nuke capability for the submarines as well as the carriers.

EM: Yeah, Bangor, Washington.

SA: Lo and behold, while we were there at Bremerton, *Coral Sea* is in there for overhaul. One of the side trips is to go through *Coral Sea* and look at her all torn apart for overhaul. By this time all of us know where we are going for our next command, and I was slated to come to Bremerton to take over *Coral Sea.* So we take the tour of *Coral Sea,* and she looks just like death warmed over.

EM: Built in World War II. [Laughter]

SA: She's all torn apart. It just looked awful. So I got free drinks that night at the O Club for a sympathy vote saying, better you than us. [Laughter]

EM: I went on board *Coral Sea.* Take-a-historian-to-sea kind of thing for a week.

SA: Oh, great.

EM: It was a good experience, a little taste of it.

SA: Who was the CO then?

EM: I don't recall. It's way back.

SA: She's a beaut. That class of ship was the prettiest class of all of them, I thought. They were low silhouette, great ships. That thing would turn.

EM: Why did you get *Coral Sea?* It just came up in the rotation?

SA: I guess that was up in the rotation. I don't know. I was hoping that I would get a more modern ship, but I was happy to have any command at that time. I was not happy to see her in the shipyard, and that I was going to have to get her out of there.

EM: How long did it take after you are CO before she went to sea?

SA: I got there in May.

EM: May of '78.

SA: Yeah, and we deployed November '79. We were out of there, I suspect, I think it was January or February, if I'm not mistaken. My first call was to go down to NavAirPac [Naval Air Force, Pacific Fleet]. Admiral [Robert P.] Coogan was the ComNavAirPac

[Commander Naval Air Force, Pacific Fleet] at the time, Vice Admiral Coogan. Quite a character, quite a rough and ready–type guy. I knew him by reputation and name only. I hadn't really officially met him at any time that I can remember. So I went into his office there in Coronado and did my formal reporting in. He said, "Well, Captain, you know what your job is?"

I said, "Yes sir. I'm going to be the CO of *Coral Sea.*"

And he said, "Yep, and you're going to come out of overhaul on time. You understand that?" I said yes sir. He said, "I mean on time. No ifs, ands, or buts." I said yes sir. "That's all." [Laughter] That was it.

I head up to Bremerton to relieve George [A.] Aitcheson. George was a fighter pilot, and I knew him from Alameda, because *San Jose* and *Coral Sea* were out of Alameda. George and I liked to tinker with cars, so we always had that to talk about while we were there in Alameda. I was comfortable with relieving George, and his XO was Kenny [Charles K.] Hutchinson, who was an E-2 pilot—little short guy. George wasn't that big either, so they were sort of a good match. Kenny had a great sense of humor—a lot of fun. George had a feisty side to him when things didn't go right, and Ken was sort of his opposite. He was not a very excitable guy, so they made a good team. We got the relieving process done. Overhaul is really a traumatic experience for a ship and the crew. I had really good people, really impressed with them.

EM: Why is it traumatic?

SA: Well, the ship's all torn apart. For the most part the process is owned by the shipyard, not by the ship. You're sort of a passive participant who is waiting for the final product, and your ability to influence the path to that final product is sometimes not very clear.

EM: They don't listen to you all the time?

SA: They don't listen to you all the time. So you see things that you want done differently, and they say hey, that's not in the package. That's not what they are supposed—that's not what they are signed up to do, so tough luck. There can be lots of friction there.

EM: Was there something you wanted done that they wouldn't do?

SA: By the time I got there it was sort of out of my hands. I couldn't influence the process other than the fact of what I could do to help ease or move toward the completion date. There were lots of things while I was there that I wanted to get done, but affordability wouldn't allow it. What had happened was, *Coral Sea* had gone in there under the premise that she would never be a fully operational carrier again. That she was going to go in there and be prepared to be the training- command carrier to replace *Lexington* [CVT-16]. So her job was going to be doing all of the carrier qualifications for the training command and for the fleet when there wasn't a fleet carrier available to do carrier quals. She was going to be a carrier that came in and out of port. Didn't do any UnReps at sea. Didn't have any place to go but to get to sea and get into the wind and bang airplanes off the front and capture them on the back. Things like the refueling stations weren't in the work package. Things like piping to the fuel tanks wasn't in the package. Things that were essential for a fleet-going carrier were sort of out of sight, out of mind. As I kept reading the tea leaves as we came closer and closer to completion, the fact was we weren't going to be the training-command carrier, that I was going to be told that we were going to deploy as soon as we got ourselves ready to deploy—I'm wringing my hands a little bit.

EM: Was Coogan aware of this?

SA: Coogan was aware of it. He knew that the package was not complete. He didn't know some of the stuff that I knew, and some of the stuff that I didn't know that I didn't find out about until the first UnRep that we had. The main thing was getting out of there, and of course now we are in to the PEB [propulsion examining board] standards for what we had gone to Idaho Falls for, to increase the readiness of the engineering department and the professionalism of the engineering department. We had to have the big inspection to see if we were safe to steam to come out of there, and we were having a devil of a time trying to accomplish that. Part of it was some of the stuff had not been overhauled because it wasn't going to be essential for the new mission. It was really getting tight relative to getting out of there. Damage control was way behind schedule. Somebody tipped off Coogan that we were going to flunk everything up there, so he came up to pay me a visit, chew on my ear.

EM: Chew on your what? [Laughter]

SA: My ear, and make sure that I understood—his first message to me was down there when we first met—that we're going to come out of there on time. I had to walk him around the ship and say, "Look, it's not pretty, but it's not as bad as you're being told. We are going to make it, we will make it out of here on time and we will be able to steam." Okay, make it so. We did. Money was tight in those days, of course. The first challenge was that once we were given the okay that we were safe to steam and we could get out of there, we were tasked to go ahead and take as many dependents as we possibly could back to San Francisco on the ship. Vehicles, kids, dogs, cats. [Laughter] You name it.

EM: Was this to be like a cost-savings thing for the government?

SA: Yeah. This was to take down the cost of moving all of these people back to Alameda.

EM: Household goods?

SA: We had some household goods. I mean, you should have seen us coming out of there. We looked like a gypsy ship. All of these cars up on the flight deck. And it's winter, so the seas are rough. I've got my family on board, along with the other thousands, and off we go to San Francisco. We get down off of San Francisco, and out there by the Farallon Islands is where our turn-in point is to come in to underneath the Golden Gate. Well, it is rough and it's fogged in in San Francisco Bay. The Coast Guard says we can't enter. Too foggy. I've got all of these dependents out there, seasick as dogs. Things are just bouncing all over the place. I mean, it's really rough. My wife is seasick and the kids are seasick. With *San Jose,* I'd been in and out of San Francisco a lot.

EM: Did your family ever say to you, "Dad, this is pretty nasty business here?" [Laughter]

SA: They were not impressed. I had been in and out of there. I considered myself reasonably good at radar navigation. Even though we hadn't had a whole lot of practice in the shipyard, we did the drills and I had confidence in my navigation team, and I knew I knew how to read a radar scope coming in to San Francisco Bay. So I told the Coast Guard, "I don't care what you say, I'm coming in. Now, I won't go past Treasure Island unless the fog breaks. I'll anchor at the Treasure Island anchorage and not go under the Oakland Bay Bridge until it clears. But I cannot stay at sea in this rough weather with all of these dependents and stuff on board." They were very unhappy with me, but it turned out okay. That's what I did. I shot an approach to the anchorage there right beside Treasure Island, and just before we let go of the anchor, all of a sudden, one of the news helicopters that had been tracking us going through the fog bank, all of a sudden he was right there. I could see him, and the next minute I could see the Oakland Bay Bridge, and I said, "Oops. We are not anchoring, we're coming on it." [Laughter] "I can see the Oakland Bay Bridge, and we're coming on in. Tell the tugboats and the pilot to meet us there, and we will turn into the channel and come into Alameda." So we did that. We got everybody in. Then we had to answer a few questions about all of that.

EM: Who, to the Coast Guard?

SA: To the Coast Guard and to the world's finest mayor, Dianne Feinstein. She made a big stink about—"they left me stranded up at Coit Tower."

EM: Who left her stranded?

SA: She claimed that we didn't tell her we couldn't bring her out to the ship because of the weather. We had. We talked to her chief of staff and said, "Hey, right now we can't see well. We are doing a radar approach and we can't see anything."

EM: She wanted to go out to the ship?

SA: We had invited her to ride the ship in under the Golden Gate and come in with us. We called way in advance and said hey, the weather is just not going to cooperate.

EM: So she thought you did it out of spite or something?

SA: Yeah, she thought we did it on purpose. That soured my relationship with her forever.

EM: So you made the papers, then?

SA: Yeah. But it was good. Alameda was a great place. Of course, we were familiar with it from *San Jose.* We had good friends over in San Francisco. The city had adopted the *Coral Sea* as San Francisco's own, so there was always some benefit or something going on where they would invite the sailors over and do good stuff for us.

EM: So, you left Bremerton on schedule, on time?

SA: Yes.

EM: Admiral Coogan was happy about that?

SA: He was happy about that.

EM: Let me ask you this: in your career, did you ever get a lukewarm fitness report or even a poor one?

SA: Yeah, yeah.

EM: I mean, you don't have to go into detail if you don't want.

SA: I probably had some fitness reports that—well, of course I wouldn't survive in today's Navy.

EM: So there was more forgiveness then?

SA: Oh yeah, yeah. The system was a little more honest then. The fitness reports, I guess—we will go back to BuPers. The first thing that happened when I got to BuPers was, my young lieutenants came up to me and said, "Hey, boss, we pulled your records for you, and you can go back and review all of your fitness reports."

EM: Your records?

SA: Yeah. "Anytime you want to see them, we've got them here for you and you can review them."

I said, "Why would I want to do that?"

They said, "Everyone wants to review their record."

I said, "Let me tell you a story. When I was in VX-1 down in Key West and I was doing a weapons project, at one point in time I came up to Washington to brief the people in NavAir, who were over in the old barracks buildings on the mall at that time."

EM: Main Navy?

SA: Main Navy. I said, "We finished up early one day. Finished the briefs, and I was ready to go back to Key West, and somebody said, 'While you are in town, why don't you go over to BuPers and read through your jacket?' I said, 'Gee, us West Coast guys, none of us ever thought about doing that.' But the East Coast guys were really tuned in to this. VX-1 was full of East Coast guys. So I said okay, I guess this is something I'm supposed to do. I go over and see this little old lady and say I'm here to review my jacket. 'You sign for it and you go over there and sit there. Don't go anywhere where I can't see you. Don't you take anything out,' and that sort of stuff.

"I started reading these fitness reports, and I couldn't make heads or tails out of them. I said, jiminy, I don't know if this is good or is this bad, and what am I supposed to do about this? I probably read for about 10 minutes, and I just closed it up and turned it back in and walked out saying, 'I'll never do that again.' You can just put it back."

EM: Well, what did you not understand? I mean there are narrative portions and then there are things checked off.

SA: These numbers, you know—

EM: The narratives must have told you something.

SA: The narratives told me something, but some of them didn't seem to match with what people had said when I was there. [Laughter] I just said, I'm happier not knowing. I'm going to go out and do as well as I possibly can. Either it is good enough or it isn't. I just can't worry.

EM: Did you ever think that a bad or lukewarm fitness report really kept you back, where you thought you were due something?

SA: I guess the answer to that is no. There was one fitness report in particular I thought would come back and bite me, and it didn't.

EM: What command was that?

SA: It was when I was XO of the squadron. It was just one mark on the scale, which normally would have been an A, but it was a B. I understood why it was a B. I thought—

EM: This was with VA-164?

SA: Yes. It was a valid mark.

EM: So routinely you're looking for all A's, right?

SA: Oh, yeah. In today's environment everything has got to be all A's, and then you've got to be marked as the number one of everybody. Big mistake.

EM: Jeez, how do you—

SA: Big mistake.

EM: Because then it tends to get inflated. You don't want to screw anybody's career.

SA: Yeah. So you can't educate with the marks. And I still haven't read them all. I've got a microfiche and I looked at it the other day, and I said, what am I going to do with this microfiche with all my fitness reports? I'm not going to read them.

I see too many people today—I saw it during the last years of my service—people worried more about their fitness report than they worried about what they were doing that day. For me, I just couldn't; I wasn't going to be able to control the hand that wrote the fitness reports. The only thing I could do was do the job that I thought I was assigned to do, and they either liked it or they didn't like it. I was never happy with the job that I had accomplished. I always thought I could do more. But I always thought I did pretty well. There were things that I wanted to do that I didn't get done. I never left any command completely satisfied that I'd done everything that I could do.

I see it in my son. He has spent more time worrying about the fitness report, and he worries more from the standpoint that, why can't people say what they really mean? The big thing that goes on in the fleet now, they stage the relief of, release of officers from the squadron so that they don't have to lump them all together. So when the skipper is getting ready to leave and he has to write the whole group of fitness reports, they will stage a couple of guys to leave before he has to write the fitness report so that they get a one-on-one observation. So that they can give the guy they really want—who they will hold for another month to ride through the change of command. So he can get a number one of three recommendations. Why do these two over here get a one-of-one and this guy gets one-of-three, when possibly you could have all five of them together? One of these guys would have been number two. The fitness report system has never been good, and I think it's worse now than it's ever been, from what I hear from the young kids out there.

EM: Interesting.

SA: Going back to *Coral Sea,* when we got out we started a workup for that deployment.

EM: How did the ship operate? You had it in for overhaul and came out of it with brand new equipment.

SA: Well, sort of new equipment. The thing that bit us right off the bat was that since we were now going to be an operational carrier, we had to do UnRep. During the first UnRep we are alongside the oiler, taking on fuel because you didn't want to take on a whole bunch of fuel while you were in port in overhaul. Too much chance of spilling and stuff like this, so you tried to do the big load of fuel when you're out at sea. We meet up with the UnRep ship and we are alongside, and pretty soon the skipper over there calls up over the sound-powered line and says, "Hey, Skipper, I'm only pumping at 20 psi. Your guys are only letting me pump at 20 psi over here. It will take us a week from Sunday to get you filled up at this rate."

I said, "Well, let me see if we've got a problem down there." So I called down to the fuels guy and I said, "I understand that we're limiting to 20 psi."

And the guy said, "Yeah, we've got leaks all over the place. None of this piping got fixed in overhaul, so we've got to take it really easy, Skipper. I'll keep pumping it, but we are finding leaks all over the place right now."

When we finally get to San Diego, one of the things I had to tell Admiral Coogan is that because of that overhaul package and thinking that we were going to be a training-command carrier, I had real problems there with refueling this beast. I said, "I just want you to know that there's probably going to be a spill here or a spill there, but we're working it and we've got a plan to methodically go through and take the primary places first and work our way down. But we will get it fixed." And we did. We finally got to the point where we could do it.

EM: Did you have repair ships come alongside?

SA: You had ship repair folks down there in San Diego from the commercial side so we could get them on board and do a lot of welding. Most of it was welding patches on the system. We had had a fire in overhaul down in our medical spaces, and we had to rebuild those. We got that all done.

EM: How did that happen? Welder's torch?

SA: Yeah. One of the typical things in the shipyard. A welder didn't follow the instructions and got down there and didn't have a fire watch standing by, and caught something on fire. It did some damage; superficial damage, really. There was no major structural damage, but we still had limited medical capability.

EM: What air wing did you have on board?

SA: My air wing was—let's see, 14, I think it was. Dave [David N.] Rogers was the CAG. Again, because *Coral Sea* was not going to be an operational carrier, they had come off of *Enterprise,* because she was up in overhaul. We couldn't take F-14s, so we had to have F-4s. The Navy didn't have the variant of F-4 that the carrier arresting gear and the catapults could handle, because the plane's gear had not been upgraded for getting more operational capability. The only model of the F-4 that we could handle aboard the ship was the variant that the Marines were flying at the time, which I think was the J model. So our air wing, instead of having two Navy fighter squadrons, had two Marine Corps fighter squadrons. We had to get them all qualified and up to speed. Even some of the skippers there in the Marine squadron hadn't seen a carrier since they had gone through their training command.

EM: Were Marine aviators different than naval aviators?

SA: Well, you know we are all naval aviators, but we sort of fly airplanes a little bit differently.

EM: How so?

SA: The Marines are a little more rustic. Now, I don't know what they are going to do now that they've got the newest airplane of everybody.

EM: Right.

SA: They're out in front leading now. They tend to treat their airplane a little rougher than we do, but they're still naval aviators, and quite a few of them had some pretty good carrier time. Some hardly had any at all, so it was a real learning experience to get them settled down to ship- board ops. Dave Rogers was the CAG and did really well. My funniest story is: Jack [John R.] Dailey over at the Air and Space, a Marine general—Smithsonian Air and Space Museum; Udvar-Hazy I guess now too. At that time Jack was a colonel, and he's the group commander there at the Marine Corps Air Station El Toro, California. It's his two squadrons that are out aboard the ship, and his photo recce squadron, which was a recce squadron from the Marines. Those guys were really colorful. Anyhow, Jack comes out to see me and introduce himself, and asks how his boys are doing. So Colonel Jack and I are out there on the bridge and we're talking, and he says, "Well, Skipper, by your leave, it's time for me to head back to the beach. Thanks for taking care of my guys. Let me know if there's anything I can do."

I said, "You bet. Glad to meet you, and thanks for coming out to see me," and all of the platitudes.

Next thing, I'm listening in on the radio circuit—now he's manning up his airplane and he's got it started, and I hear him come up on the air with his call sign; he says, you know, call sign, fuel state, and the usual stuff that we had to go through. I hear my air boss back there say, roger, whatever number it is. Are you heavyweight qualified? And there is this pause. "Yep, yep, I'm heavyweight qualified."

"Okay, we will be taxiing you forward here just shortly." I see him breaking down the airplane and taxiing up to the cat. He's on number one cat right in front of me. Cat fires. Jack goes off the end of the catapult. Not being an F-4 driver at that time, I really wasn't tuned into this question of a heavyweight cat shot. Apparently—well, not apparently for real at a certain weight, when the F-4 ends the shuttle run on the catapult, you have to go forward with the stick instead of back, to keep the nose from coming up too high. That's very counterintuitive, because normally you pull back a little bit, get the nose up, and go. Jack goes steaming out of that catapult. Afterburners are lit. Down he goes. He gets off the cat and the airplane drops like this [gestures down].

EM: Uh-oh.

SA: I alerted the OD and I said that we were going to have an airplane in the water in a second and to watch it, because we had to shift the rudder depending on which side of the ship; we wanted to kick the stern away from the airplane so the screws didn't get him.

EM: Right.

SA: He just goes like this, and I'm thinking they've got to come out of there. They've got to come out of there. He's pushing the nose over, and he's pushing the nose over, and he's getting lower, and finally he's got the nose about right where it should be, and now he's down on the water [laughter], with a rooster tail from the afterburners. I come on the air and say, "How did you like that cat shot? [Laughter]

"We got it, boss," and so the plat [flight deck camera] guys, they get a picture. They've got the picture from it. They race up to the bridge. I get it on the first airplane going to the beach. [Laughter] But everybody on the beach could see the group commander and his heavyweight cat shot. I pull his chain about that every time I see him. We see each other about every year or so. He says, "Oh no, you're not going to tell me again about that?"

I say, "I'm never going to let you forget."

EM: A perennial. Where did your ship operate? Where did you take *Coral Sea?*

SA: Back here we spent most of our time in the Southern California op areas, and then when it came time to deploy, of course my time is running short. I only got to take the ship as far as Subic Bay. I think we went to Hawaii, up to Yoko and down to Subic Bay, if I'm not mistaken. And then—yeah, we were up north, because I remember an incident coming down from up there. Then in Subic, Dick Dunleavy was scheduled to relieve me. Dick was the first naval flight officer [NFO] to get command of a carrier, and Dick and I had been in Idaho Falls together. We had been on the ski journey together.

EM: He didn't get injured on the ski slopes?

SA: He didn't get injured. I got to meet up with Dick again and give him the brief. By that time—oh gosh, don't let me forget his name—Larry [Lawrence C.] Chambers. Larry Chambers was the flag officer on board. I'll tell this story too, because I remind him every time. Bill [William E.] Ramsey, Admiral Ramsey, one of the world's nicest guys—

EM: ComCarGru?

SA: He was the first ComCarGru. Most of the time I had the ship I didn't have anybody on board, because they hadn't planned on me again being an operational carrier. They hadn't been in the rotation yet, so as things got closer to me deploying, they finally decided maybe I'd better have a flag officer on board the ship to make sure they are really ready to go. Not leave it to this loose-screw captain to do it all. Bill Ramsey comes aboard, and he had just pinned on his stars, and he'd come off of one of the nuclear carriers. He commanded one of the new nukes.

It's the first night at sea with him onboard, and I decide—it's about 2200, so before he gets too far asleep, if he's going to sleep early, I think I'd better test on how he wants this relationship between the captain and the flag to be. There is some minor thing going on down in the engineering rooms, so I say, well, this is a good thing, to give him a call and tell him what's going on and what we are doing to fix it. What the impact will be or not be. A good way to feel him out and find out how he wants this relationship to go.

I hit the hot button on the sound-powered phone, which is so he knows it's me on the other end of the line. I buzz him, he picks up the phone and he says, "Captain." He had just come off the carrier and just been promoted, so his mind is set as, here I am aboard ship, I'm a captain.

I said, "Captain? No, Admiral, I'm the captain. [Laughter] You're the admiral."

He says, "Oh, that's right."

EM: Was he asleep?

SA: No, it was just a reflex. He was used to picking up that phone and saying, "Captain."

EM: So was that a good icebreaker?

SA: It was. We got a good laugh out of that. I said, "You must think you've gone on a time capsule coming back here to *Coral Sea.*" We didn't have any fancy electronics like they had on the nukes, combat systems and all that.

He said, "Nah, it's pretty enjoyable. I'm having a good time." Then he was relieved by Larry Chambers, who went on deployment with us. Larry's chief of staff was Ming [Erh] Chang, who was a legend within the surface community. Ming and I had known each other because he used to work with Bob Kirksey.

EM: Why was he a legend?

SA: Ming had these sayings—sort of like Yogi Berra, but it's all in Chinese.

EM: Malapropisms?

SA: But it's all Chinese sayings. Ming would always mix his words up a little bit. And with that Chinese accent, people used to laugh. Ming was pretty excitable, but he was not near as excitable as Larry Chambers. Larry was on this bent of physical fitness. He was running every day, just pounding on that hard steel-armored deck that we had. He was always getting into arguments with my flight deck crew, who were spotting airplanes for the next launch and stuff, and Larry's out there tromping around on the flight deck, running.

EM: Kind of interfering with what they were trying to do?

SA: His position was, "I'm the admiral. I'll run when I want to run, and you guys will move airplanes when I tell you to." The admiral and I had a little back and forth on this running caper. Finally we reached somewhat of a middle ground on the thing. But while he was out there, he would see things. He would bring them to my attention and say—you know, they're tying down airplanes back there—the last F-4s that they spot aft by the ramp, they are putting them too close to the bead on the ramp, or where the slant comes up and meets the flush deck. He says, "I want X amount of inches between the main mount and that bead. Do you understand me?"

"Yes, sir." So I would get the air boss and his team together and say, "Look, every time you try to squeeze one more spot up there on the flight deck and put that last F-4 back there where that tire is on the bead, I'm going to get a call. And if I get a call, you're going to get a call, and we will have to do something about that."

"Yes, sir."

Well, we'd go for a while, and everybody was behaving. Then one day somebody just absolutely forgot, I guess. He was out running, and the tire was on the bead, not just off of it, but on the bead. He called me down. "*You come down here right now!*"

"Yes, sir, I'm on my way."

I get down there—and of course Ming and I knew each other pretty well. Ming is saying, "I'm not going to go in there with you. I'm not going in there with you." [Laughter] "He is really mad."

I said, "Okay, Ming." So I go in and—this is no lie—his eyes, you've seen these cartoons where the eyes start to go in spirals? And he was just out of control, yelling and screaming and carrying on.

EM: That seems inappropriate. I mean, was he worried about the airplane falling off the back of the ship?

SA: Yeah.

EM: But it was tied down, right?

SA: It was tied down. This story has a [laughter] funny ending. Not funny, really. For me it's funny, not for other people. I'm duly chewed out, and I just don't know how we are going to survive all of this. So when Dick Dunleavy comes aboard, I say, "Now listen, Dick, let me give you some clues here how to stay out of trouble with the boss. For heaven's sake, I think I've got him trained now, but I have had some chewing-outs like I've never had in my whole life over this issue, so I'm going to give you the benefit of my experience. And I'm telling you right now, those daggum guys down there in flight deck control will—if you are not watching them, they are going to move that last F-4 down there to the daggum main mount sitting on the bead of that ramp. If the admiral finds it, he's had it up to here. He gave me my last chance, and so far I've escaped, but he won't remember that he hadn't given you the same guidance, so don't let this happen."

"Got it. Got it."

I leave the ship. They spent some more time in port. They get under way. They go out. They start doing some refresher landings. One of the first A-6s that goes up to the cat, the guy forgets that it's fueled for CQ, not a full bag of fuel on board this A-6, so he was not a heavy load, he was a light-loaded A-6. There was no ordnance, no nothing.

And he gets a cat shot, and because it's so soft, because he weighs so little, his mind is awaiting a full-weight cat shot, and all of a sudden he gets what he considers a soft cat and he jumps out of the airplane. Nothing wrong at all—it was just a softer catapult, which wasn't what he was expecting. So that didn't start off the first at-sea period very well for Dick.

Next thing happens, they park an F-4 back on the ramp. I think the wheel was actually over the ramp; I don't know for sure, because I wasn't there. Some people say it was, some people say it wasn't. Anyhow, they got the thing too far back on the ramp. *Coral Sea* had a very heavy vibration at speed, when you were on the ramp. I mean, you could really feel it if you hit a certain speed or when the rudder shifted for a turn or something, and that thing started bouncing, started bouncing and broke the chains. An F-4 went over the side.

EM: So he had a reason for that. Wow. An A-6 and an F-4 were lost?

SA: An A-6 and a F-4 were lost.

EM: Poor Dunleavy.

SA: I think he recovered. It was because the admiral wasn't aboard ship.

EM: I'm sure he found out about it.

SA: He found out about it right away and got back aboard ship.

EM: Did Dunleavy go on to flag rank?

SA: He was a three-star in NavAirLant.

EM: Didn't do him permanent harm.

SA: No.

EM: Qualify for me the roles of the ship CO and the CarGru commander? My thought was the CO runs the ship, but he runs the air operations too, and the CarGru commander has the whole group, the big picture, right? But it sounds like he's getting into shipboard operations too. Or was that just an individual thing?

SA: It varied by individual guys. Of course Larry Chambers had been skipper of *Midway,* so he was familiar with the class.

EM: Same class.

SA: Same class of ship, so—in fact, Admiral [Sylvester R.] Foley [Jr.] came out one of our at-sea periods before this, and he had been a *Midway* skipper as well. *Midway* still only had the two catapults, but *Coral Sea* had three. We had a smaller flight deck, but we had three catapults. I got hold of the air boss and I said, "The two admirals are here, both *Midway* skippers. Here's how I want our launch to go this afternoon. They are going to be on board, and I'm going to have them up here on the bridge; they're going to be down on the flag bridge. I want three birds on the cats, all turning, and when you get the signal to launch, I want all three to be launched. Just *boom, boom, boom.* I want the next three on there, and I want them launched together."

That way there's only a couple of seconds between each launch, so they are almost in formation as they are coming off. So once I got that accomplished, then I called them down on the flag bridge and said, "Well, how do you *Midway* sailors like seeing all of those airplanes go at once?" [Laughter] It was something I could do that they couldn't do. Those were good days. I was really sorry to miss that cruise.

EM: That was a short time. Most of the time you were up in *Bremerton* with the overhaul?

SA: Yeah. I had, what, six months or so up there, and then another seven months, and then a couple of months going over to turn around and come back. Dave Rogers and I left the ship at the same time. The CAG and I left at the same time, which was sort of strange, but that's just the way the timing worked out. Marine squadrons—it was really tough getting them into the script to begin with—but boy, once they settled in, they were great. They just weren't used to the fuel management around the carrier. I mean they would take an afterburner launch and they would still be in afterburner clear across the horizon and out of sight. The Navy guys, as soon as the shot—as soon as they are airborne, as soon as the gear is coming up, they are coming out of afterburner saving that gas. The first thing you would hear from the Marines was, where's the tanker? [Laughter]

They settled in. What they did was, they brought a cadre of people that they screened to stay in the squadrons for the cruise. They hand-selected, down to the guys that they really wanted to take on the cruise with no penalty, to the guys that they didn't take. It was just that they wanted to make sure they took the guys who were doing the best around the carrier, because that's not—

EM: Their normal operation?

SA: Not their normal operation, and it was a no-penalty, no-foul. One of the skippers left. Another skipper was brought in who had more carrier time. I was really impressed with them. They did a fine job.

EM: Anything more you want to say about *Coral Sea?*

SA: No. Many years after, I'm making the run from Washington, D.C., up to New York City when I was working for Loral and had to fly over Baltimore. I looked down and saw them cutting up *Coral Sea.*

EM: She had quite a record, too.

SA: Yes, she did.

EM: Way back.

SA: We found—while I was on there, while we were in overhaul—we found a scrapbook that was made of her first year of service, back when they went down to the Panama Canal and everything. It was really a treasure. Oh, I'll tell you one other funny story, one other thing that I told Dick Dunleavy. I said, "Now, you know we've got this presentation silver on here. The presentation silver was presented by the City of San Francisco to a battleship many, many eons ago—"

EM: Battleship? Which one?

SA: *California,* I think it was. I said, "Now, part of the relief is that you've got to go down and see it and verify that all of the pieces are there."

EM: It's invaluable.

SA: The handles on all of this stuff are golden bears. They are gold. They are 24-karat gold. They are on the cover lid and on the chafing dishes, with the golden bear sitting on the top. It is just gorgeous stuff and I said, "It's got to be locked away." Well, when he got back from the cruise, one of the sailors decided that he needed a little extra money and snapped the golden bears and tried to pawn them at a jeweler in San Francisco. The kid must have just been crazy thinking that nobody would recognize it. The jeweler calls and says, hey, there's a sailor over here trying to pawn golden bears off the presentation silver.

EM: Could they get them back on?

SA: Yes, solder. They got them back on.

EM: We had in the museum [National Museum of the U.S. Navy] a ship's trophy cup, solid silver or something like that, and one weekend somebody broke in, and they actually carved a hole out of the glass—out of the top—went in and grabbed it with tongs and squeezed it, then pulled it through the hole.

SA: Is that right?

EM: Gone. Yeah. Never did find that one. All right, well, this has been a very productive day. Wow. It's been great. Good stuff.

[End of 19 August 2015 interview]

Interview Number 4 with Adm. Stanley R. Arthur, USN (Ret.)
Date: 20 August 2015

EM: This interview is being conducted on the 20th of August, 2015, with Admiral Stanley R. Arthur. Interviewer is Edward J. Marolda of the U.S. Naval Institute. We went through your time in Vietnam and post-Vietnam years. We are up to the '80s now. Reagan era. Let's see, you were assigned to the CinCPacFlt staff in January of 1980.

SA: January of 1980. I left *Coral Sea* there in December of 1979, and we went to Hawaii and joined the PacFlt [Pacific Fleet] staff. My job was fleet operations officer. Above me was a two-star, Assistant Chief of Staff for Operations. At the time I think it was Admiral [John C.] Dixon [Jr.], soon to be replaced by Admiral [Carol C.] Smith [Jr.]. The CinC [commander in chief] was [Donald C.] "Red Dog" Davis.

EM: Donald C. Davis.

SA: We didn't know until much later that he really didn't like to be called Red Dog.

EM: No one called him that to his face.

SA: No, we did, but then years later we found out that he didn't really like that name, but that's all we knew him by. He was Donald Cooke Davis. Red Dog was an aviator of course, and had red hair, but by this time it was mostly gray. He was a wiry, slight-built fellow. He was really good; as an operator, he was extraordinarily good. He had his ideas. He ran the staff and ran us pretty hard; he kept us busy. Every now and then, he'd send us out on missions to put out some fires that he had lit. It was a real pleasure to work for him.

EM: Quite often the PacFlt guy goes on to be CinCPac. Did he?

SA: He didn't, and we thought he would. But Admiral [Robert L. J.] Long was up there, and he stayed a long time; he did a double tour up there.

EM: Davis retired from that position?

SA: Yes.

EM: Were you involved at all on the PacFlt staff with the planning for the hostage-rescue attempt in Iran?*

SA: Yes. When I got there that was one of the first things I got involved with. That whole operation was very highly classified, so there were very few of us on the staff that were read into the program. I needed to be there because my job was sort of tracking everything that the fleet was up to on a daily basis and raising red flags if there was anything that I saw that was unusual that happened. We were moving people around and really couldn't tell them why they were going where, and this type of stuff. I was in the cell down on the PacFlt staff. There was a cell up on the CinCPac staff, and then there were some folks over there at NavEur [Naval Forces, Europe]. Ken [Kendall E.] Moranville, who was on Admiral [Ronald] Hays' staff, was over there.

We had this net with super-secret comms in the back room where we did all of the planning and sorting out of responsibilities, because there were some movements that took place that were not evident to the rest of the folks. We will cover one of those in just a minute. On the night that that rescue attempt went, Admiral Davis sent me up to be his rep with Admiral Long's staff, because all of the talk from the theater was coming in to CinCPac. And rather than have some of us down at CinCPacFlt, some of us up at

---

* On 24 April, 1980, in an operation authorized by President Jimmy Carter, U.S. military forces attempted to rescue 52 diplomats and other Americans held captive in Tehran, Iran. The operation encountered many obstacles and was eventually aborted, but not before the destruction of two aircraft and the accidental death of eight members of the U.S. armed forces.

CinCPac, Admiral Long put the war room up there, opened it up, and we came up there and were sitting listening in as the thing unraveled.

EM: Oh, real time?

SA: Real time, yeah.

EM: The things that you had done preliminary to the attempt, did they come to pass the way you had thought? Or it didn't happen in the way you had thought?

SA: No. I think part of getting that whole operation geared up were the command relationships, because we were in new territory. We had PacFlt units out in the IO [Indian Ocean] that were standing ready. Admiral Kirksey had a battle group out there where the helos were going to launch from. We had been moving those guys around and trying to figure out if they had the right configuration for what they needed to do.

EM: Configuration of aircraft?

SA: Whether we had the right carrier out there. We had a choice of—well, we didn't really have a choice, but we contemplated—do we take another carrier out there at the time? Would we need it? The forces in the Persian Gulf, traditionally the Middle East Force, a very small cadre of ships and a flagship there, that organization reported to NavEur through the Navy side of the house, with Admiral Hays up in London. In the back room, as this thing was being built up, it was determined that the Middle East Force ships would become part of the PacFlt organization.

EM: The two CinCs agreed to that?

SA: Well, that was one of the things that had not been completely disclosed to Admiral Hays. The CinCs had agreed to it and CinCPacFlt knew it, but for some reason—

EM: CinCUSNavEur [Commander in Chief, U.S. Naval Forces, Europe] didn't know it?

SA: CinCUSNavEur didn't know it. So afterward as we were returning, after the thing failed and everybody's having this revisit—you know, what did we do right? What did we do wrong? Admiral Davis says, "Okay, it's time for you to go to London and tell Admiral Hays that he doesn't own those ships anymore. I own them." They had been having a little bit of a tiff. I didn't know to what depth, but when I went to London—

EM: Davis and Hays?

SA: Yeah. When I got to London, Admiral Hays was really interested in why was I there. I'm still a captain, and he says, "Why did Admiral Davis send you out here?"

I said, "Well, Admiral, we've got to clear the air on some command relationships that I don't think you've been aware of, that the Middle East Force ships no longer belong to you in the chain of command through what happened during this rescue attempt." Those ships had chopped to PacFlt.

EM: But who chopped them? Who had the authority to chop them?

SA: CinCPacFlt just reached out and grabbed them. He said, "Is the Admiral really mad at me?"

I said, "No, no, there were certain things that were not disclosed to all of the players in this thing, and now he just wants to clean it up so that there's no more misunderstandings and make sure that nobody gets their feelings hurt over this."

EM: What was Hays' reaction?

SA: He was very gracious. And he didn't shoot the messenger. [Laughter]

EM: You breathed a sigh of relief?

SA: I breathed a sigh of relief. But that was interesting. When that thing unraveled, we could hear it on the air unraveling, with the helo blades and stuff.

EM: Did you have video?

SA: No, we just had audio in those days. As it unraveled, Admiral Long finally just got up and said to his team, "Well, okay. Get together. Get the revisit started on this thing. I want to make sure that everybody knows that we weren't the cause of what happened over there."

EM: What did he think was the cause—the people on the ground?

SA: Well, yeah. He thought, basically, that whatever planning took place on how they were going to do that refueling operation and where they were was not in his bailiwick. His bailiwick was to get everybody there that they wanted. That the PacFlt or the CinCPac part of the game had been executed.

EM: Who had operational control or tactical control?

SA: You know, I'm trying to think now how that—

EM: Did they launch from the carrier?

SA: We launched from the carrier.

EM: You had air—

SA: Then we had the Air Force tankers coming in to do the on-ground tanking.

EM: Special Ops?

SA: Special Ops. Right now I have to really dig back in the gray matter and think exactly how that—

EM: Well, wasn't that one of the problems?

SA: That was one of the problems, yes. It definitely was. And the outflow of that thing was what I ended up spending a lot of time—most of my time—with on the PacFlt staff. This was the emergence of the RDJTF, originally under P. [Paul] X. Kelley, and after a period of time the RDJTF morphed into Central Command.

EM: Rapid Deployment Joint Task Force. Admiral Long said, "We did our part and basically somebody else screwed up?"

SA: Yeah.

EM: Okay. What was your reaction to the Holloway report when it came out? He headed an investigation. What was his reaction?

SA: I don't remember what his recommendations were. I really need to go back and read that.

EM: He [Admiral Long] didn't slam his fist and say that it was a bunch of hooey?

SA: No, I don't recall that.

EM: It was probably a while later.

SA: Yeah, it was, and I don't have any reflections right now. I need to go back and sort of read that again, pick that off and see.

EM: Hard to remember the particulars.

SA: I don't, either.

EM: It called for greater jointness.

SA: Yeah, and that was evident as we stood up the RDJTF and then Central Command. But again, the Navy was not for that. Our mantra at that time—and our mantra today to some degree, but not nearly what it was in those days—was, "We operate in support of. We can go anywhere, and whoever needs us, we will end up supporting them." So the Navy was very reluctant to dedicate any forces directly to the RDJTF or when Central Command stood up. "Why is there a need to have a fleet commander out there," et cetera.? We will talk some more on that as we go through what happened here on the PacFlt staff.

EM: Was the fact that the operation went south—was that related in any way to the lack of resources? We are still—this is late '70s, in the post-Vietnam retrenchment.

SA: No, I don't think so, because we really gave the guys that were out there on the scene everything that they wanted. And if they called up and said they needed something, we moved it that way for them. My take is, talking to some of the guys who were out on that end in later years, they were all satisfied that they had what they wanted. Basically, the process was very complicated and had all sorts of potential to go south at several different points. It was really a tough operation, and it really didn't get that far because of the refueling on the ground out there on the desert floor. That just didn't happen the way it should have.

EM: What was your personal opinion of Carter giving the go-ahead, after this operation?

SA: Well, I thought it was a gutsy move. I thought that this really was a step in the right direction to show that we were not going to let this happen. Of course, from my trip over there in my ICAF days to Tehran, I thought, boy, this is going to be—this is going to be really tough to get into Tehran.

EM: And get out?

SA: And get out, because we had watched them, and their air force was good. They had the best equipment. Like I was telling you relative to the Marines that were with me when we were there in Tehran, you know. All of their air force had operating radars. The Marine F-4s, that was not always the case. [Laughter] I'm not sure whether they had the F-14—yeah, I guess they had the F-14s by then too, some of them. They were well trained from the point that we had seen there in '74, '75, when we were over there.

EM: Was part of the cover fighter aircraft from the carrier?

SA: No. Well, for a short period of time, but once they got that far inland, they were on their own out there.

EM: Just helicopters.

SA: Yeah. We didn't have anything on top of them, because that would have given them away at the time. Now, we would have had something on the exit as they came back out—pick them up.

EM: Supposed to be quick in, quick out.

SA: Quick in, quick out. Yeah.

EM: What other significant activities were you involved with on the staff?

SA: The carrier presence in the Indian Ocean, we had it out there for that particular reason, and we had been moving ships back and forth. Then the call came, we're going to maintain a presence out there in the Indian Ocean, which was really starting to stretch us pretty thin. Then we worked with the Lant side of the house to get a carrier rotation where the East Coast carrier would come into the IO, and we would swap out with a West Coast carrier, and then start that rotation. That was some of the dialogue going between the fleets, and so the ops officers from both sides would get together and try to sort out the scheduling and get that locked down.

EM: The Navy didn't think we should have a fleet commander here in the IO or in the Persian Gulf area?

SA: No. CinCPac says, "The IO belongs to me. You know, I go all the way to the coast of Africa and that belongs to me, so anybody that comes in there chops to me." That was a continuing argument. As we are doing the reconstruction and getting the fleet organized to keep the presence going out there, up pops this idea of the Rapid Deployment Joint Task Force. At the time I'm still the fleet ops officer, but about this time was the flag selection, and I was selected for flag. There were five of us on the island [Hawaii] that were selected on that board. Myself, there in the PacFlt staff; Don [H.] McDowell, who was the crypie, crypto guy on the staff; Bob [Robert] Schmitt, who was the intel officer on the staff. The three of us and Rudy Kohn, who was up at CinCPac—

EM: An old friend of yours?

SA: An old friend of mine. And Ming Chang, who I had been with before. Ming was the Chief of Staff over at Third Fleet on Ford Island, so the five of us were all selected by that board. We were the last class to go straight to two-stars. Remember the days when your first rank as a flag officer in the Navy was a two-star rear admiral lower half, and then you stayed two-stars as a rear admiral upper half? We were the last class of the instant two-stars.

EM: Lucky you.

SA: Yep.

EM: Did you five get together to celebrate?

SA: Oh, yeah. Yeah, we did. We all came back to Washington for the—what did we call that thing?

EM: Frocking?

SA: Well, no we weren't frocked at the time. We had a nickname for it. It was sort of like the tea and crumpets thing, but we all came back and went through the indoctrination. In anticipation of that, of me finally going to pin on the stars at some point in time, they moved me over to be the plans officer for the N6, I guess we were in those days.

EM: Previous to this you were what?

SA: I was fleet operations officer. And then went over to Plans and Policy. By this time Bill [William A.] Walsh, a surface guy, was over there, and he was getting orders out of there to go—well, let's see. I guess maybe Bill must have been selected the year before me. I'm not sure now. He was on his way to a new job, so that's why I moved over to be the Plans and Policy guy. He had gone to a couple of preliminary meetings with the RDJTF as a CinCPacFlt rep to the RDJTF.*

EM: What was the RDJTF? Was that in reaction to what happened in Iran? Or to the Soviets in Afghanistan? Or both?

SA: It was, I think, a reaction to both. I think from the events that happened in Lebanon—you know, the Marine barracks thing a couple of years earlier.

---

* The Rapid Deployment Joint Task Force was activated on March 1, 1980, under the command of Marine Lieutenant General Paul X. Kelley. The RDJT's mission was to deter Soviet aggression in the Middle East, discourage regional conflicts, and prevent instability that would impede the flow of Persian Gulf oil to the West.

EM: That was '83.

SA: It was '83, that's right, it was after the RDJTF was stood up. As I read the tea leaves, it was mostly in reaction to what happened in the desert and who could actually respond more to that. P. X. Kelley was a three-star down in Tampa, and Bill had attended a couple of meetings there, and then P. X. Kelley said, "We need a Navy commander as part of the RDJTF. We've got an Army component commander, an Air Force component commander, Marine component commander, and we need a Navy component commander."

EM: What were the ranks of the other services?

SA: The other services had a three-star for 18th Airborne Corps. Three-star Air Force. Two-star Marine, First Marine Division out of the West Coast; and then they named me as the Navy rep.

EM: And you were rear admiral select?

SA: I was a rear admiral select. So I go down to my first meeting in Tampa that P. X. Kelley had called for everybody to get together. He looks across the table, and my staff at that time was me and another Navy captain. But I was still two-hatted, I still had my job as the Plans and Policy guy, so I took a couple guys out of the Plans and Policy shop and put them under the role of RDJTF. Had a SEAL recalled from a bar down in Miami. [Laughter] He had been put over the beach down there and up north, and so he and I went down to this meeting.

EM: Do you recall who that was? Mike Thornton maybe?

SA: No, it was Bruce [P.] Dyer. So Bruce and I go down to Tampa, and he's a lieutenant commander as I recall, and I'm a captain. P. X. says, "Well, I see the Navy has finally shown up down here." [Laughter] And Jack [V.] Mackmull, the 18th Airborne Corps commander, a three-star, was a typical old Army fire-breather. You know, big husky guy, cigar always sticking out of his mouth. I'm sitting over there in the corner, and behind him he's got two-stars and one-stars, two and three rows. The Air Force guy is all backed up with his entourage, so we were sort of the orphans of the organization.

EM: The Navy didn't feel a need to plus-up a representative?

SA: No, in fact I said—on the way, when he had called this big meeting that was going to be the first meeting where all of the commanders were going to be gathered—I said, "I think it would be the right thing for the Navy to show that their guy is actually an admiral, not a select."

"No. No. You'll get your stars in due course."

EM: Davis did that? Or was that Long?

SA: I think basically it was the Navy staff back in Washington. They said, "No, we are not going to authorize frocking him." The request went out to do it and they said no.

EM: Who was CNO at that point?

SA: Let's see.

EM: It was before Watkins.

SA: Yeah, it was before Watkins.

EM: After Holloway—[Thomas B.] Hayward.

SA: It may have been Hayward. But the rules as I got them were sort of very mushy. I couldn't get anybody too interested . My boss, Red Dog Davis, was not in favor of this organization. Admiral Long was absolutely not in favor of this organization.

EM: How do you know that? Did he express that?

SA: Yes. He just didn't think it was necessary. In fact there was an incident later on, I think after we had already transitioned to Central Command, where he came in theater and went to Djibouti. He came and made a visit to one of the places out there and sort of pretended like it was his turf. Just to sort of dig the CinC at this time.

EM: Diego Garcia?

SA: No, well, he had made his stop at Diego, but I think it was when we came into Kenya. But in any case, he was still sort of poking around that this is not going to happen. What happened, as I got more involved with the RDJTF, was that the RDJTF was in very much of a planning mode that carried over as we established Central Command, which was really a good fit for my job on the PacFlt staff as the Plans and Policy guy, because we were updating all of our war plans and had a rigid review system set up so that we would go through each plan and update it and all this sort of thing. Of course you spend an awful lot of time on command relationships in the planning cycle, and where the forces come from, and who you have to go to to get them, and so on.

So it was becoming very evident to me, as we were working these plans, how disconnected the Navy was from where these guys were going. I started sending up the word to the CinC [CinCPacFlt] and on up to the big CinC [CinCPac] through Rudy Kohn. And the guys that worked for him up there, we'd go on exercises all the time. We'd get together and go off to Fort Bragg or go out to Fort Erwin, and eventually we were going to Egypt for the Bright Star exercises. I said, "Every time we have one of these exercises, the Marines show up and are fully integrated into this process. We, the Navy, the guidance given to me is that we are always going to operate in support of, so tell us what you need, and I'll be the connect point to go back to the Navy and go to the appropriate CinC to draw forces from them."

Well, "We don't like that. We'd like to have dedicated forces."

"You're not going to get there from here." But as the time went on, I could see that the RDJTF was going to be very short-lived, that it was going to turn into a CinC. I had already been through the battles of the CinCs—of who owns what, and do we need to redraw the lines because we now have a presence in the Indian Ocean that everybody thinks is necessary. Do we give part of that to LantCom [Atlantic Command] instead of PaCom [Pacific Command]? So the CinCs had put up all of their gates and put the locks on them, and no, they're not giving up any territory to anybody. I was continually sending the message up as, "It's not the RDJTF that you have to worry about. The RDJTF is going to become a CinC. It's going to become a CinC in a very short time period."

EM: That wasn't well received?

SA: It was not well received, and in fact nobody in the Navy side of the house believed it. They didn't think it could be done. They thought that CinCPac was all-powerful and whatever CinCPac ruled, that was going to be the decision. Well, my reading of the tea leaves was that CinCPac was going to get rolled, whether he wanted to or not. He was going; it was going to go blowing right by him.

EM: It was curious, because not only did you have things happening in the Middle East in Iran and elsewhere of concern to us, but the Soviets had just invaded Afghanistan and they are right in the operational area.

SA: Yeah.

EM: At that time their fleet was making a big push, too, into the Indian Ocean. That didn't influence anybody?

SA: It influenced the fact that we were starting to keep a carrier out there, because we started to watch their forces over at the anchorage that they used to come to on the—

EM: Island of Socotra? The Soviets?

SA: Yes. Then the French came down into the IO and so they had their Fleet Indian Ocean AlIndien [admiral in command of the Indian Ocean maritime zone].

EM: Out of Djibouti?

SA: Yeah. I was sort of a thorn in the side because they wanted me to just make sure that we kept things status quo. Then P. X. Kelley was relieved by Bob [Robert C.] Kingston.

EM: Army?

SA: Army. He was a three-star. Highly decorated from Korean War days and Vietnam. Had spent a lot of his life with Special Forces. He wasn't a Special Forces guy himself per se, but he had spent some time working with the British SAS [Special Air Service commandos] guys. A really tough, tough guy. His nickname was Barbwire Bob. [Laughter] He was rough and tough. He had a brother that was a Navy captain, an amphib guy, so he had some understanding of the Navy. I assume that he had been briefed very carefully before he took over that job, because he made it very clear that the end objective was to be a CinC. And it had to happen very rapidly, and there had to be some realignment of the CinC lines to accommodate that. There had to be a realignment of forces that needed to come along.

EML: Did any of those who thought this was going to happen, did they ever consider that it would be a Navy CinC in charge of Central Command?

SA: No, they didn't. I kept saying to the Navy, "This is going to become a CinC. We, the Navy, have got the ideal situation. We've been in theater. We've been the only force in theater continuously since the end of World War II with the Middle East Force."

EM: And lots of water there.

SA: "We've got lots of water. It's not all sand and dirt. We've got relationships already built. We should be standing in line as number-one guy to be the CinC of Central Command."

"No, no way. No." It was always this tug of war, so by the time we finally stand up as Central Command and my title changes to Commander Naval Forces, Central Command, when that happens, then I pretty much divorce from the CinCPacFlt staff. By this time Admiral Watkins is CinCPacFlt. When it's official that we are now Central Command, then my hat ends as part of the PacFlt staff, although I'm still living in the PacFlt area. They give me a suite of rooms up there in headquarters.

EM: Did you have an office or offices?

SA: I took over a set of offices there in the headquarters that they gave me.

EM: PacFlt headquarters?

SA: Yes.

EM: How many people did you have on your staff at that time?

SA: I think probably seven.

EM: Officers and enlisted?

SA: Yes. Then the first thing we thought about was, do we have a flag? Are we to fly a flag? "Well, no, you can't fly a flag at headquarters." Okay. "You can't fly a flag at my headquarters" [laughter], from Admiral Watkins.

But I'm using all of the infrastructure there at CinCPacFlt for my comms and all this sort of stuff. One of the funny things that happened is, we are really in the tooth-and-nail business of trying to sort out, once and for all, the alignment of forces. And Kingston says, "Look, your job is—I want Navy forces, but I don't want to get completely cross-wired with your boss or CinCPacFlt, and certainly not with Admiral Long up at CinCPac. But I am a CinC, and your job is to support me, and more importantly, find a way to weave through this bureaucracy of the Navy to get what I need to be a bonafide CinC with naval forces." I said I understood.

EM: Was he somewhat mystified at the Navy structure?

SA: No. He was well aware of the Navy's intransigence on this stuff. For a tough old buzzard, he was very, very smart. Very clever. But he had a temper that was just unbelievable. We were working the problem and he said, "You need to send me a message that tells me what you think I ought to go forward and ask for." I said okay. And I knew Admiral Watkins was still reading all of my message traffic.

He told the communications guys, "I don't care if it is another command that doesn't belong to me, he's using my communications facility so I get to read his messages." [Laughter] So I knew this was going on, because Dave [David E.] Jeremiah was his Chief of Staff down there, and I knew Dave real well. I write this personal message for General Kingston about Navy forces and what they should look like and what the configuration and all should be.

Three or four days earlier than that, Admiral Watkins had asked me to draft a message for him [Watkins]. I guess I hadn't turned over everything yet from the Ops and Plans, so I gave him a message that was basically the doctrine that we had agreed to on the PacFlt staff. He had read it and sent it forward and was very pleased with it. He said, "That is really a good message. I really liked it." Then I'm writing this same message for General Kingston with my other hat.

EM: You were in kind of a ticklish position?

SA: Yeah. [Laughter] I wrote it and sent it. The next morning, Dave Jeremiah calls me up—he's down at Admiral Watkins office—and says, "Admiral Watkins wants to see you," and I say, "Okay, I'm coming down."

I get there. Dave would always go in with me when I'd go in to see the CinC, his Chief of Staff. Dave doesn't come in. He just closes the door behind me. [Laughter] Well, it was—

EM: A bad omen.

SA: Admiral Watkins is up on his feet—and he's quite a bit taller than I am, and he's really—he says, "How could you write this?" He's waving this message. "How could you write this?"

I say, "Well, Admiral, you have to understand. You liked the message I wrote for you, right?"

He says, "Yes, so why doesn't this message say the same thing?"

I say, "Because I'm wearing a different hat. [Laughter] This is my other boss, and he wanted me to tell him what I thought was the best way for him to go forward to get his objective met. That's what you tasked me to do on this other message."

He says, "I just don't understand how you can do this to me."

I say, "I'm not doing it to you, Admiral. This is my world, so remember what hat I was wearing when I wrote that."

"I don't care what hat you were wearing when you wrote that message. It should have read exactly like the one that you wrote for me. You just missed."

EM: What were his big objections?

SA: The big objection was the Navy still wasn't willing, at least from the CinCPac, CinCPacFlt side of the house, to basically have a fleet that belonged to Central

Command. They said there was no need to have a fleet that belongs to Central Command. But you go through the wiring diagrams and say, well, you've got a Second Fleet that belongs to LantFlt, you've got a Third Fleet that belongs to Pacific Fleet and you've got Seventh Fleet that belongs to PacFlt, and Sixth Fleet for the European—you know, basically, the Central Command wants a Fifth Fleet.

"No, no, no, no, no, never, never, we are not going to divide up the forces that way!"

"Well, you've had ComMidEastFor [Commander Middle East Force] for all of these years."

"No, that's something different; there weren't really valid forces there. They were just sort of show ships."

EM: The fact is, this was the time when the fleet strength still wasn't what a lot of folks wanted it to be. In fact, this was the time of the "swing strategy."

SA: Yes.

EM: President Carter wanted most of the Pacific forces to go over to the Atlantic if the Soviets did anything.

SA: Yeah.

EM: So there weren't a lot of ships to go around.

SA: Yeah, because they had been fighting these battles from the other CinCs, and from the Navy perspective it was, you know, "We just don't need another CinC in the middle of this mix that we are trying to satisfy."

So there were valid reasons for the concern. Of course, CinCPac also had concerns about his area of responsibility [AOR]: "Since Central Command is going to come in on the scene, what part of my real estate are they going to try to walk away with? Are they going to take water away from me," which was the big concern, you know, "Are you going to split the Indian Ocean down the middle and say west of this is CentCom's [Central Command] water and east of that is PaCom's water?"

So there were lots of those details that were—it was real estate, because it meant who was the biggest CinC. [Laughter]

EM: Who had the biggest CinC? [Laughter]

SA: Yeah, who had the biggest CinC.

EM: Well, how often did you go out to MacDill [Air Force Base, Florida, Central Command headquarters]? You mentioned these exercises. Did you personally go on these exercises?

SA: Every one.

EM: Bright Star?

SA: Every one. I went on the first Bright Star from Cairo West. We didn't even know how or what uniforms to wear. We were trying to figure out—and we are tied up with these Air Force and Army guys and Marines, and they all have field uniforms. We don't. So I decide that to start off, we will adopt the Seabee [naval construction battalion] working uniform as ours.

EM: Sort of camouflage?

SA: At that time they were solid green trousers, shirts like our khaki shirts, but more of a working shirt rather than a dressy-looking shirt. And our insignia, "We will use the black devices for insignias," et cetera, so we had all of this stuff to sort out. That was number-two uniform. The number-one uniform, that we tried in one of the first exercises at Fort Bragg, was our dungaree working uniform. At that time I had two female sailors and three male. So we go to Fort Bragg, and my sailors are in their dungarees, and all the other troops thought they were in Levi's, so they said, "Hey, we didn't know the Navy was this forward-thinking that you guys get to wear Levi's." [Laughter]

"This is our dungaree working uniform." So they were a hit, with the strange-looking thing that hadn't been seen. The kids got creative. They took the tent, and instead of having the tent flaps, they rolled them back and got some cardboard and painted in haze gray and little black lugs on it like a watertight door [laughter], to get into the tent. It was sort of the stopping spot out on the field; everybody wanted to come by and see the kids in Levi's and their "steel door."

EM: What did you do on these exercises? Obviously you were out in the middle of the desert. Were you saying that if the balloon goes up, here's what you would provide in terms of support?

SA: Yes. What we would do is we would get out there and then we would exercise. I had a lot of interface with MSC, the Military Sealift Command, because Military Sealift Command was going to provide a lot of the lift for all the equipment. So we had to set up an organization that could talk directly through Military Sealift Command to the Army command that was down at Tampa at the time.

EM: Army Materiel Command?

SA: No, it was their big training group that also had their feathers ruffled in this—

EM: TraDoc [Army Training and Doctrine Command]?

SA: Yeah, TraDoc. That's where they had their original interface, so we sort of had to break into that interface and start talking with MSC on what they had already agreed to for lift forces for the Army and some Marines. We would exercise those comm channels and get the flow starting, and, you know, typical exercise stuff. It's all made up, but you exercised it, and then found out those ships weren't available, so where would they get the ships if the same thing happened in real time? And every one of these exercises we went on—which is why I kept sort of pinging on the system to pay attention—every exercise we went on, General [Robert H.] Barrow, Commandant of the Marine Corps, showed up. Some four-star from the Air Force would show up, maybe the Chief of Staff, or if not him one of his other four-stars would show up. Army would always have several four-stars show up, and usually the Chief of Staff of the Army would show up someplace along the way. The only guy from the Navy there that would show up was the three-star at MSC, because he knew he was sort of the go-to guy when it came to the sealift, that these guys were wanting to haul all of their stuff across the ocean.

EM: Was that going to be a tough proposition in that part of the world? Troop reinforcement of supplies?

SA: It was. And that's when we starting reaching out to find those fast sealift ships and to start developing the concept of the maritime prepositioned forces and all of that. That's what worked out of these exercises, and that's why I kept telling the Navy, "You've really got to get serious about what's going on here, because there is a big Navy impact here relative to what these guys want to do. Because if they've got all of that stuff loaded, they're going to want some kind of escorts to get it in theater." We traveled around the region out there for the first Bright Star. We were out at Cairo West, the airport west of Cairo. Still had all of the hulks of airplanes that had been destroyed by the Israelis during the Six Day War, and it was a classic.

I landed there at night in a C-141. I think I picked it up out of somewhere here on the East Coast. I didn't go west that way, I went east. Landed at night on this dark old runway and get out, and we find the tent we are assigned to. Get up the next morning, and there are still Army guys sleeping in sleeping bags alongside the runway, waiting for their gear to show up. In the meantime, the Air Force is there with their Harvest Bear equipment that they had had prepositioned in Europe, for the big push in Europe whenever the balloon went up. They've already got a canteen, air-conditioned canteen running. Their guys are all in their half tents. They've got some hard, some canvas [tents], but they've all got their own power units. They've got their own air-conditioning units. Everything is air-conditioned, and they are driving around in these Mercedes SUVs. [Laughter]

EM: All the comforts of home.

SA: All the comforts of home, and the Army guys are still in their sleeping bags alongside the runway waiting for their stuff to show up. It was really an eye-opener. But we did a lot of interface with the Air Force. The Air Force was there with some two-seat F-16s because they were helping to promote an F-16 sale to the Egyptians. They let me go flying with them in the F-16. Then we had a big air show way out in the desert southwest of Cairo. All of the Egyptian hierarchy, from their Chief of Staff on down, were out there, and Kingston and I and the rest of the component commanders were out there in this tent watching this big firepower demonstration.

EM: It was President Sadat at that point; did he ever show up to observe?

SA: He didn't show up for that one. Kingston was at another ceremony with him when he was shot. He was there, and that was one of the few times that I wasn't with Kingston. The great thing about Kingston was he took me everywhere he went. He says, "You come with me."

EM: Sounds like he really wanted the Navy to be involved.

SA: He did. When we were down in Sudan or down in Djibouti, camped out in tents on the Red Sea, Somalia. Somalia was funny. He had a meeting with the president of Somalia. The president had given us his summer castle to be housed in. It was an imposing building, with the walls around it and all. We walk in, and my Air Force counterpart who was on General Kingston's staff, a two-star who later became Vice Chief for the Air Force, was my bunkmate in this castle.

EM: What was his name?

SA: It will come to me in a second. So we go in to take a shower. Well, the shower is the commode and urinal and shower all combined. It was just a pipe coming out of the wall [laughter] that was a stream of water, and the place just smells. I guess they went in and fumigated it right before we came or something, to knock down all of the bugs that were in there. They had these old heavy drapes on the windows and velvet stuff.

EM: This is the presidential palace?

SA: Yeah. We didn't think too much of it. I'm with Kingston, and he was going to make his call—I wasn't going to go, just he was going to go in on the call. He and the two drivers and myself arrive there, and he goes in. Comes back out, hops in the car, and we are motoring back to wherever we're going at that time, and he's talking to these two young troopers up in the front in their cammies and all. He's talking to them a little bit about, you know, "How's the world treating you guys? How are you guys doing?" And they are fine. He says, "What do you got on you?"

The guy says, "We are carrying this." Their BDUs [battle dress uniform] looked just like any others. They had the buttons on them, but the buttons were just sewed on the outside. Everything was put together with Velcro, so all they did was [*crinkle*]—they had stuff packed all over them, you know, guns and stuff. He turns to me and says, "What are you packing?"

I say, "What do you mean, what am I packing?"

He says, "What gun do you have on you?"

I say, "I don't have any gun on me. I don't walk around with a gun."

He says, "Well I do," and he pats this ankle here and he's got a gun here and one over here.

I say, "Well, General, if anything happens, you'll be on top of me because I'll be already on the floor." [Laughter]

EM: They were ready for anything.

SA: They were ready for anything, yeah.

EM: So it sounds like you got around the operational area.

SA: Oh yeah.

EM: Over to Bahrain at that point?

SA: Mike [Michael P. C.] Carns was the Air Force guy. Mike, real bright guy. Out of sequence, but we traveled the whole AOR. I've done amphibious landings down off the coast of Oman. Kingston and I and Mike Carns and a couple of other guys were with the Omani Special Forces guys and the tribal guys out of Pakistan. They were such good fighters.

EM: Punjabis?

SA: No. Boy, the mind is just not working this morning.

EM: Baluchis or Pashtuns?

SA: No. They're real good; some of them worked with the Koreans at one point in time in Vietnam. They're a real small elite Pakistani unit.

EM: You're not thinking of Gurkhas?

SA: Maybe.

EM: Gurkhas are from Nepal.

SA: Nepal—maybe it was the Gurkhas.

EM: They worked with the Brits a lot. It was probably the Gurkhas.

SA: That's right. So we are way out on the desert in Oman right up against the border with—

EM: Saudi Arabia?

SA: No, it was south of the Saudi border, but with the folks there at the bottom of the Red Sea.

EM: Yemen.

SA: Yemen. So we are right there on the border of Yemen, and we are having a meeting out there, you know, we are baking the bread on the rocks underneath the sand, and all of that. They decide to have a shooting competition. We all get an M-16, or whatever at the time. Figure out how to shoot this thing, and then they put up targets. The name of the game was, we started something like 200 yards away from the targets. The game was that you fire five shots, pick up, run 50 yards, drop, fire five shots, all the way to the 50 yard line. So you shoot at 200 and 150, 100—and I'll be daggum if we didn't beat them.

EM: Really?

SA: Yeah.

EM: They were Special Forces?

SA: Yeah, yeah.

EM: Gurkhas.

SA: I think they probably threw the game. We were always out doing something. We were always exercising.

EM: What was your opinion of the Egyptians, for instance the armed forces there?

SA: They were hard to deal with. They had their own agenda. They were very forthcoming and showed us a lot. But trying to make a deal with them, tying them down to a specific "can you agree to this," their bureaucracy was worse than ours. But as far as meeting them and being around them, they were very pleasant to be with. It's just that when you had to make a decision, then it got a little bit harder.

EM: What about the Saudis?

SA: Saudis were pretty hard to deal with. They do everything on their terms. They're always working on their terms, so you need—if you want something from them, you've got to start at the high end of the scale to get them to back down to where you really want to be. If you try to start there, then you're not going to end up with much. They are very tough to deal with, the Saudis.

The Bahrainis, I call them the soft Arabs. They're just really laid-back, pleasant people. Very forthcoming. Very nice to deal with. Before I ever got to Desert Storm many years later, I had been all over that region and had interfaced with most of the forces. In Sudan they were going to let me fly one of their MiG-21s one day, and they said, "Do you want to fly the trainer first?" I said yeah, I want to fly the trainer first. They said, "Well, it's just an airplane."

I said, "Yeah, but I'll take the trainer flight first." One of their servos was out, so the trainer could only make left-hand turns. [Laughter] So I go out to make this practice flight, and I had to remember that I could never come back to the right. I always had to overshoot things to the right so I could come back to the left a little bit. The airplanes were very poorly supported, the ones that I could see. The one I was going to fly that day didn't have a battery; we I didn't get to fly it, which was probably a good thing. But over in the hangar, we weren't allowed to go to. The Chinese were over there with their version of the MiG-21s, and they were bright and shiny and looked like they were in perfect condition. But we weren't allowed to get over there.

EM: Were you getting a reputation in the Navy at that point that you were Mister Middle East? You'd gone all over the place. You'd dealt with all of the folks in the Army, Air Force, Marine Corps, who dealt with the Middle East.

SA: The Navy wasn't much interested in what I was doing; I was sort of on my own. They didn't want to pay much attention. I often thought of myself as the cork on the bottle. I was supposed to keep everything bottled in that the Navy wanted protected. I guess in retrospect, as I look back, I probably should have put more things in hardcopy than I did. I spent a lot of time talking to people, going to them and trying to make the case, especially once we became a CinC.

As soon as we became Central Command I had to be very, very careful because Kingston would have killed me and would have killed all of my people and everybody else, and we'd have gotten into a big fight that the Navy didn't need to have. But I had to be very careful that I didn't get anything in hardcopy that would be against where Central Command was trying to get to; also the hardcopy from the Navy. Things that Kingston could use against the Navy. The Navy wouldn't listen to the Marines, and the Marines really didn't care. The Marines realized that the coming wave was going to be Central Command, and boy, they put their best and brightest forward. General Barrow was really driving the equation. What a gentleman. I don't know if you've ever been around him.

EM: I've met him.

SA: The Marines would quiver when the guy walked in the room. He'd come to these exercises and have a little séance, sit down with us and talk. The guy had it all. He knew exactly where they were going. He knew what his role was, and what role he wanted the Marines to play in this. You know, the Marines were getting money for doing stuff with the RDJTF, and the Navy just wouldn't pick up on it. Just couldn't. It was just a step too far, and it just puzzled me forever. It played back to me years later—when we get to

Desert Storm we will talk more about it—but one of the things that [H. Norman] Schwarzkopf [Jr.] had against the Navy, and I hadn't realized this had happened, but the Navy had made a last-minute play for Hank [Henry C.] Mustin to be the CinC for Central Command and tried to derail the odds-on pick of Schwarzkopf. He was still reeling from that when I showed up on the scene.

EM: Did Admiral Henry H. Mauz have to deal with that too?

SA: I don't know. Hank and I never talked about that. I'm not sure that Hank and Schwarzkopf had any casual conversations. I don't think they had the same relationship I ended up having with Schwarzkopf. I may be wrong, but I'm pretty sure I'm not. It goes back to what we were talking about earlier. The Navy had a wonderful case to be made to be in the rotation for that CinC . In my mind, the Navy should have been number one pick, and then Army and Marines.

EM: Was there the concern that if they made a real push for Central Command it would have opened up CinCPac if you guys wanted to rotate over here, then give up your hold on CinCPac? Do you think that was part of the equation?

SA: Oh, I think so, sure, and that was part of the equation of why CinCPac was always reluctant to send any force—to chop any forces to Central Command, was that, hey, wait a minute. I'm the big Navy proponent to this in my AOR. And so anything that threatened the AOR relative to the Navy presence was a threat to the CinC.

EM: No matter how you cut it, you would lose resources to somebody else.

SA: Sure. So there were valid concerns on all sides of this thing, but in my mind, all of the flags were there that this region was going to be in the volatile stage—as if there ever was a volatile stage—but it was not going to be in a quiescent stage. It was going to be something that we were going to be dealing with on a much more continuing basis than we had before.

EM: If we could go back just to the time when you were the Plans guy or even the Ops guy, but the Plans guy in particular on the PacFlt staff, was the growth of the Soviet Navy a big concern to the U.S. Navy?

SA: Oh, yes, yes.

EM: Did you work any plans dealing with the Soviet Pacific Fleet for instance, or what the Soviets might do out there?

SA: Yes. We had a lot of planning. Admiral Hayward, when he went out there, had started a whole new series, a whole redo of all of the plans, and it was reflecting the growth of the Soviet Navy.

EM: When he was CinCPac?

SA: When he was CinCPacFlt. And some of the people that were his real horses that did some of that planning for him were still around on the staff when I got there. We had very, very good plans, I thought, that we had worked. All of them had been refreshed. It was a huge concern and, of course, John [F.] Lehman [Jr.]—as time went on his emphasis on the 600-ship Navy and all, was all based on some of that planning, on what it really took to meet the growing Soviet threat.

EM: Your plans when you were there, did they entail giving up a lot of resources to the Atlantic?

SA: We had a swing strategy, yes, depending on how things played out. What the swing would look like. There were several options on that play.

EM: What about the Chinese? What were you doing?

SA: The Chinese, it was, well, we've got this thing that we don't know too much about. We need to keep watching it, but right now they're not doing anything Navy-wise. During the time I was there, it was one of those places where we knew we had to devote more attention, but we didn't know how we were going to do it. Nixon's opening to China was a big step, and again—I've been out of the planning business for a long time—I don't know what kind of plans they finally picked up on for that. But we did pay an awful lot of attention to the Yellow Sea and the Taiwan Strait relative to China. I mean, we did spend an awful lot of time looking at both places there.

EM: Did you travel to Taiwan, Japan?

SA: The only time I had been to Taiwan was when I had the *San Jose* and went in there with her. Of course I had been in and out of Japan, but while I was on the staff there, I don't think I went to Japan.

EM: So you were really focused on the IO and the Middle East?

SA: Yeah, yeah. Well, Korea. We were hard into a new 1027 [CinCPacFlt Operation Plan 1027 regarding North Korea], I think, was the plan for Korea. So that one, we had to work on almost continuously because there was always some new wrinkle going on with that. I'm not sure that I did go out there. I think most of my travels were east and the [Saudi] Kingdom that I never thought I would see again, that I went to ICAF to see because I'd never go there again. Then I spent all of my time running back and forth out there.

EM: All right, well, you've covered a lot of what I had down here on the way to your NavCent [Naval Forces, Central Command] tour in July of '83. At some point, did you ever get a separate office and use your own communications facilities?

SA: That happened after I left. [Laughter] That happened after I left. I tell you, one of the exercises that we had was sort of a tragic one, really, for me in '82 at Fort Irwin [National

Training Center, California]. We were out at Fort Irwin, I think it was spring of '82. We were going to have the largest airborne drop since World War II, I think 2,300 parachutists. We had—I forget how many C-141s and C-130s. It was a big wave.

We were out there in the early morning and the wind shifted, well, the wind picked up, and we were getting different reports, in the drop zone, of the wind. Some places it was exceeding the limits, but then all of a sudden it would die. They decided to give them a green light for the drop, and from my background I'd never seen that much of a parachute drop. In fact, one of the things that I told Bob Kingston before I left was that I wanted to get a parachute jump in. We were going to set that up at someplace in time. They ended up killing six guys in that drop. I think we injured—some seriously—some 40 more. It was a real debacle.

EM: The wind kicked up again?

SA: The wind kicked up. Guys were being dragged across the drop zone. It was a drop with equipment, so there were pallets with jeeps on them and stuff, mixed in with the guys. We were lucky to escape with just the few losses that we had. It got some pretty good press coverage, I think at the time, but it was one of those things I'd thought I would never see.

EM: You saw guys being pulled around?

SA: Yeah. Saw a least one guy who didn't get a good chute. Got a streamer.

EM: A dangerous pursuit.

SA: Boy, I tell you, it was. I really admired those guys, and then watching the guys trying to make sure that they didn't get caught up in some of the heavy equipment that was coming down, you know, and getting blown off their course.

EM: Did you get your jump?

SA: Nope, nope. [Laughter]

EM: Probably not unhappy about that?

SA: Probably, yeah. Probably a good thing. [Laughs]

EM: Let's see, while you were ComNavCent [Commander Naval Forces, Central Command], January to July of '83—I think it was in the spring, I'm not sure, but that's when Beirut happened and Grenada happened.

SA: Yeah.

EM: How did that impact on what you were doing?

SA: Well, that again was the emphasis for control of some—you know, wanting more naval forces in theater to react to that.

EM: To Lebanon?

SA: Yes. Sixth Fleet is of course there, but the fact was that there was some speculation that this might not be limited to Lebanon. Trying to put the forces in the Indian Ocean on alert.

EM: Who would have been the other antagonists?

SA: This was a terrorist attack, but they were still trying to sort out whether the Iranians were involved. Who? One of the issues that we never got fully resolved during the time that I was NavCent was: How far are we willing to let the carriers operate in the region? Are we willing to let the carriers operate in the Persian Gulf? Or are we going to restrict the carriers to just the Indian Ocean, to the approaches?

The PacFlt/PaCom position had always been that we were not going to put carriers in the Persian Gulf. The threat would have to have been backed away considerably further than what we see the threat right now before we would let a carrier wander up in those restricted waters. Two issues that were always on the mind was the potential mining, which we ended up facing. The other was all of the abandoned oil wells that basically—they still have piping sticking up that will rip your hull right open if you don't understand, if they are not all mapped. We think they were pretty much all mapped. We had pretty good maps, but it really restricted your maneuverability, so if you got yourself caught in an into-the-wind situation and got in between a couple of those things and had to make a turn that you didn't want to make, you could end up losing a screw or putting a hole in the ballast tanks.

That hadn't been resolved until we got to Desert Storm/Desert Shield. Desert Storm. I think Bob Kirksey had made one short venture into the Persian Gulf when I was still Fleet Ops, with Red Dog's approval. He went in just for a very short stay and back out, but it was not something that anybody really signed up to do on a routine basis.

EM: Did you get out to Bahrain and talk to the Middle East Force commanders?

SA: Yes. Admiral [William J.] Crowe really put the Middle East Force on the map, and it was a very social network. There was a U.S.-sponsored school in Bahrain under the guise of being for the dependents. There were hardly any dependents there, so all of the important citizens of Bahrain had their children educated at the American school. [Laughter]

EM: And we paid for that?

SA: Yeah. We paid for that. A lot of the Brit kids were there as well.

EM: So it was a heavy diplomatic responsibility?

SA: Yeah. Yeah.

EM: And we had what? *LaSalle* [AGF-3] and maybe one or two other ships in the Middle East Force?

SA: Two destroyers and the *LaSalle,* usually.

EM: What is your take on the good Qataris and the others in the GCC [Gulf Cooperation Council]?

SA: Oh—

EM: Abu Dhabi? Did you meet with their military?

SA: I did. Not so much during the RDJTF days and the early CentCom days, but certainly by the time Desert Shield/Desert Storm rolled around, I was into all of those places with all of their military when I was there. Mostly from the RDJTF and the early CentCom days, we'd always come into the region and usually come into Bahrain, which was an easy place to get a commercial flight into if we were going to take commercial, and we did a little bit of both.

Sometimes with Kingston we would go in there for the military, but a lot of times if we were up in London, we would come down on a commercial flight and land in Bahrain. It was interesting. I was over there one time, and I had landed in Bahrain to do something for the CinC before I was to meet up with him over in Dhahran [Saudi Arabia]. So I get to Bahrain and do the couple of little check-ins, and I check in with Middle East Force and say, okay now I'm going to fly over to Dhahran and meet up with the CinC.

And he said, "Oh dear, are you sure you want to do that?"

I said, "What do you mean?"

He said, "We've been having lots of trouble with our people getting pulled over when they land in Dhahran, and they get the third degree quiz over there by the Saudis. Why are you coming through Saudi? Why didn't you just leave Bahrain and go to wherever you want to go? Why are you coming here?" He said, "We can't predict how they will treat you when you arrive. You may have to end up calling the embassy for some help. We will give you the numbers and all that sort of stuff."

I'm thinking, what in the world's going on? Of course nothing happened. There was somebody to meet me, no problem. They all coexist, but even though they have that GCC arrangement, they're still pretty—

EM: Independent?

SA: Yeah. The Bahrainis are always a little suspect because they are—

EM: So liberal?

SA: Yeah, yeah. [Laughter] Of course, the Saudis, you know, they wanted to build that bridge across so that they didn't have to fly or take a boat over to Bahrain, they could drive straight across this long bridge. Then they built a big mosque so that when they did come over, they could do their prayers in their own mosque and then change into their clothes and go do their drinking at the—[Laughter]

EM: That was their Disney World?

SA: Exactly. Yeah, that was where you would go.

EM: So they let their hair down when they were out of the kingdom?

SA: The thing that we got the biggest kick out of when we were traveling commercial in and out of the region was, you would leave Dhahran or even Bahrain and fly up to London or fly to Paris, or wherever your jump-off point was. You'd get on the airplane, everybody is in their robes and headgear, and soon as the seatbelt light came off, the big mad rush to the lavatories started, you know. Come out of the lavatories, go in with the face covered and head scarf, and come out—

EM: The women too?

SA: Yeah, oh yeah. Particularly the women. Gold chains up and down the arms. Silk clothes, I mean, the latest Paris fashions.

EM: No hijabs?

SA: No. The same would happen when you'd get on the plane in London to fly down into the area, and everybody's dressed, coat and tie and all nice finery and everything, and then by the time you landed, you thought you'd gone through a time warp. Very much mind their Ps and Qs when they are at home, but—

EM: Did your wife ever go over to the Middle East during any of these stints over there?

SA: No.

EM: Your family was in Hawaii at this time?

SA: For the time that I was with CinCPacFlt and Central Command we were all in Hawaii.

EM: Did they like that?

SA: They all enjoyed it.

EM: Nice place to be.

SA: It was a good place to be. The schools weren't that great, but they survived. They all had a good time. They all enjoyed. Our daughter had already gone off to college when we went out there, but when she finished, she came out and was going to just stay for summer months, and then she was going to come back to the mainland. She ended up staying there and eventually got married, the Friday before we left on Sunday to come back to the mainland. She still lives there. So we go out and visit her and her husband every spring.

EM: What do they do? Do they both work?

SA: At that time, she got a job with the Bank of Hawaii and worked for Bank of Hawaii for 17, 18 years, I guess as an executive assistant to their international vice president, and then they did away with that position and she had the ability to shift over, but she said no, she would take an early buyout of her retirement. So she took that, and then she went to work for the Marine Corps over at Kaneohe [Marine Corps Air Station] in their MWR [morale, welfare, recreation organization] world. Same sort of job that she had there at the Bank of Hawaii. Ray [Shiroma] was born and raised in Hawaii. His dad was Japanese and his mother was German. So Ray has the skin and eyes—big, tall, broad-shouldered, thin waist guy, from the German side.

EM: So what was your feeling when you went back to main Navy in—let's see, July of '83? Your experience in the Middle East—did you think, well that was interesting, but I'll never use this again?

SA: No.

EM: It was positive in terms of your career?

SA: I thought it was positive from the standpoint that I had learned a lot and had seen a lot that a lot of Navy guys hadn't seen. I was very happy with my time. I was very pleased with it. I didn't think the Navy cared much one way or the other what I had been doing. But I thought it was important work, and I knew that I had not seen the place for the last time.

EM: So you thought, I may be well placed here for later on?

SA: Well, I thought, "Something is going to happen again out this way, and maybe I'll be in a position to help; I'll be able to do something about it."

EM: Right. So you're off. Anything you more you want to add about that time?

SA: No. It was a fun time. The personalities were great. I made some great friends. Friends that will show up again in other places. I just can't express how much I ended up really admiring Bob Kingston. I don't know what it is about Army generals, but when they make four-star, they become absolutely wild men for the most part. They yell and scream and threaten to kill everybody, threaten to fire everybody, and I mean, the Army guys would just run for cover when Kingston would take off on one of his tantrums.

I said, "What in the world's going on?"

"Well he's mad at you too."

I said, "I know, but it will pass."

They said, "I don't know what it is about you and him, but man, I'd stay away from him when he's like that."

When the guys used to start running away from Schwarzkopf I would say, "Guys, this is nothing like Barbwire Bob. [Laughter] I can remember him getting up—the Army just loves these briefs, you know, morning briefs, and guys get up in front and they present stuff. I mean, he shot so many of them.

EM: Was he all bark and no bite?

SA: No, sometimes he would take action, but for the most part, just like anybody. Schwarzkopf certainly was this way, and Bob was too. If you stirred the pot and really got him going, then you had to know when to stop. Then when you knew when to stop, the next thing was you needed to know when was the right time to approach him. Right time was always when he was alone. Never when he had anybody around him. You had to get him off all by himself—both of them—and then you could gradually work your way back into the issue and have a reasonable conversation about it.

EM: Maybe the tough guy was an image—

SA: Tough guy was part of the image. For some reason the Army guys just loved to terrorize the natives and see who would stand. It's sort of a test, I think. I had lots of funny episodes, some of it just crazy as can be. Kingston always had this warrant officer who was his aide. He was never out of sight of this warrant officer, who'd been with him for years and years. Kingston liked to do a little drinking. He always had a bottle of whiskey somewhere close by. Never saw him drunk. Never saw him in any way not in control, but he always had a little sip of whiskey around someplace. For a Navy guy—I mean they even have a bar set for field use. If you go out into the field with your tent, there's a bar kit all assembled, wooden box with latches. For general officers, you know, whiskey and all the fixings and everything else.

EM: That's not part of Navy issue?

SA: No, not part of Navy issue. We were traveling a lot of times, and when we were doing mixed travel on commercial and military, we'd usually have these parachute bags we would dump everything in and tote around with us. So we were coming into Bahrain—no, I think it was we were coming into Dhahran. I think we were in Saudi Arabia. Kingston's got two bottles of Jack Daniel's in his parachute bag. Well, the warrant officer gets off the plane quick, because he wants to make sure he's out of there before the baggage handlers start throwing this stuff on the apron. [Laughter] He's not fast enough; they throw the parachute bag. Next thing we have is this whiskey seeping out on the tarmac, you know. Nobody reacts to it. We were all told, you buy a bottle of whiskey, that will get you put in the pokey in a hurry.

He's able to capture it, but in the meantime, what happens inside the parachute bag is, the thing that Kingston was supposed to wear to the next function was all soaked in whiskey. So he's beating on the warrant officer, "Why isn't that stuff ready for me to wear?" Lots of little vignettes. It was a great time, a real departure from Navy life. And I had a good time with it.

EM: All right, so you're back at main Navy, commander of Carrier Group Seven?

SA: Yes.

EM: July of '83 until July of '85, two-year stint. Went out to *Ranger* [CVA-61]. Part of the normal rotation for naval aviators?

SA: Yes. That was the desired job to have at that particular time; to get a CarGru and go out.

EM: Finally, you're rear admiral or you're still select?

SA: No, I'm a rear admiral at the time. That sort of started off funny. The CarGru was based out of Alameda, and there was training involved relative to what they wanted the CarGrus to have before they assumed command. So I'm in San Diego with the family. We are living in the BOQ over in Coronado at the amphib base. So they've got us a suite of rooms there, where we are all at. I'm going to TraPac [Training Command, Pacific Fleet] to one of their grooming schools for battle-group commanders. They are teaching us the latest—what's going on with tactics, what the threat is, and how the new systems are automated on the ships—and getting us all acquainted with what's going on. I think that was supposed to be about a 30-day course, if I'm not mistaken.

Jennie Lou has to go back to Ohio. Her mother went into the hospital for an operation, so Jennie Lou goes, and I've got the two older boys—the daughter is still in Hawaii, of course—and the youngest son, I guess maybe he was with me too. But she's back in Ohio. We are there. I get a call over at the school, and they say, "They want you to get aboard *Ranger* tomorrow morning and go to sea."

I said, "We haven't set up the change of command." I'm going to relieve George Aitcheson. George and I are back together.

EM: You relieved him once before, didn't you?

SA: Yeah. He had been skipper of *Coral Sea.* I'm going to relieve George, and they can't tell me anything. Can't tell you what's going on, but the ship is getting under way and they want you on board. I have to call Jennie Lou and say I'm not going to be at the airport to meet her tomorrow because I'm going to sea. "Yeah," I said, "The ship's going on cruise, but I don't know if I'm going on a cruise with them yet. I just don't know what's going on, other than they want me aboard the ship tomorrow morning."

I got aboard the ship and talked to George and said, "What's going on?"

He says we've got this message that we are supposed to go down and operate off the coast of Central America for a few days. I think it was Nicaragua, or one of the places was having some trouble. He says, "We don't know much about it, but that's where we are going. Since it was unknown how long we are going to be there or whatever, they wanted you to be on board, and then you and I will decide when's the best time to have a change of command."

I said okay. Off we go to what I wanted to call the Chiquita Station, but the Navy powers-that-be up in Washington said not politically correct. While I'm sitting there looking at our bananas on the table that have the Chiquita label on them, I'm saying there's something about this I don't quite understand. We can eat Chiquita bananas, but we can't say we're going to Chiquita Station.

Off we go down there, and we're going to operate in support of the Army guy who is SouthCom [Southern Command] down in Panama. We are trying to sort this all out. One of the issues we're going to have to deal with is who are we going to work with. What are we supposed to be doing while we are down there?

And so we had to get some people to fly down to Panama and start the negotiations of who we are going to talk to. Are we here for an exercise? Exactly what's going on? In the meantime, on the way *Ranger* does a refueling. I think it's the next day after we get under way. Anyhow, I'm down in flag spaces. I'm not up on the flag bridge, I'm down below, and all of a sudden I hear the collision alarm. I start running up to the flag bridge to see what's going on. Of course I have no job. I'm just a bystander at the time. I run up to the flag bridge to see what's going on, and here we are entangled with the *Wichita* [AOR-1]. We get separated and all apart.

Now since I'm the free flag officer, I'm the guy that has to investigate the collision. Jerry [L.] Unruh was the skipper of *Wichita.* Jerry was an F-8 pilot, and Jerry and I had known each other for our *Hancock* cruises in the same air wing. So I knew Jerry. Art [Arthur] Frederickson was the CO of *Ranger,* who I had not known before. So I got to do that little chore. Get helo'd over to spend some time on *Wichita* and take a look at their logs and all that sort of stuff.

EM: Did you have a staff to do that?

SA: The staff was available. I had guys there on the CarGru staff.

EM: How many did you have on your staff?

SA: Let's see, I think the CarGru staff ran around 15 folks or something like that. I had the chief of staff as a captain, and I had an ops boss that was commander. I had three commanders, I think, and then a good handful of lieutenant commanders. A couple of chiefs. A couple of radiomen. We wrapped that investigation up pretty quick.

EM: What were your findings?

SA: Basically, the *Ranger* was at fault. She just got in a little too close and got sucked in. I think there was an error by the helmsman.

EM: Who had the con?

SA: The conning officer was the CO at the time.

EM: Did he suffer for that?

SA: Yeah, he never made flag. But he was a good guy. So George and I then had a quick change of command down there, and George flew back, and we stayed down there for about—I'm trying to think if I ever did figure out how long we stayed there. Yeah, we left San Diego on the 15th of July and the collision was on the 18th of July.

EM: Any personnel casualties?

SA: No personnel casualties. No. Very fortunate.

EM: Damaged equipment, both ships?

SA: On both ships, but nothing substantial.

EM: Did the CO give an incorrect rudder order? Or the helmsman?

SA: I think what happened was the helmsman was either late responding or had started to go the wrong way and then came back, and anyhow he was already in a closing situation, and the reaction to it didn't happen as quickly as it should have. They should have picked up the fact that he was closing in on him at an earlier point in time. Of course, from your ship it always looks like the other guy is closing in on you, so the initial thing was, *Wichita* came by the—

EM: *Wichita* was supposed to be the guide ship?

SA: They are the guide, so they are always right unless you see that they did something wrong. Of course, when I got over and reviewed the deck logs and all on *Wichita,* it was obvious that they were doing their job as being the guide. The *Ranger* was the one at fault.

EM: At the time of your change of command, were either of you filled in on what the political situation was? Or were you told to go down there and stand by?

SA: By this time we knew that there was a possible coup that was going to occur in Nicaragua—Honduras or Nicaragua.

EM: Probably Nicaragua?

SA: I think it was Nicaragua at the time. There was some concern about other stability issues in the region. When I talked to General [Paul F.] Gorman, who was SouthCom, I flew in and had a meeting with him. He said, "We need to do a lot of diplomatic flag showing down here, with all of the countries here, and show them that we're here in support of stability. We are not here to impose a force. We are here to try to get people to talk a little bit better together within the region." He said, "What I want you to do is, we need to be sensitive to the possible coup up here, and be able to react to it in case we have to get people out. I'm the guy that's planning all of that, and I'll need some of your assets, helicopters and stuff if possible. In the meantime, I would like to have people come out to your ship and you sort of do the pony show that the carrier battle groups are always good about doing."

EM: A nice lunch?

SA: A nice lunch and a little fly-by. Ships coming alongside. I said okay, we will get started and let me know. "Give me your priorities, and we will make it happen." So we motored around down there. I think we were down there about six weeks, and I had all sorts of presidents of most of the countries come aboard. We had the president of Panama and his chief military guy, and General [Manuel] Noriega come for lunch. And of course we had all of the intel briefings on this guy Noriega.*

EM: What did your intel briefings talk about?

---

* Manuel Noriega ruled Panama as a dictator from 1983 to 1989. In response to his actions against U.S. interests in the Canal Zone, U.S. forces overthrew his government and captured him. Thereafter, he served many years in prison in the United States and France for murder and other felonies.

SA: This guy is as dangerous as a snake. He's been known to take his enemies up on helo rides and they never come back. Even their wives sometimes don't come back. Be careful. Watch him, and let us know if he says anything that we should know about.

EM: Anything about drug trafficking?

SA: Well, yeah, they had a pretty good rundown that he's involved in lots of things that are not right, and he doesn't let the president out of his sight, and the president doesn't let him out of his sight. So they will be a matched pair when they come, but there probably won't be much friendliness between them—keep your friend and your enemy close at hand. And that was exactly it. Noriega was a scary guy when you got to meet with him and talk with him.

EM: What did he say? Do you recall your conversation?

SA: No, he was interested in what we were doing, and he was interested in the airplanes and equipment, you know, from a military-guys standpoint. I probably had more time speaking with him than I did the president, who was not very outgoing.

EM: Did Noriega speak English?

SA: Yes. We showed him that we were flying the Panama flag from the mast in honor of them being on board, and I think we had his personal flag too that we could fly, from a military standpoint. It was a nice visit, but nothing really of import came from any of the conversations. We accomplished the task that we were assigned to do.

EM: Do you have any knowledge—I seem to remember, and it's very vague—there were some reports of us being involved in a mining operation? Dropping mines off Nicaragua?

SA: Hmm. You know that sort of strikes a bell.

EM: You didn't have any mine capability?

SA: Well, we did have mine capability. I hadn't been there for the buildup, but the air wing had a mining capability, but I'm pretty sure we did not exercise a MINEX while we were down there. There may have been some planning with the SouthCom guys as a contingency, but I don't think we exercised it at all. I don't recall it. I'll have to dig back in my memory bank a little bit deeper and see, but that would have been pretty provocative.

EM: Yeah.

SA: Yeah. I'm pretty sure we didn't do that.

EM: Were you aware of our support of the contras in Nicaragua at that point?

SA: No. We knew there was something going on, because we were getting some reports from places and we were wondering where they were coming from. We thought we were going to have to—in fact we did have to go in and extract a couple of people from Honduras, I think. There was a mini-scuffle, and they were afraid that there was going to be an outbreak of violence, and they wanted a couple of people out of there. We went in and picked them up and brought them out to the ship.

EM: Diplomats and civilians?

SA: Diplomats, yeah, civilians. But we were sort of on alert that that kind of thing was going to go on around there for a while.

EM: On the diplomatic mission thing?

SA: Yeah, we sent our C-1 into the airport, and we had it arranged that they either arrive at a certain time, or the people that they wanted out would be there and we'd pick them up and bring them out.

EM: What I meant was, did you go ashore to do the diplomatic thing?

SA: No, I didn't. The only thing I did ashore was the first meeting there with General Gorman. I flew in and met with him and his staff. I didn't go on over to any of the other countries.

EM: All right. Well, you've got a list of the ships in your carrier group, and you've got oilers and a cruiser.

SA: Yeah.

EM: Let's see, three destroyers and a frigate. How did you interact with the COs of those ships? Did they come visit you all the time?

SA: They came and visited, and I visited all of the ships. They also had their commodore who resided on board the carrier with me.

EM: Commodore of what?

SA: Commodore of the destroyer squadron. He was embarked with me on *Ranger.* John [M.] Waples was the skipper of *Camden* [AOE-2]. John had a lot of Vietnam time. Real character of a guy. [Laughter] But he had a supply officer that was a young lieutenant commander who absolutely had more energy than the law allowed. We get alongside Waples on the *Camden* there for the first time, and the next thing we know this bright, spiffy lieutenant commander from *Camden* is on board *Ranger,* and he's bringing me ice cream up on the bridge. "The skipper wanted to make sure you had your ice cream for the day." He had a rag with him, and wanted to make sure that the windows were clean there on the side of the bridge. "Fast Eddy," we ended up calling this guy. He kept us entertained for almost the whole deployment. He had one scheme after another going on over there on *Camden.* Of course Jerry Unruh with the *Wichita* was a good steady hand, and of course he had to get some repairs done from that collision. So he had those done while we were in port at Subic after we were released and went straight to Subic. We had to go directly into the Indian Ocean.

EM: After Subic?

SA: After Subic. This was in reaction to the Lebanon events then, '83, I'm pretty sure.

EM: Well, this says '85 though. No, you are right.

SA: Yeah, so we are in a presence role out there in the IO starting right away, and *Wichita*'s in Subic finishing up her repairs, and we have to release her to go up north for the KAL-007 shootdown [Soviet aircraft destruction of a Korean Air Lines passenger plane]. So she goes up in support of that until released from that. Then she comes back out and joins us in the Indian Ocean.

EM: Where did you deploy to in the IO? Up toward the north?

SA: Oh, we set up an orbit outside the [Persian] Gulf. We were probably 150 miles down south of the entrance to the Gulf. We called it Gonzo Station. We operated around there, and that's all we did on cruise. We went out there and we stayed.

EM: Nothing really happened in Lebanon after the bombings.

SA: Right. We were on alert out there for all this time. Had to find a way to relieve the crew. By this time John Lehman has passed the deal that after 45 straight days at sea, each sailor was authorized to have two cans of beer.

EM: Steel beach party?

SA: Yeah. Steel beach party, and you just couldn't hand out two cans of beer. You had to have food and all of that other stuff with it, and figure out the rotation so everybody had a chance and nobody was on duty drunk or nobody was bartering their beer away to other souls. I had tried to figure out if I could run a scheme with the ships up in the Persian Gulf to see if there was anybody up there that had anything other than Budweiser, so maybe some of the ships might have pulled something from port. But nobody had anything different than what we had. So we had to manage that. We had a lot of fun with that. We had a good picnic up there on the flight deck and all.

EM: In regards to the sailors, are they allowed to bring instruments on board? A band?

SA: Yeah. The sailors have always got a band of some sort. Sometimes we would have boxing matches and volleyball matches. But what the guys really like to do is get the barbeque cans up there and start barbequing some hotdogs and hamburgers, and then it's amazing—just the thought of having two cans of beer, you know. Guys would swear they were drunk just because they—[Laughter]

EM: They wanted the feeling?

SA: Yeah, they wanted the feeling is right.

EM: Well, what was the reaction of folks in your carrier group to the Marine bombing? Was it a shock?

SA: Yeah, it was. Everyone thought well, if we are going to do something about this we will be ready to do something, but that never came.

EM: Going back to your posting off Central America, was there a thought of, oh no, we're going into another situation here in the jungle?

SA: We didn't think it would be anything like a Vietnam thing, but we did think that there was going to be some possibility we would get involved in some humanitarian effort, or maybe some effort of holding people back from moving forward to take over a government, or something like this. But I don't think any of us thought it would blossom to the extent that it would be another Vietnam.

EM: General Gorman, when you met with him, didn't say, we've got big things coming?

SA: No, he was pretty much on the side that says we need to reassure everybody that we are here, and we are here more in a mode of helping to reduce the friction that's going on and not in the role to participate in it. He was very definite about keeping the profile of—we're not going to do any show of force–type things. We are going to do the diplomatic role and play that as hard as we can.

EM: Back to the IO. Did you have any contact with the Soviet Navy ships because their ships were out there?

SA: Yeah, we did. I think there was *Fife* [DD-991]. I think *Fife* had an encounter with one of the Soviet ships there who tried to nudge—who tried to come in a little too close to her, and she barked them away, but they didn't approach the carrier. They didn't get close to the carrier. The one thing that we had happen to us there, with all of those days at sea, was a major fire down in main machinery room number four. That was in November.

EM: How did that happen?

SA: I think we were transferring some fuel at the time. Two firemen down in the space were supposed to be taking readings on the tanks, and what you would do is you'd take the reading on the tank and then you'd cap it again, and one of the guys left the cap off of one of the sounding tubes. Then fuel came up out of the sounding tube and lit off, and the place, the main machinery space, was enveloped in flames and a lot of smoke immediately. As the people tried to get out, some of them tried to get out without donning their smoke masks. We had OBAs [oxygen breathing apparatus], and we also had the smaller masks for quick emergency use.

Some of those weren't readily available, or some of them, they just forgot they were there. But we ended up losing six people in the space and about 35 injured. We were at GQ for that for a long period of time. I went down on the hangar deck to watch and see how everything was going. The ship really did a wonderful job of containing it.

EM: What happened? The damage control parties responded pretty quickly?

SA: Damage control parties responded. We used a ton of the fire-retardant, the foam. We used up almost everything we had on the ship, and then started helos lifting it over from *Camden* and *Wichita.* I forget how many hours we fought that thing.

EM: It was contained?

SA: It was contained to that space, and so what we ended up doing, once we got everything down, stabilized, got the injured to the sick bay, and found out how many people we had lost and got them out of the space, we sent to have a team from Cubi [Naval Air Station] fly out to Diego Garcia and then come up to the ship to assess what we could do. We were operating on three shafts and with the rest of the plant, and we could continue. We could continue operating, so we could still launch planes and everything. You know, the decision from on high was if you can, stay out there. If you've got to come to do the repairs in port, you can't do your mission, then let us know. But if you think you can still do the mission, we want you to stay.

Once we got it fully assessed, we said, well, not only can we continue with the mission, but if you give us the right people and we can get the right parts flown out to Diego and then back, then onto the ship, the survey team thinks it can be repaired while we are under way. That's what we did.

EM: Was Frederickson still in as CO?

SA: Let's see. Art was, yes. He was still CO.

EM: Did *Ranger* have a reputation for being—

SA: *Ranger* had always had a reputation as a hard luck ship. I can remember an earlier cruise, one of the *Hancock* cruises I think it was, and *Ranger* was out with us, and she ended up having to get pulled off the line and go up to Yoko for repairs, and I forget what the episode was. But, yeah, *Ranger* was sort of a hard luck ship at several different times. But this was a really a tragic, tragic event. We spent 121 days straight days at sea out there in the IO.

EM: No liberty ashore?

SA: No liberty ashore. Now, what we did was, Masirah Island off of the coast of Oman up there had been a resupply place for years. For a long time it was classified that we were using Masirah, but by this time most people knew we were using Masirah for resupply. You'd fly stuff in and lift it out to the carriers. We'd pull into an anchorage there and the planes would land. We would send the helos in and drag stuff out. I wanted to put the whole battle group to rest for a few days and play like we were in port someplace. We got an anchorage off of Masirah, and we had to be careful about its use. The Omanis were very restrictive, very leery of us being too aggressive around there, so we had to limit the number of helos we would send to the beach and how much operating we were doing off of there.

EM: Turn a blind eye. Was there any alcohol ashore?

SA: No. We wanted to make sure that we didn't impose on them too much, so we had the anchorage enough separate that it didn't look like we had Masirah surrounded by the fleet. But we found a nice place for anchorage, and we decided that we were going to have the Great Gonzo Bakeoff. The CarGru staff was going to be the judges of the bakeoff. We had an op order and the whole kit and caboodle, and what the rules of engagement were and what we were going to, you know, ambience and taste and the criteria for the grades. We would come around each ship with a team and have a tasting. And the judges were going to be me and my submarine officer, master chief, cook, and I forget, there were about five of us on the team. Then I needed a barge. We didn't carry a barge with us. So the cruiser had the captain's gig. No, he had a barge on board, that's right. So he said, well you can use my barge.

EM: What ship was that?

SA: *Horne* [CG-30]. So on the appointed day, the barge shows up. We get on board, and we are going to go from ship to ship. Water is a little rough. My submarine officer is the first one to get seasick. [Laughter] Then the master chief. We all made the mistake of—instead of sampling, we took a full measure of the dessert at each place we went. So we were on a sugar overload by the time we got back to the flagship.

But it was more fun, and we had costumes. I had a big chef's hat, and I had a necklace made up of Rolaids and Alka-Seltzers. So we would board each ship, and the guys took the presentation part real serious. Some of them had side boys, with the cake rolled down on carts by side boys. Over there on *Camden,* they rigged up a high-line transfer in the mess hall—out of one door with this high-line came this pallet, with the cake coming through. So we went through all of those things.

I had let it be known that I had a special place in my heart for Baked Alaska. The skipper of *Shields* [DD-596], the small frigate that was with us, he took that really serious, so boy, he had one of the best Baked Alaskas I'd had in a long time. We had a good time. And then they had some games, you know, guys from other ships could come over to the carrier, and back and forth there for a little basketball or something like that. But the Great Gonzo Bakeoff was a big hit.

EM: What other morale boosters did you have? Did you have communications back home? What was the state of communications at that point?

SA: State of communications back home was still pretty much nonexistent. But we did have good mail. We had good mail delivery from both Diego and out of Masirah, so we were able to do that. We ran out of ice cream right before Christmas, I think it was. Let's see, 29 February we got back, so it was a seven-and-a-half-month cruise. Right around Christmas we found out that we had run out of ice cream. There was no ice cream. Again, from my earlier days as RDJTF in NavFor CentCom, I remembered that Bahrain had a nice dairy up there, so I called the guys up at Middle East Force and said, "How much ice cream do you think I can get out of Bahrain? I've got a battle group down here and we've just run out."

The pork chops [Supply Corps personnel] all got started talking together, they said, "Well, we think we can fill your needs. How are you going to get it?"

I said, "I'll get clearance for the C-1 to come up Bahrain, if you guys can make the arrangements to get the ice cream out to there to the airport."

EM: How do you refrigerate it?

SA: They had to pack it in dry ice. It was a big race. What we had to do was, we had to know exactly when that airplane was going to leave Bahrain, and then we were going to have to take the carrier up just as close as I thought I could without any diplomatic issues, and as soon as the C-1 trapped, I had the helos all manned and ready. Each one had their destination for the ship that they were supposed to go to. So we had this big ice cream fly-off. [Laughter] That helped a little bit with morale, especially mine.

EM: I get the picture that throughout your career, ice cream was a factor.

SA: Ice cream—well, one of my jobs as a kid was pedaling one of these ice cream carts, so it goes way back.

EM: That's great. Were you under the OpCon [operational control] of Seventh Fleet when you were out there? Did the Seventh Fleet commander ever come out to see you?

SA: Not when we were in the Indian Ocean. It was a long haul out there. All of those days at sea and pretty boring.

EM: How was the weather, hot and humid?

SA: Yeah. But it wasn't as bad as it would have been if we were up in the Gulf. It was pretty pleasant. We had some good flying. Again, with knowing the CentCom business and remembering the Omani Air Force, we were able to make everybody happy. I was able to lash up some exercises with the Omani Air Force and use their bombing ranges for our guys to go in and bomb.

I flew with the A-6 squadron. We ended up being able to do a practice raid on one of their bombing ranges. So we would come in on a certain day and see if we could get into the target without their intercepting us, and drop. I don't think we had some exercises with live ordnance—this one, I don't think we used live ordnance. I think we used dummy ordnance on this particular one. And I got to fly with one of the A-6s. I was acting as bombardier/navigator with the skipper of the A-6 squadron. It was a good experience. We got in there and sure enough, the Jaguars—the Omanis and the Brits mostly were flying the Jaguars. They jumped us. We got in, but they got us on the way out.

EM: Did you have fighter cover?

SA: Yeah, we also had we had some fighters with us, and so it was good. From the aircrew standpoint they did get some good flying. Not as much as they would have liked, but they did have some bombing ranges and some exercises with the Omanis.

EM: Did you have foreign military come out to your flagship? Any visitors?

SA: Oh yes, the French, AlIndien was the French Navy guy, the Indian Ocean commander. He had flown the French F-8s. One of my F-14 skippers on *Ranger* had been a JO in Air Wing 21 flying F-8s. I sent a message to him, and I said, "How would you like to come over? I'd like to host you on board the ship for a couple of days, if you would like, and show you what we are doing and introduce you to the ship and the air wing. I note that you're an old Crusader pilot," and I said, "I happen to have one of the really old—he's not old; he's really young—great young Crusader pilots in Vietnam with me, and he's now the skipper of the F-14 squadron. He'd be very happy to take you up in an F-14 and show you the difference between what you all are able to do with the F-8 and what he can now do with the F-14."

He jumped on that. He said, "Yes, I'll be there." I had him come over, and he loved it. He had a great, great session. Dick Evert was his pilot. Dick was really a good guy, and he gave him a good demo on the F-14.

During the process of the RDJTF and NavCent and all, it was again how to deal with the French in the Indian Ocean because they did have a presence there. We now know all about their Réunion [Island] down in the south Indian Ocean. But they had had a presence, and every now and then, PacFlt, PaCom, and the French would cross swords a little, wondering, you know, can we count on you? Can't we count on you? And this type of thing. I had remembered that and reached out to this guy; I thought it would be good for all of us if we could sort of wrap him into our arms.

EM: So you had a good relationship with him?

SA: Yeah. Very good. Again, that paid off. It's funny how the building blocks go. You do a little of that and the next thing you know, here you are in Desert Shield/Desert Storm, and one of the first things I had to do after Hank left was—Hank had set up a meeting with all of the different navies, and we were going to have a come-to-Jesus meeting with who was going to be the next. There was this big furor from the Europeans that this thing ought to be part of the European Union.

EM: Western European Union?

SA: Yeah. And we weren't ready to buy into that. This meeting was going to be held on the French flagship. And I hadn't met any of these guys before. I was just sort of thrown in there, and this was the first big meeting that I was going to go to that had some political and diplomatic ramifications. So I was able to let it be known that yeah, I had worked with AlIndien a few years back, and I had the *Ranger* battle group out in the Indian Ocean. We had a great relationships, and I'm hoping that we can continue this type of thing. These little things that sort of—all of a sudden they just—every step of the way, one of them pops up and you can play back.

EM: In their preparations somebody probably told them, Oh, he's a good guy.

SA: Yeah. Yeah. He's not going to—it's funny how that all works out.

EM: Any last thoughts on your *Ranger* experience?

SA: No; it was the CarGru, and then we came back, and then we got involved with switching flagships and working other ships up to get ready to go for their cruises. It's a busy time, but it's also a time you can have a lot of fun. That was the CarGru. You get to go around, visit the ships and see the people.

EM: That's the real Navy.

SA: Yeah, yeah. It was good.

[End of 20 August 2015 interview]

Interview Number 5 with Adm. Stanley R. Arthur, USN (Ret.)
Date: 27 October 2015

EM: All right. We are off and running. This interview is being conducted on the 27th of October 2015 with Admiral Stanley R. Arthur. The interviewer is Edward J. Marolda, U.S. Naval Institute. Good morning, Admiral. How are you today?

SA: Good morning, Ed. It's good to see you again.

EM: It's good to be back. Let's see, I guess you wanted to say a few things that you had remembered from last time about race relations, if you want to elaborate on that.

SA: Yeah, you asked the question, I think at our last session during that last period with the carrier command and CarGru, and any incidents that came about. I said no. I thought about stuff, but we were pretty lucky. I got to thinking about it later, and yes, you know, I was lucky enough that in the commands we didn't have a big incident. But I got to thinking about how much time and effort we spent trying to make sure that we were doing everything we possibly could. It was just an inordinate amount of time.

EM: What did you do to handle the racial tension that was present?

SA: We tried to work with all of the different commands that I went through about that time, and we did an awful lot of dialogue. We talked to people about the importance of what the job was, and that it took everybody do to the job. So we tried to get people to believe they were in a system that everybody was in together. The other thing was, we tried to identify people who had access into the black community to find out what the real issues were. For example, one of things I did on the carrier—we had one of the carriers there on the West Coast that had a death down in the correctional custody unit.

EM: Brig?

SA: Brig. And when I took command, the one thing that everybody would talk to me about was how great the correctional custody guys were, the Marines down in the brig. I thought it odd that everybody was focusing on that. I mean, if I talked to the supply officer about what was going on in the supply department, somewhere along the way he would talk to me about the brig. And so it got me a little suspicious, and finally we had a wonderful chaplain, whose father had been a bootlegger from Louisiana. He was a colorful character. I said, "Look, there are some things happening down there. We end up with folks going to sick bay more than I think they should." I said, "You're doing your daily visits?"

"Yes."

I said, "Are you seeing things?"

He said, "Well, yeah, there's, you know—"

EM: Hemming and hawing?

SA: Yeah. So I finally decided I would just put a white rat down there and send the guy down there to sort of cover at the time. I started saying that everybody who comes out of correctional custody, the last thing they do before we release them is they have to come up and have a personal interview with me. So it wasn't unusual that this guy would come up, and so he came up and said, "Yeah, the guys—if they really got a tough guy down there, they will make sure he falls down a ladder or something like that."

EM: The Marine guards?

SA: The Marine guards. I ended up having to bring the whole cadre of Marines who ran the brig up to Captain's Mast, and that solved the issue. It might have added a little more tension between sailors and Marines, but we were able to work through that. It was those kinds of things. I had forgotten how much time and effort was put in trying to make sure things didn't go wrong.

EM: Over time do you think it was worthwhile? Or did you waste too much time?

SA: I think it was worthwhile.

EM: Yeah.

SA: I think it was worthwhile, and I think we were able to get some folks who were in the fringe elements to come back inside the boundaries. I know we didn't get them all. I'm sure we still had a lot of things that went on that we weren't ever able to see. Since we were in the San Francisco Bay area there at Alameda, with Oakland and all of the stuff going on in San Francisco, we were in a prime area for outside influences that our sailors could get to. We had one sailor, I think he was a second class, black, who was pretty aggressive in his feelings about what he didn't like. So one day, chief master at arms, who was just a wonderful guy, came up and said, "Hey, I've got information that this guy is going to try to come aboard with a .45 caliber pistol," and we were able to intercept him.

EM: What was he going to do with that? Did you have any idea?

SA: I don't know. Whether he was just going to have it to have it, or whether he was going to do something or not. It was an interesting time.

EM: Did you have groups that were demonstrating outside the gates of the base?

SA: We had some of that. Not as much as you would have thought for that period of time, but every now and then there would be a little demonstration. For the most part, we were sort of able to contain things within our own bailiwick.

EM: What was your reaction to the increased racial tension in the Navy? Did it surprise you, or did you see it coming?

SA: It surprised me about how hard the two sides were, you know. In my youth I traveled so much with my dad in the service, and we went from place to place, so we would end up in Virginia, and then all of a sudden I would realize that the black kid I was playing war games with, it wasn't acceptable for him to go to somebody else's house and play. It was strange to me. And then the time we lived in Philadelphia. In Philadelphia it wasn't so much focused on race, but it was focused on religion. Between the Jews and the Catholics and the Protestants.

EM: Really? You lived in South Philly?

SA: Yeah. For a while we lived in North Philly, and we finally moved out to the suburbs. When I was in the public schools in central Philadelphia, I came across an incident one day walking to school when these three guys were beating this little kid up, and they were all white.

EM: The kid was black?

SA: No, he was white. I should have kept my mouth shut and kept going, but I didn't, and so I said, "Hey, what are you guys doing?"

He says, "Oh, he's just a Jew."

That didn't equate to anything in my mind at the time and I said, "What difference does that make?"

They shoved—"Get out of here."

I said, "No, just let the kid alone."

So we walked away from that, and boy, by noon hour I came out, I came out to the play yard at noon, and they were all ganged up on me. We got that fight broken up, and then I'm ready to walk home that night from school, and a guy slips up to me and says, "Don't go home the way you normally go."

I asked why. He says, "Because the guys are down in the alley with a knife waiting for you."

EM: Oh, jeez.

SA: I thought, jiminy Christmas, what in the world's going on. I knew people got pretty disturbed about different things—in that case it was more religion than it was race. So it did catch me by surprise, because I thought that people could get beyond that. I had in my memories of the black community, especially in Philadelphia. I would go through the black community every Sunday morning. I would ride down to the airport and stuff like this, and I had to ride right through there on my bicycle. Nobody ever said any bad word to me or anything. They were all dressed in their finery. They had all been to church and they were out, the dads were out shining the cars, and the kids were still dressed up. It caught me by surprise.

EM: Later on, you were there during Linebacker, 1972. How were race relations on board your ship at that point? You were heavy into combat operations, but did you notice any friction?

SA: Very little. Very little. Everyone was working so hard that it didn't show up as much as a lot of people thought it was going on from the Army reports coming out of the field, and fracking [attacking officers with hand grenades] and all of that stuff. No, it was pretty good, and I don't remember any real liberty incidents that had the whites and the blacks at each others' throats.

EM: You were aware of the trouble on board *Hassayampa* [AO-145]?*

SA: Yeah.

EM: Did you think it was an aberration?

---

* In October 1972, fights between black and white sailors on board the fleet oiler *Hassayampa,* only days after a similar outbreak on board the aircraft carrier *Kitty Hawk* (CVA-63), reflected the parlous state of race relations in the Navy at the time.

SA: No, by that time I realized this was going to be something that really had to be dealt with. That it could happen in an instant, you know. Everybody started picking up their ears a little bit, listening for where the agitation would start from. A lot of people said it all started from the black side. Well, that was not necessarily so. You know, the redneck philosophy. It was a very unsettling time.

EM: It just jogged my mind, last time we had talked about women in the service and you mentioned Coogan, does that ring a bell?

SA: Yes. Vice Admiral Coogan.

EM: If you could tell that story. I don't know what it was about, but I had a note.

SA: When I had *Coral Sea* and came out of the shipyard when we were down out of San Diego doing our workups—or out of San Francisco, but we were doing our workups and came in and out of San Diego. He pulled me into the headquarters and said, "You're going to have midshipman cruise this summer. You are going to have two sessions of midshipmen come on board. And there's going to be females." He says, "There's not going to be any trouble. There is not going to be any publication. There's not going to be any PR stuff about female midshipmen—you know, officer of the deck training."

EM: Don't make a big deal out of it.

SA: "Don't make a big deal of it. I don't want to see anything in the press about female midshipmen on the ship." And he said, "Where are you going to put them?"

Well, as you know, with *Coral Sea* you had a very unique arrangement down in the captain's import cabin. You came into the import cabin from the starboard side, and you walked right into the cabin, and there was a little office, and then there was a short passageway leading out to the port side, fore and aft passageway, and there were three or four little staterooms there. So I said, "That's where I will put the females. They've got their own head. They've got their own passageway with a watertight door there, so they are sealed off, and on the other side of them is my import cabin, and I'm not down there during sea time, so that won't be a problem."

He said, "Good. I'm going to assign one of my officers to be sort of the den mother for these female midshipmen." He had a lieutenant, female, that he assigned to come out to the ship, and we were able to do what he wanted done. It ended up that we probably could have taken his lieutenant to mast, because she got a little too cozy with one of the guys on the ship.

EM: The fox watching the hen house?

SA: The female midshipmen were really good, and it was interesting for me to see. For both sessions that I had out there, the top midshipman ended up being a female. By all measurements, they both were just far and above their counterparts.

EM: They weren't aviators at that point?

SA: No, no. They were just straight midshipmen. It was funny, we had one girl from Georgia, and she was exactly what you would think of when people say Georgia peach. She was a beautiful young lady, and she knew it. She had a real pleasant personality and this big southern accent, and so when she was up on the bridge, she just sort of was like queen bee. Instead of her doing the work, she would say, "Oh Joe, can you get that signal book for me to break this signal?" She had them all running around the bridge, and her counterpart, who ended up being the number one midshipmen on board that session, was just exactly the opposite.

I mean, she knew exactly what she needed to do and what other people needed to do. She assigned everybody tasks, and she just ran the bridge like she had been there forever. So at the end of this period we were at sea, the lieutenant came up to me and said, "We've got trouble with this female from Georgia."

I said, "Well, what's the problem?"

She said, "She just tried her charms a little bit too long, and everybody realized that she was sort of using them, and they've gone into giving her the silent treatment."

EM: The other women?

SA: Everybody. All the midshipmen. They all just wouldn't talk to her. Just sort of isolated her.

EM: Did it affect her morale?

SA: Yeah, it did, it did. We had to go in and say, "Look, you've got to realize it's a team effort, and you were sort of playing—'I'm the cheese and everybody else does the work,'" and I said it just doesn't work that way. "So it's something that's got to be changed, or else you are going to find this happening throughout the rest of your life."

EM: How did the male officers take this?

SA: You know, everybody was a little nervous about having them aboard, but as it turned out, everybody enjoyed having them aboard.

EM: Were they distracted from normal routine?

SA: The only time they were distracted was with the Georgia gal. The rest of the time, as long as they were pulling their weight there was no problem at all. So it went a lot smoother than I thought it would.

EM: There was no PR that accompanied this?

SA: No PR. Not a word. [Laughter] In fact, a lot of my counterparts never knew that I had them on board.

EM: What year was this, ballpark? You said it was *Coral Sea,* right?

SA: *Coral Sea,* yeah. So let's see. That was probably the summer of 1979, or '78.

EM: All right, anything more you wanted to say on race or gender issues at this point?

SA: No. They all come back.

EM: All right, well I wanted to start off today with a discussion about your time in OpNav [Office of the Chief of Naval Operations], Director of OP-50, the first one, Aviation Plans and Requirements. This was July '85 to July '86. First of all, where did you settle your family when you came to the Pentagon?

SA: We had a house there from when I had come back to town to go to ICAF and then BuPers. We had bought a home out in Fairfax, and so we rented and then came back to that house. Most of the time I had a little carpool out there that I could come to work with, or my Marine. I had two deputies when I was OP-50, a Navy captain, Don [Donald V.] Boecker, later Rear Admiral Boecker, and the Marine was Colonel Norm [Norman E.] Ehlert. Norm later became a general, and we crossed paths again in Desert Storm. So we lived out in Fairfax, and the kids were in school out there, so that worked out pretty good for us.

EM: You had to carpool in the Washington area.

SA: Yeah, although it ended up that I would carpool just on necessity sometimes. I've always had sort of a tricky back. I was coming to work one morning and I got down to the corner to put on my brakes, and I couldn't lift my foot. So I sidled off to the side of the road and got out of the car; I'd had this happen off and on.

So I'm out with my hands on the roof of the car, and I'm trying to stretch out and see if I can get going again. This guy comes by and says, "Can I help you?"

I say, "Well I think I'm going to be all right, but I'm not sure. I should be able to walk right now, and I can't."

He says, "Where do you live?"

I say right around the corner. He had a van, so he says, "Why don't I take you home?"

I said I didn't think I could get in. He said, "Here's what we'll do. You stay there, and I'll get as close to you as I possibly can, and we will open up the van door"—slide it open, because it didn't swing out—"and maybe you can get in from there."

So it worked out. I'm laying back on the floor of the van, and he takes me home. I get out, and now I'm on the doorstep and still can't walk, but I can crawl. I ring the doorbell, and Jennie Lou comes and gets the door open and I say, "It's my back again."

The guy says, "Can I help you?"

I say no, I'll be fine. I crawl into the house and call good old Navy medicine, and they say just call for an ambulance. "Well," I say, "I don't think I want to do that."

EM: I know.

SA: Spent a couple of days on the floor, and then Norm Ehlert and his buddy that he carpooled with, they carpooled me in, and in those days you could still drive in underneath the Pentagon.

EM: Oh, right.

SA: You know.

EM: Oh yeah.

SA: So they would drive in and drop me off at one of those freight elevators. [Laughter]

EM: You couldn't straighten your leg?

SA: Just couldn't move.

EM: Did this happen when you were flying?

SA: No, never happened when I was flying. But I had a couple of episodes during that period that were interesting.

EM: How did that happen?

SA: I had hurt my back as a kid, probably when I was 13 or 14 years old. I had gone to a summer camp, and we were taking diving lessons and doing backflips off the diving board. The guy was saying, "Oh, you're all flipping way out, you need to move it in. You need to go straight up and flip and come down." Well, I finally moved in and went up, but when I came down I ended up coming down with my back right across the edge of the diving board, so it gave me a tricky thing back there, that every now and then would—I guess would get inflamed, and the muscles would lock up.

EM: Do you still have the problem?

SA: Yeah. I'm doing physical therapy three times a week right now. But it's one of the funny things, because I couldn't sit, and so I would lay on the floor after I got up in the office, and you know how many briefings you have to do in the Pentagon. Everybody would come in, and I'm lying on the floor. [Laughter] It was a little comic relief.

EM: All right, what were your responsibilities in OP-50? You're a rear admiral at this point?

SA: I was, yes, and of course if you remember, I'd left Washington when I was at BuPers, and I was forbidden by the OP-05 [Deputy Chief of Naval Operations (Air)] guys to ever come back in the building. I thought it was funny that I was now back as OP-50. The job was basically, as the Plans and Requirements guy, how to get the current Navy and the future Navy programmed into the POM [program objectives memorandum]. So it was building a POM and getting it—

EM: Program objectives memorandum ?

SA: The old McNamara thing, which, when I was at ICAF they were teaching it at that time.

EM: Systems analysis, right?

SA: Yeah, so I didn't pay much attention to it. So now I was involved in feeding the beast as far as getting all of the numbers together and lots of issues that we were dealing with at the time. We were just starting to talk about what are we going to do about the F-14. Are we going to continue to try to finally upgrade it to the big engines that it was always meant to have? Or are we going to go to a new airframe?

We had the A-12 going into the system at the time. Lots of debate about what that airplane should look like. Secretary [John] Lehman, of course, thought it ought to be side-by-side seating. Most of us thought it should be fore and aft seating.

EM: Wasn't he an A-6 B/N?

SA: He was an A-6 B/N [bombardier/navigator, the second crewman in an A-6 Intruder who sat side-by- side in the cockpit with the pilot]. He was also into UAVs [unmanned aerial vehicle] at the time. There was an Israeli UAV that he was very much interested in looking at. Ordnance issues. We had a lot of new ordnance coming along. Smart weapons were in the programming stages. I'd been in the "black world" to a degree on the CinCPacFlt staff, especially with submarine stuff, from the operational side of the black world. Now I had to get briefed into all of the black world stuff of acquisition things, you know, on the drawing boards, and were they going to come into the fleet or not come into the fleet. So it was very interesting from that standpoint to see where the technology, the future technology was really—where it was going to be at.

EM: The F-14, at that point it had been operational since '75?

SA: Yeah.

EM: So it was getting old at this point?

SA: Yeah, it was getting old, and we had already got the A-4 out of business. The A-7 was—it was time for something for the A-7, so it was for the F-14 too. The F-14 drivers were still upset that the airplane never came with the engines that it was designed for. So the idea was to get the engines, the big engines, in the F-14.

EM: Did you get the engines into the F-14?

SA: We did get some upgraded engines in the F-14 eventually. And then, of course the F-18 was now on the drawing board as a replacement for the A-7 and for possibly the F-14. It was, do we go to the next generation of the F-18? We had the A's, B's, and C's—do we do the E/F? So that was all stuff on the drawing board that we were playing around with. The V-22 was big for Marine air, and they were trying to get that program.

EM: Wow, that goes back a ways, doesn't it?

SA: Yeah. It was a long, long program.

EM: Can you talk about the A-12, which of course was never operational and ended up being pulled back. Why? What's your opinion about the A-12 program?

SA: Well, it was one of those programs where they really were stretching the technology. It just dragged on too long. The costs were going up. I guess it was [Richard B.] Cheney who finally—

EM: I think [Henry L.] Garrett [III] was Secretary of the Navy. I'm not sure who SecDef [secretary of defense] was.

SA: Maybe it was Cheney. I will have to think about that some more. But it's sort of like the same drumbeat that you see with the F-35 today—you know, too expensive, we can't afford them—and the thing just kept slipping. I guess finally the SecDef made the decision to cut the program. The Navy was still behind it. He just cut it, and I'm still surprised that I guess the settlements have finally been made, the money that was lost by the contractor and all. By that time it was a fairly mature program when they finally cut it.

EM: Where was the fault? The contractor was not doing its work?

SA: You know—

EM: Or did the Navy want too much of this airplane?

SA: All of the above.

EM: [Laughter] Right.

SA: Let's look at the presidential helicopter replacement program. Like the A-12, OSD [Office of the Secretary of Defense] canceled the program. It just so happened that I ended up on two different panels to look at what happened. Lockheed Martin put me on an internal review, since I was in a business area different than the one who had the program. Later on John Douglass headed up an industry panel to report out to the SecNav on what went wrong from the DON program side. From what I saw, it looked like a mirror image of the A-12 program, although it had one more partner in crime—the White House, the ultimate customer.

The program started very simply, find a "green"—an in-production helo capable of carrying the cube and weight of the mission package—and marry it with a system integrator. That is how the contract was signed. Lockheed Martin integrated the system into the MH-101. NavAir was not happy with the basic MH-101 and wanted it modified to meet NavAir specs; the mission package grew due to new White House inputs, and the net result was too much weight and cube for the basic airframe. So new engines and main rotor needed to be sized up to haul it around.

There's a philosophy in business that we've heard ever since we were young guys: the customer is always right. Well, that doesn't work in these big acquisition programs. If the company believes the customer is always right, that is a mistake. And if the customer believes he's always right and those charged with the delivery listen to him, you're bound for disaster. Somebody has to tell the truth. Somebody has to say, "Look, you are asking for something that we can't give you in the timeframe and cost that we were contracted for."

EM: So finding that individual is key?

SA: Yeah, finding that individual is key.

EM: Where would that individual be? On the Navy side or the contractor's side?

SA: It has to be, in my mind, it has to be on the contractor's side. The contractor has to end up telling the customer no. And that's not a very comfortable place for the contractor to be because maybe the guy that's providing is right; maybe he's wrong. So there's always this belief that, hey, we know you can deliver, just get busy and get serious about it and do it. That's a big mistake.

EM: Who did you work for in OP-50?

SA: The OP-05 was Vice Admiral Ed [Edward H.] Martin, one of the POWs that returned. Very pleasant guy to work for. Ed was sort of an elitist. He was very trim, good looking guy. He was just sort of a pretty boy, and he was very politically attuned to what was going on around him, so he knew the politics of the building well, and he didn't particularly like to get his hands dirty on issues. But he had a quick mind, and he let us do our job.

EM: You never had any real run-ins with him?

SA: Never had any real run-ins with him. I wasn't there long, and his deputy, 05B, Cookie [Paul T.] Gillcrist, was retiring. Bud [Leon A.] Edney was coming back. Ed thought he was going to—Ed thought Bud would come back to be OP-50, and I would move up to 05B. He told me this is what was going on. I said, "You know Bud and I, I'm maybe a couple of numbers senior to him, but that doesn't make any difference. We're good friends." I said my preference would be to stay as OP-50 and not go up to be 05B.

He said, "Why is that?" I said I'd looked at the past OP-50s, and none of them had put a POM together and seen it go through the whole process and end up in the president's budget way down the road. I said, "So therefore, I think there needs to be some more continuity in OP-50, and I'm willing to stay as OP-50 if you are happy with what I'm doing."

He said, "Well I didn't want to bring anyone in and put him in front of you."

I said that didn't bother me at all. I said, "You know, I think it's best for the system if I stay as OP-50," and he said okay.

EM: And yet you were only there for a year.

SA: Yeah.

EM: Other things came in later?

SA: Well, yeah, completely.

EM: We will get to that in a minute. The CNO at the time was Admiral Watkins, submariner. Did he involve himself in aviation matters to a great degree, or not?

SA: He didn't, except there was one thing that happened during that period of time that was sort of close to the end of my time as OP-50. I had worked for Admiral Watkins, of course, on the PacFlt staff.

EM: When he was CinCPac?

SA: CinCPacFlt. So I knew Admiral Watkins quite well. Paul [D.] Butcher was a surface rear admiral at the time. Paul and I had been in the Rapid Deployment Joint Task Force in the standup of the RDJTF to CentCom. Paul had been the plans officer there for P. X. Kelley and then Bob Kingston. So Paul I think was deputy of OP-06, and he comes down, hair on fire one afternoon and says, "You've got to help." He says OP-06, I forget who it was at that time, was gone, "and I've got to go to the tank session this afternoon, and they are going to vote on the air component commander issue." He says, "I need your help."

We had talked about this long ago, back in the days when we were in the RDJTF/CentCom, about the Air Force, which had been working this ever since World War II to have an air component commander who would own all the air. And of course there's a lot of flaws in that process, as far as I'm concerned, and he was concerned, and everybody else should have been concerned.

EM: This was to be the air component commander for which command?

SA: This was just endorsing the concept, so that then you could start changing all of your war plans—

EM: In the JCS tank?

SA: Yes. So I gave him the argument for not doing it, and he went off. But of course we had Admiral Watkins who was the Navy guy for the tank, and Admiral Crowe was the Chairman. Neither one of them understood this issue. Sounded fine to them.

EM: Crowe was surface as well, right?

SA: Submarine.

EM: Submarine?

SA: Submariner. [Laughter] That's when we lost. That's when the JCS said, "Oh yeah, we are going to have air component commanders come in."

EM: We know what happened during the Gulf War, but what were the reasons for your opposition?

SA: Well, the reasons for the opposition were, why should carrier aviation be out at sea and may or may not be involved with the air war over the beach but might be involved with access to the area, and why should naval aviation have to be ordered by the air component commander to either do or not do the mission of protecting the fleet?

Obviously there had to be better coordination, and there are ways to do that, but if you had an air component commander that said all the air belongs to me, in my AOR for example, not just a little theater dustup, the carriers were going to be neutralized. If they couldn't launch their airplanes without an okay from somebody that's 2,000 miles away, it didn't make sense. So we just thought that there were other ways to accomplish the task of how to blend the airs when they were both focusing in on a precise target. We just thought it took away a lot of the flexibility from a Navy standpoint.

And it does even today. From guys on the scene, we've had carriers go down off the coast of Africa outside the Gulf thinking they had to have the okay from the guy in the Gulf to fly their airplanes down off the coast of Africa. Crazy, absolutely crazy. So we made this argument, and Paul took it back down. It passed. Years later the Air Force invited me down to Maxwell to speak. I'm not sure if I was retired yet, or I think I was just about ready to retire. I was Vice Chief.

EM: Into the belly of the beast?

SA: Yeah. So I jumped aboard this little Lear jet, and there's an Air Force major sitting right across from me. No sooner did we launch in the air and headed down than we started talking. He said, "You want to know how the air component commander issue was solved down in the tank," back in whatever this year was? I said yeah. He said, "Well, I was part of the Chief of Staff of the Air Force's little staff." Each Chief of Staff of the Air Force always has a little cadre of people that he assigns to do certain things. He said, "I was part of the team that was tasked with doing all the reading of the Navy periodicals, especially the Naval Institute *Proceedings,*" and he said, "of course you know that the concept of the air component commander started with the bombing campaigns in Europe, and then the Air Force tried to bring that out in the Pacific. The Marines and the Navy just fluffed them off." He said, "so I'm reading all of this stuff, and I finally go to the chief and say, now is the time. The Chairman is a submariner, doesn't know a thing about aviation. [Laughter] The CNO is a submariner and doesn't care a thing about aviation. As long as you are not going to fool around with the submarines and how they do their business, it's a slam dunk. Take it to the tank now."

I said, "Well, you made a good call." [Laughter]

He said it was the truth; "that exactly how it happened."

EM: On its face it seems logical that one person should be running the air show, but when you get down to the details, maybe it's not as clear.

SA: No, it's not as clear. It's just like the big argument during the Gulf War was why did I have airplanes for self-defense of the fleet, and "How many of your assets are you using for this?" I said as many as I want to use, and as many as I think I need, and it's nobody else's business but mine, because we've got a threat on the east side of us here that nobody knows anything about and you're not watching it because you don't care.

EM: Iran.

SA: Yeah, and they've got this mountain ridge that sits over there, and a valley on the other side of it, and I said, "I've got about a five minute look at these guys if they ever decide to come down the valley and bop over the top of the mountain. That's not enough." I said, "I've got, I cannot have all of these ships out here exposed to something that you don't think is a threat or could be a threat, so I'm obligated to take care of it." So it was one of those things that we had to fuss about.

EM: We are jumping ahead, but let's get to that issue. Did he give you delegated authority to take action right away if that threat appeared, or did you have to go to the air component commander to get authority to do it?

SA: I would have probably had to go through them to do it. To get to shoot.

EM: Yeah, right—with five minutes you don't have much time. All right, Secretary Lehman. Now, he liked to get into the details of everyone's activities, but especially aviation. So did he come down to your office?

SA: No, I got up there a lot.

EM: All right.

SA: With him and Mel Paisley.

EM: Did you brief him when you were on the floor?

SA: Yeah.

EM: Really?

SA: Yeah.

EM: I mean, did you lay down? You said earlier—

SA: Oh, no, no. [Laughter] I almost did. When you presented the POM to the Secretary of the Navy, he had a little blue room, and we would bring in these volumes of books, and in the blue room he would have all of the chief members of his staff. The CNO, Commandant of the Marine Corps, and my job and 090 [Director, General Planning and Program Division] and the CNO were all there—and my job was to brief the POM to the SecNav [Secretary of the Navy].

Well, one year—I'm not sure whether I was OP-80 or 90, and I was having one of these periods of time where I had the spasms in my back and I couldn't sit, so I had to stand and be propped up against the wall [laughter], while laying out these books and reading excerpts from them to him. But no, he was very active.* John Lehman was really an interesting guy. I don't know whether we talked about him, my meeting with him when I was in BuPers.

EM: I don't recall that, no.

SA: Well, when I was there as the head of the detail shop, I got this call one day that said, "You've got to go over to the old office building, right by the White House."

EM: Oh, the executive office building, EOB.

SA: And meet this guy John Lehman, who is on the Arms Control Board or something like that at the time, and find out what he wants to do on his active duty this summer in the Naval Reserve. I'm thinking, why am I, a commander, going over to meet this guy to figure out his summer active duty. So I get ahold of the little old lady down in active duty orders shop and say, "I've been told I've got to go over and talk to this Lieutenant John Lehman."

"Oh," she says, "not again." [Laughter]

I say, "What do you mean not again?"

She says, "This happens every summer. [Laughter] He wants some of the most outrageous things."

I say, "What am I supposed to do?"

---

* John Lehman was one of the most energetic and active Secretaries of the Navy in modern times. He championed the 600-ship Navy and the Navy's Maritime Strategy.

"You go over and find out exactly what it is he wants to do."

EM: Were you told that he was politically connected, or why did he get special treatment here?

SA: It was sort of indicated that it was important to do—to take good care of this guy. So I went over, and John said, hey, let's go out and have lunch. I said okay—

EM: Did you go to the Metropolitan Club?

SA: Probably.

EM: The former Navy Secretary told me one time it was one of his favorite lunch spots, where Alfred Thayer Mahan and Teddy Roosevelt had lunched together.

SA: Well, it may have been because we went to the place and we had a nice little lunch. A very pleasant guy at that time. He says, "Here's what I want to do. I want to get on a carrier for a week or so, here and there." I said okay.

So I come back and give the little lady the itinerary that he would like, and she shakes her head and says okay, we will make it happen. So then again, when I got out and ended up out at PacFlt, he would come out and see the CinC and see the big CinC up on the Hill, CinCPac, and I finally got the clue that this was an important guy. I don't know who—I never did discover who the cadre of people were that had seen in him, at an earlier stage, that this guy was going to be very influential, somebody that they ought to take care of. Some of the old-hand aviators had identified this guy.

EM: He's a comer.

SA: Yeah, he's a comer and we've got to get close to him. So we did, and he liked to stir the pot, and he liked to get into the nits and grits, and he had his own ideas about certain things. Like anyone with that kind of vision, part of the vision is great and part of the vision is a little flawed, but for the most part he wanted a 600-ship Navy, and I was able to brief him when I was OP-09, OP-80 or 90. In that POM brief I was able to show him the number 600 ship. That's what he wanted, and we were able to do that.

EM: What did he want in naval aviation in particular? What types of aircraft?

SA: Well, he was interested both in the types of aircraft and in the weapons. He was one of the first guys who really saw the value of UAVs. He wanted them for gun spotting, and while the rest of us were saying we didn't know about that, he says, "The Israelis have got a system. Why don't we just buy it off-the-shelf and try it?" Of course, I used it out in the Gulf War.

EM: Did they buy it from the Israelis?

SA: Yes. He was fun to work for.

EM: What were some screwy things?

SA: Well, most of the stuff that got him across the breakers [at odds] with a lot of people were his personnel issues. He liked to dabble a little bit.

EM: You mean like promotion boards?

SA: Promotion boards and stuff like this.

EM: Did you think that was proper? Improper?

SA: I thought it was probably something that he should have stayed away from, although he did identify some real talent that could have been bypassed, and I'm sure from his standpoint he would say, "I know they were going to be bypassed, so that's why I stepped in and made a difference."

EM: Do you remember any individuals in particular?

SA: Well, I think that one of the guys he promoted very heavily was Joe [Joseph W.] Prueher, and that was a good, good pick, you know. I don't think Joe would have been bypassed in any shape or form.

EM: Guess who my next person I'm interviewing is?

SA: Joe?

EM: [Laughter] Yeah.

SA: I think Joe was handpicked by Secretary Lehman to be the standup to the outfit out at Fallon. He was the first CO out there. He [Lehman] was a great fan of [James A.] Ace Lyons [Jr.] back in the day. That may have been not as good a choice. But all in all—I know conventional wisdom now sort of says that a lot of the guys he thought were his buddies turned against him once he was gone, but I think for the most part, most people thought he did good things. I think one of the unintended consequences of one of the things that he did was, he could see that we were eventually going to go to a fleet of all nuclear carriers. He said, "We've got to have an alternate path for aviators. We've had an alternate path for aviators to make flag rank with conventional carriers, nuclear carriers, so we've got guys who may not be so technically oriented but can make flag rank because they were COs of conventional carriers. We are going to all have nukes, and so we are going to end up with a cadre of people to select from the aviation community that have all technical education so they can be the CO of the nuclear carrier. And we are going to miss out on the guys who have had a more liberal arts education who maybe have a little bit better ability to look at the big picture, instead of the carrier types. Technologists. And so I'm going to introduce this thing as super CAG so we can make a CAG a captain. Before, he lost the air wing because he made captain. We didn't have captains as air wing commanders."

Now, a couple of years later, we are going to have captains as CAGs and we are going to have a deputy for the CAG, and that will be the alternate path to flag rank. I think Jay Johnson was probably the first of that group who was able to do that. So that's sort of him thinking out of the box and looking forward. He did that. Now, the unintended consequence of that today is that the aviators will do anything to not be the CO of a nuclear carrier. That is looked at as second class. You need to be the super CAG. That's the job.

EM: Huh, interesting.

SA: I think that is entirely wrong.

EM: And the last conventional carrier went out with what, with *John F. Kennedy* [CV-67]?

SA: Yes.

EM: So it's not career enhancing to be the CO of a carrier?

SA: Not necessarily. Guys will do almost anything to get them out of being nuclear CO.

EM: Well, that would be captain rank as CO of the carrier and captain rank as super CAG?

SA: Yeah. So they are co-equals. There's so much more difference now. There are guys today who will argue, "As a CAG, now I'm a warfare commander," and all this other stuff that we've developed around that over the period of time. I still believe that the CO of the carrier, who has to know what the air wing is doing and what the air wing is capable of, and who has all the management issues of running that ship with all the technology, plus the fact that he is the centerpiece of any afloat business. Although you've got warfare commanders out there of every ilk, the air warfare guy and the surface warfare guy, they all still have to look at the carrier to get to the where-are-we-going, what-are-we-doing routine. A good carrier CO is certainly involved in all of their decisions as well, so I think they end up with a much broader look at the world.

EM: So now they are just running the ship?

SA: They are supposedly just running the ship. So I don't think John Lehman thought that was where we'd go; that it would end up taking the carrier CO down a notch. I think he thought he was going to be able to bring the CAG up to the same spot, and both would be appealing, but now the aviators are not so strong on it.

EM: All right, back to the OP-50 year that you were there. The Maritime Strategy was all the rage.

SA: Yes.

EM: Were you involved in development of that strategy? You mentioned the 600-ship Navy, which pretty much went hand in hand. Did you think there should be a naval strategy or not?

SA: Oh yeah. Absolutely, there had to be a Navy strategy, naval strategy. Although that was up in the 06 side of the house for the most part, when I moved over to OP-90, then that's where we sort of started looking at how these things played into the strategy from the standpoint of the OP-05 or OP-50 perspective. Were the aircraft and the weapons compatible with the strategy? Could we use the strategy to make a case for a more sophisticated weapon?

At that time, of course, Tomahawk [land attack missile] was on the drawing boards, and we were trying to bring the Tomahawk into the fleet. There was some speculation whether it should also have a Tomahawk for the aircraft. At the time, it was too big to put on an airplane. Some of the stuff that we were trying to do for precision weapons for the air, we had to compete against Tomahawk for funding. So I was more interested in some of the air systems, Harpoon and some of the other stuff that was coming down the pike.

EM: The overlying concept of the Maritime Strategy was to hit the Soviet Union around the periphery, not just in Europe, so Asia was important, the North Atlantic was important around Norway, and that sort of thing. Did you think that the aircraft that were then in development and the weapons that were in development, were they appropriate to that strategy?

SA: We were trying to get there, so what we had on hand was not, for all intents and purposes. We were still trying to get out of the Vietnam iron dumb-bomb routine and into the smart bombs. The Air Force was working much faster at it then we were.

EM: Why did it take so long? We had PGMs [precision guided munitions] in '72, and here we are '86.

SA: It has always remained a mystery to me why we couldn't get the Navy to move faster in that. You are always fighting the concept, "We've got plenty that work. We've got other things to do." It was a tradeoff. We missed some opportunities.

EM: There were still plenty of iron bombs around, maybe that was it. Use up these resources.

SA: That's right.

EM: All right, any particular challenges or accomplishments when you were in this OP-50 job?

SA: It was really a busy time, and I know there were some victories that we won and some things that we lost. It's really hard now, this whole evolution from OP-50 to OP-90 to OP-04.

EM: Runs together?

SA: It all runs together because the same issues—I was seeing from a different lens as I moved along. In building the POM for the aviation where we were headed for the future, we did, I thought, a pretty credible job. We had lots of issues with spares and stuff, and this really came later when I was OP-04, that I really had to get more involved in that. You sort of measured in that world at the time, the surface guy and the submarine guy, the aviation guy, as we built our POMs, and then we had to get it all blended in by the OP-90 guy and OP-08, you sort of got graded on whether it was a tight program or a loose program. Basically we started getting some credit for having a pretty tight program, which had not been the reputation on the aviation side of the house. I was pleased that we were able to start getting some of our own stuff together, so that what we really needed got noticed.

EM: Was that one of the reasons you were pulled up to OP-90?

SA: That job, I would have thought I was the last guy in the world that they would have sent up to OP-90.

EM: Why? Let's see, General Planning and Programming?

SA: Yeah.

EM: Well, you were just saying you did a tight job with your POM, so maybe that impressed some people.

SA: I guess maybe it did. [Laughter] Because that's where I ended up. But you know, I was really a novice at this game. I sort of thought OP-50 was going to be talking about airplanes all of the time.

EM: When you were in OP-90, was it more than planes, or were you into surface and submarine?

SA: In OP-90 I built the POM, the whole POM. Basically under the auspices of 090 or 080, or whatever the three-star was.

EM: Who did you work for at that point?

SA: Dan [Daniel L.] Cooper and then Bill [William D.] Smith, both submariners. Submariners always had a tight hold on the programming. Trost had been in 090, you know.

EM: Why is that? Are they detail oriented?

SA: Well, they are detail oriented, and they always had a power structure that ensured there was always a good submariner in a spot that could make a difference.

EM: Are they good guys to work for?

SA: Yes. Dan Cooper had a brother who was an aviator that I knew well, and I didn't know Dan at all, but we got along fine. Bill Smith was a wonderful guy, just a wonderful guy.

EM: Were there issues that, if he was an aviator, he understood what they were about when the submariners might not have? Did you have to persuade them? Or were they hard sells on certain things you wanted to get across?

SA: You had to have your act really together. The submariners, the one thing, the one wonderful trait about every submariner in this business that I met, they didn't give a nod to any fluff. It had to be hard. It had to be hard facts. It had to be hard numbers. There was no—well, it will come. I mean, you had to have a path. And if you couldn't describe that path, and if you couldn't weave the story to show that you had all the alternatives covered, you weren't going to get very far.

EM: Do you think that had anything to do with Rickover's preparation for nuclear power?

SA: Yes.

EM: Could you elaborate on that?

SA: Well, between *San Jose* and *Coral Sea,* that's when they started that Senior Officer Material Readiness course, so I got a chance to go. I was in the second class or the third class to go out to Idaho Falls and get some of that nuke style training. I was really impressed with the fact that they started with basics out there to make sure you got a good understanding, and I think I probably mentioned this during an earlier interview. Out there with guys that—my counterparts who didn't understand algebra.

EM: That would have been me. [Laughter]

SA: And they had a wonderful civilian instructor out there who taught algebra. I mean, it was algebra for dummies. Of course, I had a lot of math and stuff by this time, and I said, "Jeez, I wish somebody had taught me algebra this way when I was a kid."

He said, "Look, you go into the grocery store and they've got these 'buy two, get one free,'" and all of this other stuff. He said, "How do you know it's a good deal? You sort of have to understand algebra to do that. If you get three of them, you get two and one's free, well, what's the price of the two? And so now you've got a price for two, and you add a third one, and you've got to figure out, okay, now what's the unit price for each of them? Now, if you go to the store next door that's just selling them one at a time and the price is this, which is the best deal? To buy two, get one free, or just buy one at a time?"

I said, that's algebra. All of us could figure that out, you know, if you can figure that out, just remember there is a formula for that, and that's how easy it is. If you want to know how much gas mileage you are getting on a car, you've got how many miles did you go. How many gallons of gas did you use?

EM: Right.

SA: X is miles over gallons, miles per gallon.

EM: I wish all of the algebra teachers would get it down to the basics.

SA: Yeah. And so that was everything—that's how his training went. Everything built on top of good, solid reasoning. The nukes had that instilled in them from the start, that you start here and you get to there, and everyplace along the way there's a checkpoint that says, are all of the bb's accounted for when you get there? He did the Navy a great service.

EM: So you prepared the overall POM in—it went from OP-90 to OP-80. Is that just a numbers change?

SA: It used to be OP-090 in the early days. Then they did a renumbering of the staff at that time and said, no, 090 is going to be 8, and so that made me 80 as the next step down.

EM: Just the letters changed, it didn't matter if the numbers changed.

SA: Yeah, the jobs didn't change; 81 still became the budget shop, so we had the planning shop, the budget shop and the—

EM: So you were in that job for two years.

SA: Yeah.

EM: How did you like working in the Pentagon? [Laughter]

SA: You know, people think I'm crazy because I'm the last person that people would think that would enjoy the Pentagon, but I did. It was very frustrating. There were so many frustrating things that—you know, you had to find a way to get rid of the frustrations.

EM: What were the frustrations?

SA: All of the briefs—you'd think you'd solved a problem, the issue was dead, and two months later it was right back up in front of you again with another flavoring to it that you had to deal with. Just seemed like if there was something that you didn't really want to deal with, you couldn't get rid of it. It just kept showing up, showing up, showing up. The political games that went on between the OpNav staff and the SecNav staff and the SecDef staff, and then as the Joint Staff became more powerful, and the waste of talent and time. There were just lots of those things that were very frustrating.

There was so much wasted motion, but it was motion that had to occur. So I had my own ways of handling that. I had little electric cars, and I would race them up and down the hallways [laughter] and crazy, crazy stuff, you know, but it was a stress relief thing. I spent long hours, I was always there early and late, but you did accomplish things. It took a lot longer than it should have, but you were able to accomplish things.

EM: Who were some of your key subordinates in both your OP-50 and OP-90 jobs?

SA: In OP-50, like I said, my two deputies were such great guys, Don Boecker and Norm Ehlert, and then I had my counterpart down in the readiness side of the house, John Weaver, Rear Admiral John Weaver. His claim to fame was he was the great punter on the Navy football team back in—wonderful guy. He and Don Boecker played golf every Saturday morning. I used to kid them since I wasn't a golfer, you know, I'd try to make them think that they were going to have to stay and work a little bit longer so they would miss their golf game. Chuck McGrail, who had been my deputy in BuPers, was now across the hall from me and had the carrier desk for OP-05, and then Bud Edney was up there as the deputy to Ed Martin. We had a really good group of guys there when I was in the 05 organization. Then when I came to be OP-90 and then 80, my civilian deputy was an SES [senior executive service] by the name of Irv Blickstein, who was one of the best SESs around in those days. He now works for RAND along with Charlie [Charles P.] Nemfakos.

EM: Oh, yeah, I met him.

SA: He and Charlie are over there. I did a lot of work with Charlie when I was also down in OP-90. Don [Donald L.] Pilling was my bullpen guy who was the—where the worker bees for OP-90 that put all of the paperwork and the POM together and brought it up to me with the briefing books. I used to be reading books stacked to the ceiling. One of his top guys that worked for him was Jonathan [W.] Greenert, the CNO that just left office.

EM: Oh, yeah, right.

SA: There was just an extraordinary amount of talent to work with and be around. That was another real plus.

EM: Who were big supporters of your advancement, your career at this point? Did you have someone who was your mentor, or somebody that you knew who was looking out for you specifically?

SA: I don't really know. I had guys who I had worked for. Bob Kirksey, who had been a CO of mine back when I was in OP-50 or VA-55. He was a three-star in the building at the time and his career was about winding up, and he was going to go out to be working in the civilian world. He wouldn't have been in a position in the organization to have championed me. Admiral Trost knew me pretty well from my days in BuPers. I think Admiral Trost was the guy that was probably, he never said, but I sort of assumed that he was probably the guy that got me on the PacFlt staff after I came off the carrier, because he was a deputy at PacFlt and he left just as I got there, or maybe even—no, I think he was there a week or so before I got there and then left. So I think he knew me, and I think he probably had a chance to put in good words or bad words, either way, I don't know which, but I assume good for maybe one of these moves.

EM: Yeah. What about the aviators?

SA: You know, when I was leaving the battle group, Bill Lawrence was the chief NavPers [chief of the Bureau of Naval Personnel], and Bill had sent out a memo to all of the battle group commanders and other guys of that ilk at that stage of their careers, and said, "Look guys, we are having too many of you retire after having a very plum job. We expect you to come back to Washington and do some payback. I don't want to hear anything. I'm not going to tolerate any of you saying you won't come back to town."

Well, that memo had been on the street just a very short period of time when all of a sudden I got called and they said, "You're going to have orders to be ComFAir Japan, fleet air at Atsugi [Naval Air Station]," which was—it was flying jobs, staying close to aviation and all that sort of stuff.

I said, "But he just told me that we were supposed to be coming back to Washington," and so I wrote for the first time in my life. I wrote him a letter and I said, "Admiral, I read your memo that you sent out, and you are absolutely right. I think we do owe the Navy to come back and spend time in Washington. I know most of my counterparts don't want to come back and some of them are going to retire, but I feel obligated that if you want me back in Washington, I want to come back to Washington."

EM: So that didn't hurt?

SA: That didn't hurt. So I ended up coming back to Washington for the OP-50 job.

EM: What was his position at that point?

SA: He was Chief NavPers.

EM: Did he retire from that job?

SA: Yes. Yes. So I think—now Admiral Watkins of course knew me, and we had our ups and downs because of the RDJTF and NavCent when I was there in Pearl Harbor, but I think he respected my positions on things. I don't think he thought I was the brightest bulb in the world, and he is also pretty much of a Lehman man too. But nobody talked to me. I mean, nobody ever came up and said, hey, do you want this job?

EM: I'm going to do this for you? Well, Lehman, what about him? Did he have a positive opinion?

SA: I think John Lehman had a positive opinion of me, yeah.

EM: That didn't hurt either.

SA: No, that didn't hurt. And I think later on Larry Garrett, he and I got along very well, and he would share a lot with me when we were in different meetings and stuff, and talk to me about things. I think that Larry Garrett was probably—

EM: Didn't he take a major hit for the A-12 thing, is that correct? I interviewed him years ago, but that's my recollection.

SA: Yeah.

EM: All right. Okay, then you are off to the DCNO for logistics, OP-04.

SA: Yeah.

EM: Let's see, from January to November—so it's a short time in that job, and you were vice admiral select in February?

SA: In February of '88, yeah. I went to OP-04. Now, that was another surprise.

EM: Yeah, how did you get this logistics job? Maybe because you were good with the POM?

SA: They came down and told me I was going to be OP-04, and I just absolutely—it was the furthest thing from my mind.

EM: Who told you this?

SA: I think it was—

EM: Smith?

SA: Bill Smith, yeah. I didn't understand it at all. I was convinced, and I knew that would be my last job—by this time I'm getting pretty old and long in the tooth. So I was thrilled to get a third star, that was nice. I didn't think I was the guy to go to OP-04, but it was a new challenge and like always, you say okay, it's—you know, put on a new, better pair of shoes—

EM: Did you do anything to prepare for the job?

SA: No time for preparation. Of course, when you build a POM, you know where all the pieces of the Navy fit. There was no doubt in my mind that there was a lot of work to be done in the OP-04 business, because OP-04 sort of had his tentacles out in the whole world. I didn't quite understand all the things that OP-04 did, and I was totally amazed once I got engaged how much more the job was than I thought it was. So it was a very good learning experience for me, and I got to dabble in things that I had seen in the

POM—now I had to try to understand why they were as important as some people thought they were, and some people didn't think they were, and try to sort out what were the real, real keys to success in the business.

It turned out to be a real challenge and also very entertaining. Of course, during the period that I was OP-04, we went through this period of defense management review through the DMRs [defense management review] and then the defense management review decisions, and this was really a period of time when lots of changes were starting to happen in the building.

EM: Generated by whom?

SA: By OSD. So it turned out that now I was sort of the center of activity for the Navy staff for all these management decisions, and so my team—we ended up interfacing with Dan [J. Daniel] Howard, who was now the Assistant or the Under Secretary of the Navy. He was SecNav's designated guy. He had some people put away in an office there, and so I and the team that I had, we were sort of the OpNav input to the process to the SecNav, and so we spent an awful lot of time on these management things, which covered the spectrum of how business was going to be done in the building and who was supposed to be in charge of what.

EM: What were the big changes? Do you recall any?

SA: This is where I get a little fuzzy, because there was some monumental stuff that was going on. We were going through the process of where the maintenance facilities, especially in the aviation side of the house—should all aircraft maintenance be done through the Air Force logistics system, or should the Navy still have its NRFs [naval repair facility] and overhaul facilities and stuff? And so there were going to be tradeoffs between Navy and Air Force of closing NRF Alameda. And work from North Alameda would go to Hill Air Force Base, and close NRF North Island and put that work someplace else.

EM: Did that work out from your perspective? From the Navy's perspective?

SA: From the Navy perspective, we held on to the pieces that were most key to us. The argument was that the Air Force had these huge, huge repair, overhaul and repair facilities. Warner Robbins [Air Force Base] and all around in Utah, and I forget the one in Texas right now. But, they were big, big, big, big, and the Navy had close to the fleet a place in Jacksonville, a place in North Island, a place in Alameda, and the recruiting commands, you know, for recruits. For a while we were going to have combined recruiting commands.

EM: With the Air Force?

SA: With the Army or Air Force. That was one of the possibilities. Finally, from the Navy's standpoint through the BRAC process, because these things sort of fed into the BRAC processes.

EM: Base realignment and closure.

SA: Yes. The Navy finally ended up, we had too many, and of course we closed the newest and best here in Orlando and kept the oldest and worst in Great Lakes.

EM: Why?

SA: Tradition!

EM: Oh. [Laughter]

SA: So we were going through all of these different exercises that really had significant inputs to what would finally happen with the BRAC process and also significant differences of what was going to happen with the way we did business in the building, and that's when the JCS ended up with a Vice Chairman, and then the Vice Chairman became part of the JCS that sort of was the program guy. All the services had to bring their programs into him and get the blessing of the Joint Staff.

So it was really a very fast-paced time. The Hill—there was lots of money trying to be brought, pulled out of the Department of Defense budget. The Hill had gotten on this kick that you still had spares for World War II airplanes sitting in buildings. Yes, we did. But we didn't know how to get rid of them. And if we bought too many spares—and I had to go over to the Hill and testify on that kind of stuff and got beat up bad over there.

EM: Who were the—not the foes, but who gave you the hardest time in these briefings? Senators, congressmen? Do you remember?

SA: Yeah. John Glenn was after me pretty much on the spares. Kennedy was onto me about sealift and money that should have gone to build up the Merchant Marine that's for the future, and I got into a real argument with him on testimony. One of his aides had given him a piece of paper that said the Navy hadn't spent hardly anything on sealift in the four or five years.

EM: Was that accurate?

SA: No, and I challenged him with it right on the spot. I said, "Senator, that's not true." I said, "In the last POM that we built, there is $1.5 billion in that for sealift. So I don't think that's an inconsequential number." Boy, *whew,* red-faced.

EM: He didn't agree?

SA: He started yelling and screaming at me. The OLA [Office of Legislative Affairs] guy was running out the backdoor. Mike [Michael L.] Bowman. They gave me a picture someplace along the way about that incident. He was just—

EM: Well, if you had the facts and figures in front of you and you told them?

SA: He said, "I want to see those numbers."

I said, "Yes sir, I will have them on your desk tomorrow," or today or whatever, but that didn't stem the anger. I had gone over to see Senator [Robert C.] Byrd.

EM: West Virginia?

SA: West Virginia. At the time he wanted a third atomic clock put into West Virginia. At that time, we had two. This is after I'd stopped building the POM.

EM: Why do you need three?

SA: So Mike Bowman, who was OLA at the time, says, "You've got to go over and talk to Senator Byrd. He wants to talk to you about some projects," and of course MilCon [military construction] was a big thing from the OP-04 standpoint. "He wants to talk to you about some MilCon in West Virginia."

I said, "Oh, I know he does. He's not going to like my answer."

He says, "Well, he wants to talk to you." I said okay, we will go over. Of course, you know, he was a gracious southern gentleman when you first met him.

EM: Yeah.

SA: We go, and we sit down and talk. We finally get around to the subject that matters. He says, "Well now, how about this atomic clock that you are going to build in West Virginia?" [Laughter]

EM: Foregone conclusion?

SA: Yeah. And I said, "Senator, I'm afraid I'm going to have to tell you that we are not going to build that atomic clock in West Virginia. We've decided that we don't need three. We are going to stay with the two we've got. That's all we can afford. I'm sorry to be the bearer of bad tidings."

Mike Bowman was watching him. Byrd started to shake. We were scared to death that he was going to have a heart attack. So we got a not-so-gracious departure from him.

EM: Did he ever get his atomic clock?

SA: No.

EM: What was the purpose of the atomic clock?

SA: It's a backup; they need it for the GPS [global positioning system] and all that stuff.

EM: It has to be what, precise?

SA: Very precise. And so we had two, and they were well separated from each other. And the other, not-so-subtle selling point, non-selling point for him, was that third atomic clock would be very close to the first atomic clock that we had, which was in the Washington area. So having it there, and then 100 miles east or west, another one didn't make much sense. The second one was down south.

EM: Well if you go to West Virginia, every building is named the Byrd something-or-other. [Laughter]

SA: Then, let's see, I had another notice that was sort of funny.

EM: Let's get back to Senator Kennedy—what did you think? Was he helpful to the cause of defense, or what was your opinion of him? He had a huge following; he still does.

SA: Yeah. He had a big following. He didn't do his homework. He was controlled completely by staff people, as far as I was concerned. He just sort of grandstanded when he was doing business. So I didn't have much respect for him from that standpoint. Of course, they are all sort of controlled by their staffs. I mean, they get in these hearings, and they've all got papers that they do. But I didn't think he was particularly good at what he should have been.

EM: Who impressed you on the Hill? Who were some of the good senators and representatives supportive of defense and Navy?

SA:[Daniel] Inouye and [Ted] Stevens. I spent time with both of them in private briefings for them. They were very—they were very supportive of the military. I had met both of them. They came to the flagship after the Gulf War, and so they knew who I was. After I retired, I ran into both of them in airport settings. They always remembered me and talked to me. They were both very good. We were not always on the same page, but they were very respectful of my opinion, or whatever I was giving them, and then if they didn't agree, they weren't mean or vindictive about it. They said, well, that's not in the cards.

The guy from South Carolina, for so long Strom Thurmond, Strom was, he was a Byrd. He was like Byrd, and he had a program that fell under logistics called RAMP [Rapid Acquisition and Manufacturing Parts], and the concept was that with all of our old ships and stuff, we had a lot of old iron sitting around that had to be supported, and yet there was no industry now supporting the spares for that old equipment. But we had all of the drawings, we had all of the technical—the specs for what the metallurgy had to be, and all of this other stuff for bearings and stuff like this. His concept was that they would have a facility that had all of this stored—the drawings, the technical stuff, and if you lost a bearing to a fuel oil pump out on a ship, and that fuel oil pump was no longer being supported by the manufacturer, you could send a message into this center, and the computers would find the documents that the manufacturer needed, and the manufacturing facility there, which could handle most of the stuff that you would think about, they would just send it over. Metal would come in, lathes would turn, and they would manufacture a bearing, and out it would come. Of course he wanted this facility to be built in South Carolina.

EM: Naturally.

SA: We didn't think South Carolina was exactly the place. We loved the concept, but we weren't sure that South Carolina was exactly where we wanted to build this thing. So we had lots of negotiations back and forth, and finally we acquiesced to the fact that we were going to build the RAMP facility in South Carolina.

EM: Charleston?

SA: Charleston. And so I'm the designated guy to go down to be the Navy presence for the dedication of this building. The senator and I are up on the stage, and I do my little part thanking the senator for his vision, and how proud we are to be here today to be a part of the dedication, and sit down. Strom gets up, walks up to the microphone, and he says, "All of you folks here in South Carolina, you wonder how I got this facility to be brought down here, dedicated on this special day? Well, you know, John Lehman wanted his 600-ship Navy, and I wanted my RAMP facility." [Laughter]

EM: Right to the heart of the matter.

SA: TV cameras rolling.

EM: Say no more.

SA: It was classic.

EM: He's talking to his voters.

SA: I tell you, nowhere, nowhere could anyone do this but him.

EM: Yeah, oh yeah.

SA: The press would have been, you know—

EM: He had clout.

SA: But it was one of the special days, that's what made the Pentagon fun. You had these little excursions that—out of nowhere comes something. He was just classic.

EM: I forgot when you are talking about Lehman, but you knew Melvin Paisley. You were his assistant. What was your opinion of him, and what do you think he was involved in?

SA: Quite honestly, I always knew there was something wrong over there, that he played a little loose. Mel was a guy, I briefed him lots, especially on the V-22 [Osprey]. From the time that I was OP-05 to the time I was OP-80, and until the time I was OP-04, there were issues with the V-22. The program had finally been approved, and this brief to Mel was to make the case for a bigger engine for the thing, because the Air Force wanted a little bit more performance than the Marines did. And of course Lehman and Paisley were selling the program to be a very joint program. The Army was going to need this thing. The Air Force was going to need it. The Navy was going to need it, and it was going to be the ASW platform instead of other aircraft.

They were going to get rid of the S-3, and the V-22 would be a replacement for the S-3, and it would be the replacement for the COD, and all this other stuff. So that was their vision. They had to have that big macro number to make the program affordable. So the final piece that needed to really be done was decide on the engine. And so I went down to brief him on the engine decision. Mel was a guy who always read the last page of the brief first.

EM: The conclusion?

SA: Yeah, the conclusion. He read the last page first, and then if he didn't like it or if there were some questions, then you could go back to the rest of the brief, but you went right to the back page and said, okay, here's the decision we're asking for. We knew for some reason, he and John Lehman did not want the bigger engine. We never could really understand it, because we could see that the need was there, and of course the experience was that every airplane you built always ended up having an upgrade.

EM: A need for a bigger engine?

SA: A bigger engine. The A-4 went from J65 et cetera up the line, so every airplane had done that. So this was a chance to do the first increment, while we were doing the manufacturing. Wouldn't even listen to it. Wouldn't even listen to it.

EM: Cost, was that the factor?

SA: No, the cost was not the factor. The factor that he hung his hat on, and John Lehman hung his hat on, was it was too hot. We were using much hotter combustion chamber temperatures, and they said, "We don't think we are ready for that." It seemed liked a very simple thing to do for a program that young to not make mistakes, because all of the indicators were you are going to need more power—absolutely, you're going to need more power. That decision by Mel.

EM: Was it just off the seat of his pants? Or did he have studies to back this up? He was probably following Lehman's lead, his boss?

SA: That's why I always thought, I always wondered—I thought John Lehman could possibly be involved in some of that stuff. But apparently, none of it ever led back to him. So Mel must have been on his own. Yeah, there were lots of things when you would brief Mel that didn't make sense, but he knew the business. Boy, he knew business. He knew how to work the contractor.

He would say, "Look, I can bring them to the table and we can fix this." You'd go down to the briefing and get to one of these things, and he says, "I can fix that." A lot of times he could. He was an interesting guy.

EM: Did you suspect there was some hanky-panky?

SA: Yeah, when that all started to break, I was pretty sure that there was—

EM: Substance behind it?

SA: Substance, yeah. There was just always this uneasy feeling that he knew something that you didn't, and where was he getting the information from?

EM: What happened with the Osprey? Did they eventually get the larger engine?

SA: You know, I don't know. When it finally got to the fleet, I guess I was already gone. I think I went down to see the first testing of it down in Pax River, but I don't know what they've done with the engines.

EM: Just the Marine Corps has it now, right? Or the Air Force does too?

SA: The Air Force Special Ops people have some. I don't know how many, but they've got a few. Now we are going to replace the C-2 on the carrier with the Osprey.

EM: Oh, I didn't know that.

SA: Yeah.

EM: In your job in OP-04, did you get out to the repair facilities and the other logistics sites the Navy had?

SA: We had a group, joint service group, called the Joint Logistics Commanders [JLC], and this was Air Force Systems Command guy, four-star general; Army Materiel Command, four-star general; Marine Assistant Commandant for Logistics, three-star;

Navy OP-04 three-star. And we met every quarter at one of our facilities. So it was sort of a dog and pony show to showcase yourselves. We would go out, and of course now that we were starting to talk about combining facilities and who could do what where and do this there, we were really involved in testing—big, big push to close down Pax River and move everything to Eglin AFB [Air Force Base] or to Edwards AFB. So we would try to take each other to facilities that the other team members hadn't seen. These became sort of "got you" moments every now and then. Like Mike Carns, who was—let's see, no, that was another trip. That was when we were the Vice Chiefs. We also had the "vice squad" when we were Vice Chiefs. But in this episode, we were at Edwards AFB and he was showing us the B-2 [stealth bomber].

EM: He's Air Force? Mike Carns?

SA: Yes. This was something that came up when I was OP-4, but we were seeing it now when we were Vice Chiefs. In the OP-04 days, we always had this argument about what facilities were paid for under MilCon and what facilities were paid for under programs. Well, the Navy had a very hard line. You could not build a hangar based on the airplane program. The hangar had to be MilCon; the airplane was procurement. In the Air Force, when the B-2 came along, the Air Force decided to hide a lot of their MilCon into the procurement.

EM: Okay. To hide the B-2?

SA: No. Well, yeah.

EM: Hide the cost?

SA: It was to show that they didn't have to have new MilCon to support the B-2.

EM: Because MilCon goes to Congress, right?

SA: Yep. We get out to Edwards AFB, and I'd been out there as part of the Joint Logistics Commanders just a couple years before. So I knew what was there, because I'd been there. And Mike flies the Gulfstream. He goes up and flies us out there; Mike and I had been together in the Rapid Deployment Joint Task Force in CentCom days. So we go through this big operations center, I mean, big test operations center. They had all the video stuff, radios, and all the data collection and everything for testing of the B-2. Of course, now most of the testing for the B-2 was complete, and so they reconfigured it to test as part of their regular test program stuff.

They had these hangars for the B-2s, and Mike is looking around, and he says, he's asking the colonel out there, "When did we build this building out here? When did we this build this big ops center?"

The guy said, "We built it three or four years ago for the B-2."

Mike said, "I sure don't remember this on our MilCon bill." I used to go through that, and Mike was a detail guy, I mean, he should have been a nuke submariner.

EM: Yeah.

SA: I knew the answer, you know. He's starting to pull the string, and you could see the colonel's not comfortable with what he's doing. Finally I said, "Mike, you bought this with the B-2."

He said, "What do you mean I bought this with the B-2?"

I said, "It was in the program dollars, not in the MilCon dollars."

"How do you know that?"

I said, "I was out here when they did that, and as part of the Joint Logistics Commanders." He couldn't get over the fact.

EM: Well it obviously worked.

SA: Yeah, it worked. But—

EM: He was upset because the two were not separated the way they were supposed to be?

SA: Yeah, because as the Vice Chief, because we all got beat up on MilCon when we had to go over to testify on the Hill, he would've remembered right away if there was a big project like that that he hadn't seen. So that's what we did with the Joint Logistics Commanders. We would take each other around to our different facilities, and especially to facilities where it looked like in the future we were going to go and combine, like the big, big air depot that the Air Force had there in Texas that I was talking about. We went out there, and I got to go through all the engine shops and see how much capability—they even let me fly the C-17.

EM: Really?

SA: Yeah. I took them all out to the carrier to see how we did our maintenance on the carrier, and the engine test cells.

EM: Did you pilot the aircraft into the carriers?

SA: No, I didn't. I let the fleet guys. They all had a good time, and the Marines took us up to Cherry Point—their facilities, and of course in those days we were very much into Six Sigma stuff.

EM: Oh, yeah, God.

SA: So, we were all—I was sort of the Six Sigma guy.

EM: You were?

SA: Dan Howard driving the train from the SecNav standpoint .

EM: Well, let's talk about Six Sigma, because I was involved in that down at even my level. You were supportive of the program?

SA: I was.

EM: Describe what it was, first of all.

SA: Six Sigma was sort of an industrial program that said quality matters, and the quality starts the first time you put a piece of pencil to the paper. You start the process, and there are ways as you go along that you can always find efficiencies. As you find these efficiencies, you need to find a way where they can be inserted without changing the whole thing at once. You can just gradually keep moving and refining and refining and refining, until you get to the point that you've got the right quality at the end. To do that, you have to have very good data, you have to observe everything that you are doing, and you have to measure it. It gets a little out into the ether there, saying, you know, why do I have to see how many pieces of paper I have and how many pencils I have at the start of this thing? That's sort of what the process is about.

EM: This was based on the Deming thing? The guy out in Japan that developed this?*

SA: Took it out of Japan. Toyota and some of the Japanese manufacturers jumped right on it. That's how they got their quality to their cars so improved so fast. So I believed in the concept. Admiral [Frank] Kelso, when he got into it a little later than I did from the standpoint of when he became CNO, he wanted to add the leadership piece to it. TQL [total quality leadership] was his thing, instead of TQM [total quality management]—instead of management, which works too.

I was a fan of it. Parts of it I didn't think were readily adaptable to everything that we did, and people were trying. It became a theme song, so you had to have TQM or TQL associated with everything that you did; that's where it did what Deming didn't want to do. It became overburdened with too much oversight and too much process. Deming was pretty clear on where he thought it should go. It became Lean Six Sigma as

* American W. Edwards Deming's lectures throughout Japan during the 1950s significantly influenced that nation's successful adoption of the total quality management process with regard to industrial output.

it finally got to the manufacturers. They all added lean to it. It got rid of some unnecessary stuff that was lying around, and processes that we had been using and just thought that's how you had to do it.

EM: Was it still in place when you were VCNO?

SA: Yes.

EM: It did not last—well, it went away after a few years.

SA: Yeah.

EM: Why?

SA: I don't know; after Kelso left, I think, because Kelso was very much on the TQL. He kept emphasizing TQL, especially the "L." I think when he was gone, of course, Howard was a big pusher from the Under Secretary of the Navy standpoint. So when Howard left, and Kelso, I think that's probably when things started to fade.

EM: When you were implementing the program, did you get pushback?

SA: Yes.

EM: From where?

SA: Well, everywhere. [Laughter] Nobody thought this was the right thing, especially because they had to go to school.

EM: Right, training.

SA: You had to go to training, and some of it was pretty fluffy, I have to admit. Because we really hadn't gotten it to the point where we were sure that we could take a manufacturing process and bring it into an information process.

EM: From a personal standpoint, I can say that was our big gripe about it, that naval history is not a manufacturing process. I mean it's a lot of fluff, if you will, and it was just a hard fit—that was the reaction we had.

SA: Yeah, yeah, and I would agree.

EM: I mean the concept was well thought out, and it made sense to have a process to what you do. It was just that that the training ate up a lot of time, and it seemed to drag things out. I don't know, that's my personal opinion.

SA: And quite honestly, it's hard to say this, but it's the truth. People who had not had as much success in their career paths at a point found this was something you could really jump on and sort of make your mark. So you had a lot of people who didn't have the knowledge of what the original process was all about—why were we going things this way. Now, being the teachers of this is how you got to do it; they didn't know what you had done before to begin with. But they were sure they knew how to tell you how to do it better.

EM: Good point.

SA: I tell you, there was a lot of effort put into it.

EM: Is there anything comparable to it in today's Navy?

SA: You know, I don't know.

EM: Because we went through a number of Lean Six Sigma, TQL, TQM, a few others in there too.

SA: Yeah, I don't know. I've lost touch.

EM: When you went out to the various bases, did you ever decide that this facility had outlived its purpose? Naval facilities, I'm talking about that: "Yeah, we should BRAC this one or put it on the list." Any bases that come to mind that were not as efficient as some of the others?

SA: Yes, we did. Trying to think if I can get a good example.

EM: Did you go to submarine bases and the naval air rework facilities?

SA: We went to some of the naval rework facilities. Went to one at North Island, and went to Cherry Point. Let's see, I think we went to Warner Robbins with the Air Force. Didn't go up to Hill, although I knew a little bit about Hill. Let's see, I'm not sure that we went to Alameda. Alameda was one that we did away with; it had outlived its usefulness, because the carriers weren't coming into Alameda anymore, which was a big thing in those days. We could offload an airplane right there.

EM: Why didn't they? Would they go to San Diego?

SA: Yeah, the carriers were going to San Diego, then up north to Bremerton. I'm trying to think.

EM: Treasure Island in that same category?

SA: Treasure Island, yeah, that went, but that wasn't so much. I'm trying to think of something joint that would come to mind right now. Let me see, I scribbled some things down here from ammo.

EM: Ammo?

SA: Ammo. We spent some time out at Crane, Indiana—ammunition facility out in Crane, which was an interesting place. I remembered it well, because when I came back from the Gulf War, I remembered what we had seen out there [at Crane] relative to 16-inch projectiles and gun barrels. We still had a stock of 16-inch gun barrels and a whole lot of 16-inch ammo for the 16-inch guns. We also had some really nifty facilities out there for testing deep-water sonobuoys and stuff like that that nobody had a clue about. The thought being, are there places where we do lots of ammunition-type stuff that can be combined with all the rest? So we went to the Army facility up there in Maryland.

EM: Aberdeen?

SA: Aberdeen and their testing ground, and quite honestly you would think that was going to be an easy sell. It wasn't, because the Army tested differently than we did and manufactured their ammo differently than we did, and so the thoughts of taking Crane and the Army depot at Aberdeen together, it just didn't play right. Besides, Crane had learned a way: they harvested their forests out there.

EM: Oh, they had live oak?

SA: Live oak.

EM: Well, the *Constitution* used some of that wood.

SA: Exactly, exactly. Now, there's a command that I would have loved to have had. They are out there, you know—that place was all built up during the '30s in the WPA [Works Project Administration], with stone houses and cottages for the staff to live in. A lake to fish in. What a gorgeous place it is.

EM: Is that in southern Indiana?

SA: Yes. It is south of Indianapolis and very close to Kentucky. In fact, some of their workforce is Kentuckian, and of course, being both big basketball states, the Kentucky workers are under a little—during the basketball season.

EM: Were you involved in when *New Jersey* [BB-62] went off Lebanon in '83 and fired the 16-inch ammo? They hadn't mixed the powder in many years. It was stored, so their shots were going all over the place. Were you involved in reworking that ammo or restoring it?

SA: Yeah. That was all being done out at Crane. They pretty well had that controlled. When we brought the battleships out for Desert Storm, didn't have those problems. We will talk about this at a later time, but that's why I was so convinced, or so bound up out there to fire every round that we had, get them out of the guns, and get them going, because that was going to be the last hoorah. Might as well use them for something; but they shut me down.

EM: Your time in the Pentagon then, before you went off to the Gulf, what stands out in your mind as an accomplishment during this '86 to '90 period? Is there some program about which you say, "I really helped ram this through and it's paid off," or just a number of things that worked out?

SA: I think there were a number of things.

EM: What aircraft and what weapons came to fruition?

SA: Harpoon came along. Tomahawk came along, and as OP-04 I got more involved in Tomahawk than I had as OP-05, because when I was OP-05 I was very biased against Tomahawk. Back in my VX-1 days, I had been involved with a project of putting Bullpup missiles on ASW airplanes, and I shot an awful lot of Bullpups, and then I shot some during Vietnam as well, off of attack planes. I was very unimpressed with the Bullpup A warhead, in that it didn't have enough punch. I mean it got its punch from the Bullpup, because it was coming in basically supersonic. But the original Tomahawk basically took that warhead as the basis for the missile, so I thought it was a whole lot of expense to deliver a very small warhead. Now when I got to OP-04, of course, we are now in the production of the Tomahawk and trying to get the cost down and trying to get an inventory, and I still had this queasy feeling about—gee, the first time we've used these things in a serious manner; I'm afraid people are going to be not impressed with what it can do.

EM: Its hitting power?

SA: Yes. So I was a little worried about it, but obviously we had decided that we really needed to have that capability for our submarines, and that was part of the program we worked during those tight years when I was there.

EM: Well, surface ships too, right?

SA: And the surface ships—the surface ships first, then we went to the submarines. So that was something that I used later on and I was very happy with.

EM: Who were the supporters? The submariners and the surface guys?

SA: The surface guys were big supporters. I don't think the submariners were to begin with, but became that way. Interestingly enough, one of the things I was most satisfied with that we did was the fast sealift ships, the SL-7s. That's sort of where I got in trouble with Kennedy and all of that. The SL-7s were just magnificent ships. During my days in *Coral Sea,* I was coming up from San Diego one time and there was an SL-7 coming down under the Great Circle Route, from the north. We were both going to hit the turn-in point at the—out beyond San Francisco Bay to head for the bridge and come in. I'm watching this guy tracking, and the ship is coming at 30 knots. I thought, who in the world is that, because I was making good speed coming up from San Diego, and I'm wondering who in the devil is this guy. We are approaching from opposite directions, so I'm really interested to see what this is. They had beautiful lines, the SL-7s. They had really rakish lines, and it was an SL-7 coming into San Francisco, and he made that turn and just left me in the dust, and man, that is really something. Then of course we got into this sealift and prepositioning ships, maritime prepositioned ships.

EM: Were you involved in the prepositioned ship business as 04? You were supportive?

SA: I was supportive. I tried to get the Navy more interested in it because I thought this business was one of the classic coups by the Marines. The Marines found a way—I mean, they sold this program and said, we need it. We've got to have it. And here's why we've got to do it this way, and of course it was all built around, it was originally all built around the issue out in the Gulf, you know, which began from my Rapid Deployment Joint Task Force and CentCom days. I understood and realized that, yeah, that's a long way to get a lot of stuff there in a hurry, so you should find a way to preposition things nearby that you can get in there if you need it in a hurry. I was a believer in that concept. The Marines took that and ran with it. The Navy didn't.

EM: What was the Navy's objection?

SA: It was money. We needed money for today, not tomorrow, and the Marines were, "I'm getting a second set of equipment; I might need that today and I might need it tomorrow."

EM: A MEB [Marine expeditionary brigade] set?

SA: Yeah.

EM: Entire MEB support.

SA: And basically everyone bought into the program and bought it for them. The Navy bought it for them.

EM: Who in the Navy was opposed; I mean, the CNO at the time?

SA: Yeah, you just couldn't get it through. First of all, you couldn't get enough people to argue what it was that they wanted to have forward, and so they couldn't. When the Air Force came on board and saw the Marines, the Air Force wanted their own ammo ships.

EM: Was there any particular antagonist, though, on the Navy side that you recall?

SA: No, I don't recall anyone who was just absolutely against it. I just couldn't find anyone other than me who thought it was a good thing to do.

EM: How about the Merchant Marine community?

SA: Oh, they were all on, they were all over it. That was one of the politic jobs of OP-04, was keeping the unions and the Merchant Marine group happy because they were always wanting me to come to meetings and going here and there, always bending my ear, and that is what they were feeding to Kennedy also.

But it was that there were other, higher priorities from the Navy standpoint. That was very successful. The work that I did as OP-04 when I got out to the Gulf War, I was very pleased with what I had seen. There was a lot of angst when Desert Shield kicked off, and some of the SL-7s, we couldn't get them ready in the five days that they were supposed to be. It took us 14 days, and so lots of stones being thrown. But once we got it moving it really moved, and then when I got out in theater and saw what was there, it was even more impressive than what I had seen back there. I just couldn't believe that we moved all of that stuff out there that fast. And again, we were out there in very good shape from the Navy and the Marine Corps standpoint. The Air Force was begging us for weapons and bombs, and we had plenty of bombs.

EM: Iron bombs?

SA: Iron bombs.

EM: Going back to the Maritime Administration [MarAd] and the Merchant Marine, how were your dealings with them? Did you get a good audience with them, or were they hostile?

SA: No, they always had their hand out for more. They wanted more work. They wanted more ships, but I think we did pretty well. You know, we had one of our vice admirals that retired and went over to MarAd.

EM: Was he at TransCom [Transportation Command] at some point?

SA: Well, let's see, was he TransCom? No, he'd been MSC.

EM: Oh, Donald?

SA: He also had been in BuPers.

EM: Donovan?

SA: No, not Donovan. Herberger, Al Herberger. We did well with each other then. I had pretty much an open door. They could come and see me. They knew they could come and talk to me. They knew that there were certain things I could and couldn't do, and so we got along good. I had Jonathan Kaskin, who was my SES [senior executive service] that worked that shop in the OP-04 business and was well respected by them. He knew the business really well. He had been a long time mariner. So we had good dialogue.

EM: Talk about what happened during your watch in the Pentagon there, the strikes against [Muammar] Qaddafi. Were you in any way involved in that? Against Libya in '85, '86?

SA: No. There was a lot of feedback from that, of course. John Lehman was right in the middle of that, and that's why the Strike U out of Fallon came about, out of that Libya strike.*

EM: Lehman pushed that?

SA: Yes. Said, "Get the fighter guys, you've got Top Gun down there. There were a lot of mistakes made. We've got to get to a place where people can sit and think and work out some of the do's and don'ts when you are in this kind of stuff." He thought there were basic errors made.

EM: Interesting, because Top Gun came out of the Vietnam experience. We weren't happy with the ratio of wins to losses and all that. Everyone was happy with strike operations during Vietnam, so it didn't lead to the same sort of thing until what, '86, '87?

SA: It didn't.

EM: What were the problems in the Libya strikes? They just weren't hitting the targets?†

SA: Tactics, some tactics and some weapons not on target, or the need for some better weapons to do some targeting with. I think the thing with Strike U was to get everybody's head in the game. That there were some relative do's and don'ts. You've got to have some preplanned stuff that you can execute and not just draw up at the last

* In 1996 the Fighter Weapons School, formerly at Miramar Naval Air Station, was moved to Fallon, Nevada, and combined with Strike University to form the Naval Strike and Air War Center.

† On 15 April 1986, U.S. Navy and Air Force aircraft, in Operation Eldorado Canyon, bombed Libya in retaliation for Muammar Qaddafi's support for terrorist attacks that killed American and allied civilians.

minute. You ought to be able to reach back and pull something out and say, okay, here's the framework of what we are going to do. Now, how do we tailor it, the specifics. There was a little bit too much—like before we had NATOPS for airplanes, each squadron had their own way that they flew an airplane.

I was reading an article the other day, and I didn't realize that some squadrons, when the early jets came along, some squadrons flew their passes with the speed brakes out, and some squadrons flew the same airplane with the speed brakes in on their carrier pass. Speed brakes out meant you were further up on the power curve on the thing, and you probably had less lag time to get the extra power if you needed it, whereas if you had the speed brakes in, you were further back on the power curve and you had to let more lag time to get the full power. So once they figured out which was the best and NATOPS said this is the best way to do that, we took out the alternative way of doing it.

EM: Of course, NATOPS could be amended if enough opposition developed.

SA: If you found that—

EM: It didn't work?

SA: Yeah. That's what Strike U was like, bringing NATOPS into the world.

EM: The tactical philosophy that came out of Fallon, wasn't it to go in low, which had been the thing earlier?

SA: We still had the low and the high guys, and of course there is a place for both. But you've got to know when is the right time. Basically, low only works if it's a surprise. If it's not going to be a surprise, then you might as well come in high and figure out how you are going to deal with the threat that is at that altitude.

EM: Yeah.

SA: Because if you come in low, you're going to get slowed someplace in the evolution, unless you can go zipping by at 50 feet and drop, which requires a special kind of weapon to do that with.

EM: In the larger picture, what was your view: Was the Navy in good shape at the end of the '80s?

SA: Yes.

EM: Were the '80s positive, in your view, toward the Navy?

SA: Yes. Those were good years. I think we were good.

EM: Personnel-wise?

SA: Personnel-wise. I thought a lot of things had come together, and were working pretty well.

EM: Relations between the races and between the genders, where were they?

SA: They were good. I think things had quieted down, and things were moving along pretty well. There wasn't as much talk about it, and I think things were reasonably good. We will come back to that, I guess, during my Vice Chief years. But no, it was sort of a good time. Of course you only remember the good things after so many years.

EM: Right. Well, that helps us get through it, I guess. Before we break for lunch, anything you want to add about these pre–Desert Storm years? Anything is game.

SA: Well, some of the fun issues—like in OP-04, the vice president's house at the Naval Observatory.

EM: Naval Observatory, right.

SA: It needed a lot of rework, and then we had our favorite vice president from Indiana.

EM: Birch Bayh?

SA: It was the vice president.

EM: Oh, Quayle, Dan Quayle?

SA: Dan Quayle and his wife were living in the place at the time. They became a thorn in the side for us trying to do what we really needed to do with the house, and what they wanted to do with the house.

EM: In what capacity were you at that point?

SA: I owned the bill for maintaining the house, as OP-04.

EM: What did they want to do to it?

SA: The normal things, "We want to do this, we want to do that."

"No, you can't do this, you can't do that because it's historical."

EM: Historical preservation.

SA: We had to get this kind of work in to do this and, you know, "Why can't you just do that?" It was always one of those interesting things as MilCon, always is. The MilCon issues that went on during the OP-04 days were exciting. One of the other things, OP-04 is looked at as—he's always here, and he's always available to do something, so we had a crisis in Navy Medicine at the time, and the Vice Chief now was Bud Edney, and Bud became chairman of the Navy Blue Ribbon Panel. He was designated by the SecNav do to a Blue Ribbon Panel on Navy Medicine, and of course I became the action officer for Bud to round up the facts and all. So here I am standing at the head of the operating table while they are doing open heart surgery on a guy so I could get the experience of what happens.

EM: What was the issue with Navy Medicine?

SA: Well, we had lots of issues. Again, they were starting this emphasis on—can we combine medicine? Why do we have Navy, Army, and Air Force medicine? What's unique about Navy medicine? Why can't you combine with Army, et cetera. We got all of these hospitals. Do we really need all of these hospitals if we combine? Then the argument came about teaching hospitals over just a straight operational hospital. Now we are talking about field hospitals. Are we going to preposition hospitals.

EM: Mobile hospitals?

SA: Mobile hospitals. We got the hospital ships.

EM: Were you supportive of having those hospital ships? Because there was opposition to having two.

SA: Yeah, well that was sort of a SOP to the industry to have two, and yeah, I thought the hospital ships were a good idea. I wasn't sure how we were going to keep them functional, but then they do these mini-deployments, which are very costly as a matter of fact.

EM: Really?

SA: Taking a lot of people out of the hospitals and servicing people that you normally wouldn't service—but they were good. Of course, Desert Storm, we rolled them out and we didn't need them that much, but they were there. But the Blue Ribbon Panel on Navy

Medicine was another one of those—who in the world would think you'd be sitting in an operating room and watching open heart surgery, unless it was you on the table, which made me a believer. I never want to be on that table.

EM: Oh really?

SA: No.

EM: Why?

SA: Well, this is in the days when they still split the breastbone and pulled it apart. When they put it back together, if you've ever seen an aircraft mechanic do the threading, where they wire nuts, tight. In other words, you put safety wire in the nut and you spin it, and that holds the nut and the nut will never back off.

EM: Right.

SA: So they had special pliers to do that, like the old Yankee screwdrivers, you know, that *szzz, szzz, szzz.*

EM: Oh yeah, right.

SA: Well, it's the reverse, so they get the tube and take the wire and feed it underneath, and pull it up and pull the breastplates back together, and they put on this big pair of pliers with this rotating thing, and *szzz.* The body comes off of the table. It's all pulled together real tight. I said, oh my goodness. All I could think about was somebody tightening a noose.

EM: Never me, yeah.

SA: So it was good. It was good. We verified the need for a teaching hospital. Why a teaching hospital was necessary.

EM: Did you combine any Navy hospitals with other services?

SA: No, but they eventually did out there in Bethesda. The thing was we found, again, taking a look at TQL and TQM, we found that the medical department had not gotten involved as much as they should have. You would go into a hospital with five or six operating rooms, and you would have three operating rooms that were usable. The other three were stuffed with supplies and junk that—it would've taken you a week to make them ready to ever take anybody there for an operation. So we found a lot of unused space, and then we had to task the medical department to figure out why they needed that many operating rooms. If they didn't need them, then shut them down and turn them into—

EM: Storage rooms?

SA: Storage rooms, or if they couldn't find another place to store stuff, admin offices or whatever. But to use an operating room with all of that equipment in there as a storage room for junk was just crazy. We found that everywhere we went.

EM: Can you talk a bit about—I know during Lehman's time, he established the Naval Material Command.

SA: Disestablished.

EM: Disestablished it, okay. So you must have been involved in that evolution.

SA: That happened before I got there, and so that was one of the things that I found lacking when I went to the JLCs. When the JLCs and I meet and the Air Force guy is a four-star, the Army guy four-star, and now the Navy guy is a three-star and the Marines a three-star. The Marine would always be a three-star, but the Navy guy should have been a four-star, under the old Material Command. Basically, when Lehman did away with Material Command, the Navy only had the OP-04 staff office, nothing on the scale of the other services' logistical commands. So I had all of the same issues that these guys had, but they had commands and they had all of the stuff you had to deal with when you sat on top of a big organization.

I was already embedded in the organization. So I was questioning what had made Lehman do away with the Material Command. My basic view is that he did away with the Material Command because he believed it had become the "master of no." That he [the chief of the Naval Material Command] was the voice that said yes or no, and he had become a voice that said no too often. That's a very simplistic view, but I think he saw it as a bureaucracy that was now just serving itself and not serving the greater need.

EM: Acting as a stopper for programs he wanted?

SA: Yeah. Now, because I wasn't there when he first got into that, I don't know what the real issues were. What ended up missing was somebody who could have made NavSea [Naval Sea Systems Command] and NavAir look much more alike than they did. My bias was that NavAir was much more progressive than NavSea. NavSea still had the attitude, maybe we will return to sail someday. [Laughter] NavAir was, we are going to the moon. There were just a lot of things I thought that a really powerful Material Command could have done to make our two procurement activities look like they were in the same era.

EM: No one stepped in, or no one was given that responsibility after the disestablishment?

SA: No, they were left on their own.

EM: And it's still the same?

SA: Yeah.

EM: Are they still perceived in different ways, NavSea and NavAir?

SA: Well, yes. They do business entirely differently. There's good and bad about both of them. NavAir still chases tomorrow, and NavSea is much more comfortable with today or yesterday. From an aviator's standpoint, when they had to do the Senior Officer Material Readiness instruction out at Rickover's place [in Idaho], they had to bring that into the system. So those of us aviators who were going to be ship drivers went out there and we saw this. We understood what the Rickover thing was. We were sort of familiar with checklists and checks and balances. The surface guys weren't, and when you went aboard ship and started seeing how ships were maintained compared to how aircraft were maintained, it was a different world entirely. The training was different. The quality control was different. A pump on a ship would crap out and you would go find the bearings, or maybe have to send ashore for a new pump to be brought out. Nobody paid any attention to why it broke.

EM: In the first place.

SA: You know, it broke. That was just heresy for the aviators. Why did it break? I want to know why it broke. Is there something I should have heard before it broke that would have saved my fanny if I had just heard that noise? Sad to say, I'm told—now, it's been a long time since I've been out on a ship, but I'm told that there is still a great vast void between the two, and I can see it happening in the Army Materiel Command and the Air Force Materiel Command. They were driving processes throughout the organization.

EM: Who is on this same level now, the Army Materiel Commander and the Air Force, these four-stars? Does the Navy have OP-04? And, it's a three-star billet, right?

SA: Yes.

EM: It seems like the Army and the Air Force have all of these four-stars running around for everything.

SA: They do. [Laughter] But they are big commands. So I think Lehman threw away some good pieces that could have been reconstructed some other place. I'm not sure how we would do it, but I also recognize, and I talked to some people that were in the Material Command in those days, and yes, there were lots of layers, and it just became a bureaucracy.

EM: All right, any last thoughts on this period?

SA: No. I have to say that the job that I didn't think I would enjoy at all, I really enjoyed OP-04. You touched everything. You touched everything in the Navy.

EM: I guess so.

SA: I wish I could remember all of the skirmishes. I mean, there was a skirmish every day with somebody somewhere.

EM: Well, we can dip in, if it comes to your mind. It doesn't matter where in the interview.

SA: I had good people that came and went from the organization. It was always entertaining. The JLCs was a fun organization just from the standpoint that you got to see how the people tackled the same problems that you had, and it was fun. The people that I got to meet in those organizations came back when I was Vice Chief, some of those same people showed up other places, so it was good.

EM: Okay, Admiral, one thing that I forgot to mention, they've named an award after you. The Stanley Arthur Logistics Award, right?

SA: Yes.

EM: How did that come about? Was that connected to your time as OP-04?

SA: I guess it was something that happened after I retired from the Navy, and I didn't have any input to it, and I was very surprised. But yeah, it had to do with my time as OP-04 and the Gulf War.

EM: When did they make that award?

SA: I think it started in, let's see, I retired in '95, and so I think it was a couple of years after that it was brought about. I remember Bill [William J.] Hancock [Jr.] was the OP-04 at the time, and it was under him. He was the one that started the ball rolling.

EM: What are the criteria for the award?

SA: What they do is, they usually give out four awards each year, and they are for excellence in something to do with logistics. They have a panel from supply and OP-04, and I don't know who else is on the panel, but they have a group that takes all of the nominations for the awards, and then they select them down to individual military, individual civilian, a team award, and I think a team award military and a team award civilian from the civilian organization. I think I've only missed one or two of the ceremonies up in D.C. They have them at the Pentagon. The last few years they've been doing them by VTC [video teleconference] to save on travel.

EM: Right.

SA: It's always fun. They do great things. Most of the awards are for people who have introduced a new system—in the supply piece of the new system, bringing it online or savings, where they are able to go in and drag a bunch of savings out of the process. A lot of times the team awards are for something like—I think the last one had to do with improvements they made over in Naples, a service to all of the fleet units over there, on time and delivery and the cost of delivering. So it's pretty impressive, what some of the folks have done. Some of them have designed new programs that looked for places to make savings, pinpoint them. So some of them are things that will continue into the future.

EM: It's impressive that they associate you with the award.

SA: Yes, I was very honored. The first one I went to, they gave the honorees a plaque, and they got a bronze relief of me, and I said, jiminy Christmas, normally you don't see one of these until you're the guy that is underneath it. [Laughter]

EM: Well, that's quite a kudos. All right, we are off to your time as Com7Flt [Commander Seventh Fleet] and ComNavCent. You didn't have much time to be Com7Flt out in Japan, right?

SA: No.

EM: Did you get there at all, I mean—

SA: Yeah, I went directly there. Hank Mauz was already down on the flagship in the Gulf, but I thought it was important to go to Japan to at least introduce myself to the locals, especially the Japanese Navy there. Then went basically straight out to the Gulf. Went down and stopped through the Philippines. Met my old buddy Norm Ehlert, the Marine that I talked about that was with me in OP-50, and he was on his way. He didn't know it quite then, but shortly thereafter he was on his way to be on Schwarzkopf's staff in Riyadh. So we hooked up again. Went out and got aboard.

EM: Was it you that was the critical factor because of your previous experience in the Gulf with the RDJTF, or was it because the Seventh Fleet was the best fleet to come out there and take charge?

SA: What had happened was when Desert Shield kicked off—well, when the invasion kicked off and Schwarzkopf started to get things rolling for Desert Shield—the Navy component commander, the NavCent, who was still in Hawaii at the time like I was when I was NavCent, they deployed him out into Bahrain to be the Navy guy out there.

EM: Admiral Foley?

SA: No, it was Bob [Robert] Sutton, and Bill [William] Fogarty was Commander Middle East Force. So they had the Commander Middle East Force flagship out there, and then sent out NavCent to be there with Fogarty, and then the decision was—well, they really needed somebody with three stars out there in the Navy—so the decision was, from back in Washington at the time, to send Com7Flt out there to send him down. So Hank went there from Japan and then the flagship *Blue Ridge* [LCC-19] followed him.

EM: Well, of course, the alternative would have been to take the guy from Sixth Fleet.

SA: Sixth Fleet, yeah. And I don't know exactly why they made that particular choice of Seventh instead of Sixth, although Seventh Fleet traditionally had more access to the AOR ship-wise, because Seventh Fleet had been sending ships in and out of there for some time. There was also thought of sending Third Fleet.

EM: Into the IO too.

SA: Into the IO and support, although Sixth Fleet had sent ships down too. And I was surprised back in Washington; from my standpoint, of course, I was again saying, well, if we had done it right with ComUSNavCent [Commander, U.S. Naval Forces Central Command], we wouldn't have had to do this. Now they are making an accommodation because, obviously, the one-star was not going to be a guy that was going to be in charge of fleet operations. So I sort of felt vindicated from my days in the RDJTF.

EM: Right, you had recommended—what was the reaction of CinCPac and CinCPacFlt at that time, when Mauz went out? Did they readily give him up, or were they not wanting to do that?

SA: There was quite a bit of dialogue as I remember, but of course, I was 04 and I was flailing around with getting stuff moving out that way and wasn't too much involved with the plans and policy guys out there. So I don't remember as much about that piece of the action as I should have. I think at one point in time, they were also thinking of Third Fleet as a possible. In fact, that issue came up again, I think, when it looked like it was going to be extended longer, that maybe they would send Third Fleet out.

EM: Who was Third Fleet?

SA: I don't know who it was at that time—an aviator, if I'm not mistaken. I think [Huntington] Hardisty was thinking maybe that would be the right move, but anyhow, Hank was out there, and he got the flagship down there. So by the time I got there, we had the change command on the first of December, and they were well ensconced in Bahrain. And she [*Blue Ridge*] was alongside the pier with a thousand phone lines attached. [Laughter]

EM: Wow. Communications at that point.

SA: Communications was phone lines.

EM: We might have talked about this before, but if you could go over it again. Why was Mauz not extended to cover the war? Be in charge of the war, instead of bringing you in at that time?

SA: I had been announced to be his relief back in early summer, to be Seventh Fleet, and Mauz was coming back to town to be—I think he was coming back to be 08, the big programmer. Then I think Kelso had in mind that Hank would go right on down to be

CinCLantFlt. Hank and Kelso were and had always been very close over the years, and they got along well. They were both good golfers. They golfed a lot. Frank, I think, recognized that Mauz had tremendous intellect and knew the political terrain back in Washington pretty well, and he had a lot of confidence in him.

EM: Well, wouldn't the first priority have been to run the war?

SA: Neither one of them thought the war was going to happen, is my belief.

EM: Yeah. Yeah.

SA: I'm sure that Kelso didn't think it would happen. I'm pretty sure Mauz thought there would be a lot of staging and maybe some small skirmishes, but it would all shut down before anything really bad happened. Hank and I never really talked about that, but as I go back and sort of review the bidding and all, I think Hank was of the mind too that there was going to be a lot of posturing and a lot of stalemate, and I think he was sort of surprised at the end, when all of a sudden it turned into a real shootout. If he thought it was going to happen, I think he would have made more attempts to stay.

EM: What had he done before you got out there?

SA: I'll tell you, I think Hank has been criticized unnecessarily on what he accomplished during that period of time. Hank was a very good policy guy and plan guy, and he understood, he really understood some of the dynamics of Washington and also the dynamics that kept the Navy in play. So he did a good job of making sure the Navy was not an outsider to the process. He was very much engaged with the preplanning that went on. He looked out for Navy interests very well—maybe sometimes too well for the other component commanders.

EM: They got a little antsy? Felt he was intruding?

SA: Yeah. There was some evidence that they were a little worried about that.

EM: Which other component commanders?

SA: Well, all of them. All of them had some little concerns, but from what I could see, Hank did a really great job of setting the stage. This was not a world that he had dealt with before, and he very much was tuned into what Seventh Fleet's roles and responsibilities were, and he tried to carry out those roles as, "I'm Seventh Fleet and here's the way I operate and here's how my ships operate and here's how I want to operate with you." He did not have the same advantage that I had; I had been on the other side of the fence with the RDJTF and NavCent before, bringing other people in to act in support of us. So he was very much, "I'm here as Seventh Fleet, I don't care what you want to call me, I'm Seventh Fleet and here's how I operate and here's how I operate my units."

EM: Not so much as the component commander for Schwarzkopf or CentCom.

SA: He had never done that before, and of course Seventh Fleet is the component commander of CinCPacFlt to CinCPac, that was all fairly clear. As I used to say when people were talking to us about NATO, a lot of the NATO influence was out there with some of the teammates and all. Us PacFlt guys thought NATO was a Japanese admiral. [Laughter] You know, we didn't speak NATO.

EM: Right. One of the things he had set up clearly was the multinational force out there with the embargo patrol.

SA: Yes.

EM: When you got there, was that in good shape?
S: It really was. That was a spectacular operation, and it was a well-oiled machine. There were lots of things that went on, but he had the procedures in place. Everybody knew

what they were expected to do and how to do it. I think it was a really, really well-organized effort, and I will tell you, the guys that had to board those ships and stuff, that was really dangerous work. We had a couple of episodes out there and they really, they really did it well. Going up and down the sides of those merchant ships is tough.

EM: Up a ladder or being dropped by helicopter.

SA: Or dropped by helo, and you are not sure what you are going to see until you come over the gunwale, or what you don't see when you get dropped off of a line there from the helo.

EM: When you got there, was there much rejiggering that you had to do?

SA: Yeah.

EM: The force?

SA: I did some things. I renumbered the force a little bit to make it clear where the lines of responsibility were, and we had some little diversions there that made it sort of, who really was in charge of this, this, and this. I did some renumbering that the communicators weren't too happy with, but I just wanted to simplify the system a bit. By that time NavCent took over the job as a logistics guy. The old NavCent had the logistics job and I, from my OP-04 experience and my *San Jose* experience, I looked at it and wanted to do a couple of little things that I wanted to simplify from the fleet user standpoint. And so I made a couple of changes there that put some of the responsibility back on the guys at sea and took a bit of his responsibilities out, except he was the guy that had to get it all coordinated.

But I thought there should have been more fleet input on when and where they were going to do their resupply efforts. So I changed a little bit of that. Other than that, that's about all I did. I was very uncomfortable with the fact that I was on *Blue Ridge* and Schwarzkopf and the rest of the team were over there in Riyadh, because that's not the way I had envisioned NavCent operating. I wrestled with that and kept wrestling with it. Through most of December I was very, very uneasy.

EM: Did Schwarzkopf at some point say, "Come out here to Riyadh, I want you here"?

SA: No, but we talked about it. One of the first things I told him was this was not the way I had operated in my days as NavCent. I said, "Of course it's much different now, because we've got all of these ships and stuff out here."

He said, "Well, why do you have all those ships out here?"

EM: Thought there were too many ships?

SA: Yeah. I said, "I would be much more comfortable being in Riyadh with you, and I think that's where I should be, but I don't think I can get there in time without causing some disruptions that could actually be pretty fatal mistakes if I did it."

He said he agreed. He said, "If we had it to do all over again, I sure wish you were in Riyadh, but I understand and, you know, it's your decision. You let me know."

EM: If you had done it, who would have controlled fleet operations?

SA: I would have designated one of the carrier group commanders as the guy in charge. We could have been made to go back and forth with Washington.

EM: Since you had two separate battle forces, Zulu and Yankee, one guy would have to run the whole show, wouldn't he?

SA: Yes, and that was part of the problem, and that's why Hank's fix was, "Well, they all work for Seventh Fleet." And that's why I always thought there should be a Fifth Fleet. There should have been a Fifth Fleet hat. We'd nominate somebody as Fifth Fleet. But like I said, a thousand phone lines [from *Blue Ridge*] may be an exaggeration, but I don't think by much. We were doing everything by secure phone. It was just constant chatter,

chatter, chatter, chatter, chatter, and communications weren't that good, and so that's why as the time got near, I told Schwarzkopf, "I am not going to be in port when this thing kicks off. Whenever we decide what D-day is, I don't want to be in port. I don't want to trigger anything by going to sea that you don't want triggered." So I've got to have enough time for me to make a good excuse for going to sea and not look like I'm going to sea because we are going to war.

EM: Everybody knew it was the command ship, right?

SA: Yes.

EM: What kind of excuse would you come up with?

SA: Well, you know, underway training.

EM: An exercise?

SA: An exercise or something like this. Of course we had been doing that with the amphibs. We had taken one of the amphib groups out, taken them down off the coast of Oman, and they had done an amphibious exercise. That did two things as far as getting them some room doing what they like to do and also to send a message. And so I said three or four days, something like that, I can get under way and get everything shaken down so there's no drop in connectivity and stuff. I had no idea how much comm activity we lost when we went to sea, dropped all of those phone lines.

EM: It was significant?

SA: Oh yeah.

EM: So how did you communicate?

SA: I had more satellite comms than almost anybody in the fleet in those days by that time, and we starting using InMarSat [International Maritime Satellite], the commercial British system for some of our comms that could go unclassified.

EM: That's voice comm?

SA: Yeah. So we had a little shaking down to do there to get all the circuits back up, and we could talk to everybody that we had been talking to on phone lines. The problem was that the staff in Riyadh had gotten to the point where they were so used to picking up a phone to talk—a lieutenant over there could pick up a phone and talk to the flagship on matters that should have been dealt with by their colonel and my captain.

I can remember times—in fact I had to call the deputy for Schwarzkopf one time, and I said, "You've got lieutenants and Army captains calling my battle staff out here for information for your morning briefs. Tell me what you want for the morning brief and I will get it to you." But, I said, "I picked up the phone one day and I heard this guy calling in and using your call sign, and it's some lieutenant colonel wanting to know, or major wanting to know what time we did X, Y, Z, or something." I said, "totally administrative thing, nothing operational to it. You've just got to knock this stuff off, because I don't have much bandwidth out here as I need, and we can't be dealing with this kind of stuff like that."

So he took it on board and tried to cut down some of the nonsense that was going back and forth between the staffs. But it was an awkward period. We could have certainly done a lot better if we would have had the NavCent job as someone that Schwarzkopf was comfortable with, sitting beside him like the Army guy and the Air Force guy was.

EM: Who was your liaison with Schwarzkopf?

SA: At the time over there was Connie [Conrad C.] Lautenbacher. Connie was there. He had been preceded a little bit I think. We had Tim [Timothy W.] Wright over there earlier as a rear admiral. But that was early in the game. Connie was there for most of the time I was out there.

EM: Did that work well? That contact?

SA: Yeah. I had good relations with Connie. He was sort of captive, again, to the communications setup. He didn't have as much direct comms as I think he would have liked to have had with me, although he could always get ahold of me almost instantaneously. But I think that Connie was a little unsure on some of the day-to-day stuff that was going on out in the fleet. He had a good chief of staff, D. J. Wright. D. J. had been a young pilot in VA-55 when I was there as the ops officer. He was one of our lieutenants, so I knew him quite well. Good warfighter. He knew the aviation game pretty well.

EM: In what capacity was Lyle Bien; wasn't he there?

SA: He was there earlier to be in JFACC [Joint Force Air Component Commander]. But again, Schwarzkopf wouldn't pay much attention to those guys in there. I think Connie was a one-star at the time, maybe a two-star, but I think one. Of course Schwarzkopf didn't pay much attention to one-stars or two-stars. He wanted to talk to three-stars and four-stars. And the personalities didn't mesh quite as well. Connie was a really bright guy. Always thought through a problem and put a lot of thought into everything he did and had it pretty well laid out in a nice format. Schwarzkopf wouldn't give you much time to warm up to the subject. He wanted—

EM: Get to the bottom line?

SA: Yeah. Connie was the type of guy that wanted to make sure you had all the information before he jumped on the bandwagon. So I think there were some times there where Connie thought the general wasn't giving him due regard to what he wanted.

EM: All right. You talked about this in the earlier interview, but how did your Vietnam experience relate to the Gulf War? What did you want to do differently from Vietnam, the way you perceived it being run back then?

SA: First thing, when I got there and made my rounds and introduced myself to the other players, it became obvious that we were all Vietnam veterans. We had all been through the thick part of Vietnam, and we all had some grave reservations about what the ROE and the political climate from Washington had done to what we thought should have been done in those days. And so we had this discussion amongst ourselves that we weren't going to do Vietnam all over again. We were going to do our best to make sure we did the things the way we wanted to do them, and not the way somebody else wanted us to do them. Schwarzkopf was on that same sheet of music, as you know from many of his writings and talking. He really disdained any influence from Washington.

EM: But did Washington respect that perception?

SA: No. But they ended up having to tailor it pretty much to a degree; obviously Colin Powell and Cheney were not about to let Schwarzkopf go running off on the ranch.

EM: Was there anything comparable to the Tuesday luncheon meeting to pick out targets at the White House?

SA: No. The Air Force tried that. The Air Force formed a team back in Washington that came out and briefed Schwarzkopf on the way the war should be run, and this is the way it's going to go. He had them ushered out of the theater. [Laughter] And told them nobody comes out here without my permission. Told [Colin] Powell and the rest of them back in Washington, "I'm the final guy who comes in theater." Of course, the first time I came across that was when the Commandant of the Marine Corps wanted to come out and visit the troops.

EM: Al [Alfred M.] Gray [Jr.]?

SA: Al Gray. Al had been out in theater several different times, and every time he came and left, Schwarzkopf, I guess, was saying that's the last time that SOB gets out here. [Laughter]

EM: What was his gripe against Gray? That he was just there?

SA: Well, that he wanted to talk to the Marines and tell the Marines how to fight the war, you know, and he wanted to lobby Schwarzkopf to make sure the Marines get this and the Marines get that, because they are not going to be second players in this game.

EM: Al Gray wasn't in the operational chain of command, was he?

SA: No. He came out right before the war started, and he came in theater, and Schwarzkopf wouldn't let him on the ground in Saudi Arabia.

EM: What was your reaction to Al Gray? Did he come out to the flagship?

SA: Yes, he came out to the flagship. Al had been in the Marine detachment aboard that flagship as a captain or first lieutenant. In fact, one of his sentries had fired a round, or even I think maybe Al did, cleaning their gun at one of the watch stations. [Laughter] So Al comes out and he says, "I've got a big favor to ask of you."

I say, "Yes sir. What is it?"

Walt [Walter E.] Boomer had warned me that Al Gray had something in mind. And Walt knew what it was, and Walt said, "I don't want that guy here. So you've got to do something for me."

EM: What guy are you referring to?

SA: Jack [John J.] Sheehan.

EM: In what capacity was Sheehan going to act?

SA: He was going to be Al Gray's spy.

EM: On Boomer's staff?

SA: Well, he wanted to put him either in Riyadh or put him with Walt. Walt says, "I don't want him." Of course, Schwarzkopf wasn't going to take a three-star Marine on his staff that he wasn't controlling.

So Al came, and he said, "I would like to have General Sheehan out here."

I said, "Absolutely. As you know, this flagship has dedicated spaces for Marine staff," and I said, "I haven't violated that. Every bit of that property is vacant, ready. Everything is in there. The radios are all set up. He can come and in an instant be online. He's got a general's quarters just like my quarters, right across the passageway. He's got a flag plot," you know, a command center. "Everything's there. We haven't invaded that at all. Even though I'm about twice the size of my normal Seventh Fleet staff, we've kept that pristine." So I said, absolutely send him. Send Jack out and I'll be glad to have him on board. So that got me lots of favors from Walt Boomer. [Laughter]

EM: So he came out? He was on your ship?

SA: For the whole war.

EM: What was his duty?

SA: His duty was to be there and take a look at the process of how things were going, and if he saw anything that didn't look right, he would come to me.

EM: Kind of difficult to report on what Boomer's doing because Boomer's in Riyadh, right?

SA: Well, no.

EM: He went up front?

SA: See, Boomer did the exactly the same thing I did. Walt didn't want to be tied into Riyadh, so it ended up that the Marines and the Navy were both absent without leave from Riyadh.

EM: Right.

SA: Walt was up on the front.

EM: Right.

SA: Right south of the border of Kuwait, and I was out at sea. And Jack was a great adviser to me. As we were doing the amphibious landing preparations and thoughts and stuff, I had a great time. He was a great, great help to me.

EM: It's fairly well known that Al Gray was really pushing for a Marine amphibious assault. Did Sheehan reflect that too?

SA: Sheehan wanted to make sure that we could do it if we had to do it. I never got the impression that he would have done it any differently than I would have. That if the CinC said we had to do it, or if Walt Boomer and his guys got in trouble coming up the shoot and needed a diversionary coming in from the side, we were willing to do that, and we had plans for doing that. And it was going to be pretty risky, but it could be done. Jack was good about making sure we thought of all the alternatives and what we really had to do in case. I found it valuable to have him on board. We solved a couple of problems. We made Al Gray happy and we made Walt Boomer happy, and Schwarzkopf was happy and I benefited from having—

EM: The best win-win, right?

SA: Yeah, I had a good smart guy on board that could give me a sanity check every now and then.

EM: In terms of Washington's involvement in affairs, did Admiral Kelso come out at some point?

SA: No. Frank Kelso left me totally alone.

EM: He didn't communicate with you? You should do this or that?

SA: No. There was no pressure. The only pressure I got from Washington every now and then was Tony [Anthony] Less, who was the Ops and Plans guy, and Tony would ask me questions about this and that, and sortie counts and stuff like this. Basically I told Tony, "We are not into that game out here. We are in the game that the CinC wants to play, and we are going to play it the way he wants. If it's looking bad, if it's looking bad back there from your standpoint, I can't help it. I'm doing what the CinC wants, and I'm also doing what I need to do." I said, "Yes, we are getting beat on the sortie counts sometimes by the Air Force."

Even the Air Force said they were not going to play this game of counting sorties and all this. Of course back in Washington they were, and they were looking at my flight schedule and how many I had for fleet defense and how many I had for strike. Tony would call me every now and then and I said, "Tony, we are meeting the numbers for strike that the CinC wants and the air component commander wants, and the rest is what we need to do to make sure we don't get surprised out here in the Gulf."

EM: So was Less reflecting the view that the Navy ought to have more of a showing? Naval air had a greater part?

SA: Everyone wanted the Navy to have a little bit more showing, but as we did in the reconstruction of things out there, basically we matched the Air Force in almost every category as far as ships—you know, airplanes available. Sorties flown. Basically we did the same job, and we had some extraneous stuff that we had to just do to keep ourselves able to sleep at night. Tanking was a big argument. We had all sorts of tanking complaints. We never did win them all, you know. The agreement when Watkins and—

EM: Goldwater-Nichols, are you referring to?

SA: Well yeah, but when we went to the air component commander and then we finally went to the Air Force, we said, "Okay, the Air Force was wanting the tanker business to be all their tankers. We will support you when you need tanking."

Well, we called their bluff out there. We said, "Okay it's time, we want our tanker support."

Well, okay, yeah. That's fine. Then the guys show up with JP4 [aviation fuel] and we were JP5 guys. We could burn JP4, but to bring it back aboard ship, then you have—

EM: Higher danger of fire?

SA: Fire, yeah. JP4 on those steam catapults, you drip a drop of JP4 on the steam catapult track and they would burst into flame right away. They couldn't honor their agreement to give us JP5 all the time, so we ended up having—a lot of times we brought airplanes back that had JP4 in them, and we would have to defuel them and put JP5 back in the airplane.

EM: Was there a problem with the fuel transfer equipment too?

SA: The Air Force likes to use booms more than drogues, and so we preferred to get the KC-10s, which had drogue stations on the wings and probe on the centerline. It's much easier for the Navy guys, once you get used to the drogue—plugging into the drogue and sitting back there waiting for a guy to put a drogue on the back of the boom. And what happens is you've got a very short hose, and you've got this hose hooked into the back of the boom, and the basket is—I forget how long, 6 or 10 feet or so, maybe even less. It's hanging back from the booms, and the boom comes down close, and then the basket is back there. But it's a much stiffer plug now, because there is not so much slack, there's not that long hose, and so when you hit it, it tends to be a harder hit, and if it flexes or bounces at all, it will bounce right off of a probe and hit your canopy and shatter a canopy.

It's a much more dangerous operation than having the long hose drogue. So we tried to emphasize that we wanted to have more than our fair share of the KC-10s rather than the KC-135s. But we eventually worked it all through, and even though we complained sometimes that we weren't getting as much support from the tankers as we should have, in the final analysis when you got all the data down, we came pretty close to getting our fair share of support. It was just one of those irritating things that was always popping up.

EM: Going back to the air component commander business—or JFACC was the term used—and this relates to Goldwater-Nichols. There was more of how joint operations should go. Did you notice a change in operations that resulted from that legislation that came out in '86? How did it impact on your activities?

SA: We had the JFACC to deal with and they sent out the flight schedule for the whole shebang.

EM: Air tasking order?

SA: Air tasking order, the old ATO. You see referenced it in most of the publications, but the Air Force made sure that most of it was sort of buried in the fine print. The Air Force went in there with the concept of the air component commander, and the operating cell that controlled that air component commander was the JFACC. That was the operational entity that produced all the documentation of who, where, when, how and with what kind of loads and what the targets were.

Well, this thing was going to be electronically passed every night, and they had this computer, CAFMS [Computer-Aided Force Management System] I think was the correct term. CAFMS was the computer-assisted system that put all of the stuff together and put it into a format that could be electronically transmitted from computer to computer. So it was going to be digitally sent out. The fact is the Air Force made it seem like everybody in the Air Force had a CAFMS terminal, and everybody had worked with CAFMS for years—not so. Not even close to being so. It was just a small part of the Air Force that had CAFMS and had worked with it. The rest of the Air Force had not, but they didn't—

EM: Play that up?

SA: They played up the fact that we couldn't connect with CAFMS. Well, you know, Jerry [O.] Tuttle was the big communicator in the sky back in Washington. When I got out there, this was a burning issue before the war started. I said, why aren't we on CAFMS?

"Well, we don't have enough bandwidth to get on CAFMS."

I said, "Before I came out here, Jerry Tuttle assured me that we had more bandwidth than we ever could have utilized, and so I bet I can get us another satellite link where we can dedicate it all to CAFMS. And it's got to be a two-way, it's not a one-way. It's got to be a two-way thing." So I said I'd fix this. Well, that was too simplistic, of course. I couldn't fix it. We couldn't do it.

EM: Couldn't get the satellite?

SA: We couldn't get the satellite linkage, and plus that we didn't have the interface that would play with CAFMS at all. So that failed, and CAFMS came to us in an airplane. They would produce the document to the Air Force units that could copy it. It went out manually to the Air Force units that couldn't copy it, it went just like it did to us at sea. Again, very few people really know that.

EM: So this document—which was huge, right?—it would be flown out to the carriers via a couple of stops in between.

SA: Yeah. The Air Force would fly it in one of their Lear jets. I mean they sent their own Lear jets down to Djibouti and all the other places for the Air Force guys that couldn't [get it digitally]. They had a fleet of Lear jets out there, all first lieutenants and second lieutenants flying them, and they would fly these stacks of paper to the units that couldn't copy CAFMS. They would fly mine to Bahrain, and I would have helos at Bahrain that would bring them out to the flagship and to the carriers. The Red Sea guys would go to the field over there on the east side, and a helo would take it out to the carriers from there. So that's how we had to deal with this thing.

EM: Now, the ATO would have everyone's mission? Whatever service?

SA: Yeah.

EM: So could you frag [fragment] part of it, which was your part, or—

SA: Yeah. We would pick up that number and assign that number to whichever place it was to go. Then the intriguing part of the ATO was the back piece of it, which was called the SPINs [special instructions].

EM: Information something?

SA: Yes. What it was was the amplifiers to the information. Okay, if it was a CAP [combat air patrol] station or something you were going to—you cannot fire without authority from X, Y, and Z. Always at least two people. You can't go into this area unless so and so provides X, Y, Z beforehand. So it had all of the information that then now started to give you rules of engagement and special instructions that might not have been agreed upon someplace else. So what we were finding was, there were things written in the SPINs that prevented our airplanes from doing certain things that they would have normally been able to do. Such as the F-14s taking up—

EM: Intercept positions?

SA: Yes. So then it became gamesmanship of trying to read the SPINs fast enough to see who had dealt you a bum deck of cards.

EM: Did you have an appeal process where you would go back and say wait a minute, this doesn't make sense?

SA: Yeah. The telephones used to be so handy, because you would pick up a telephone right away and start arguing with some colonel on the other end. Why do I have this dumb thing to deal with? So that process was fraught with all sorts of issues that, again, all stemmed around writing the ROE ahead of time and everybody agrees, which was one of the things that didn't happen here. And also, let's find a way to deal with this information in real time, so we can have the back and forth when the process is under way instead of all of a sudden we've got this 1,000-page document to look through and assign our missions and go off and do them. So it was very cumbersome. It had a long lead time, which the Navy was not used to. This thing had a 24-hour lead time to it, and when you're in a fast-moving situation, you can't deal with 24-hour lag times before you can get a mission into the game. It was the new things of lessons learned from Vietnam, "We are never going to do this again." The lesson learned from this was, we've got to find a better way to do this CAFMS ATO thing than what we had.

EM: How did they assign targets to the carriers? How did the JFACC assign targets? You certainly didn't have route packages like Vietnam, right?

SA: No, we didn't have that.

EM: What did you have?

SA: We had specific targets. In other words, they would line them up ahead of time and you would have this specific target to go to.

EM: Were they all coastal, or did you have some that were farther inland?

SA: No, we went downtown. We were all over the place. We went almost everyplace the Air Force went if we could reach them, if they were in our ability to get there. So the target assignments, obviously, the Air Force had the ability to select the better targets. We got our share of good targets. The thing that was the most frustrating was the popup targets that you could get to that all of a sudden you would see, and the AWACS [airborne warning and control system], which was now the big daddy in the sky that was seeing everybody—it's like the airline-FAA relationship, you know, getting your radar controlled up there, and everyone had their own squawks, and this sort of stuff—so the AWACS knew whether you were going to the right place or the wrong place. And if you were going to the wrong place, they would call you on it, say you had to do what you were told to do on this piece of paper.

But the Air Force had a way of changing that. If the Air Force had a target of opportunity, they had a way that they could get an approval to go for a target of opportunity. If the Navy asked to go to a target of opportunity, the Air Force said no, that's got to be put into the ATO. We had one instance where one of our guys had a brother flying F-15s for the Air Force over there, and he calls him back to the ship one day and says, "I want to thank you." He says, "You just got me a DFC [Distinguished Flying Cross]."

The guy says, "What do you mean?"

"Well, weren't you the guys that called in that popup target down in so-and-so?"

He says yeah.

He says, "Well, I got to go get it." [Laughter]

EM: I wonder what his brother thought of that.

SA: He was not happy. [Laughter]

EM: Interesting. These are targets over land, right? At sea did you have more latitude?

SA: At sea I had more latitude. Again, within the ramifications of—did I have a flight up there that was supposed to be up there? From the fleet support or fleet perspective, I had some opportunity to do things, and sometimes that got me in trouble. For example, the way I had signed off on the ROE was, if a ship is out there and is contributing to the effort of the bad guy, then he is free game for me. And at one point in time, in front of Secretary Cheney and Colin Powell and everybody else, the CinC got me up on the stage to say this was what he wanted, to sink ships out there anywhere in the AOR.

EM: This was in a public forum? He said sink any ship you say is a threat?

SA: Yeah.

EM: Okay.

SA: When the war started the ROE changed, and due to the immense amount of information that was going over the circuits, the new ROE didn't get to me until about four days after the war started. In the meantime, I sank two ships.

EM: What was the ROE; what did it say?

SA: That I had to have permission.

EM: To sink ships?

SA: Yeah.

EM: All ships or any or just particular ones?

SA: Particular ones. And so the next thing I get is a call from Schwarzkopf, saying I'm going to court-martial you.

EM: Because you sank some tankers, right?

SA: Yeah, I sank some tankers [laughter], the tankers who were broadcasting the raid count of our airplanes going over the beach. Because he positioned his ship off Bubiyan Island and was just counting our raids coming off the carriers in the Gulf, and basically we could hear him saying, "Okay, there's a flight of 20," or 25 or whatever it was, "coming north where I think they are headed for so-and-so. They are over us right now." And so we heard them reporting the traffic and all, so the guy said we want to go sink them. I said go sink them.

EM: Were they carrying oil?

SA: Yes. But they were carrying light. They were not carrying crude oil. They were carrying light oil, which we could burn off of the surface.

EM: Did you burn it?

SA: Yeah. Basically, from those two ships we didn't have anything environmentally catastrophic, but when he [Saddam Hussein] started dumping oil himself from the oil wells out there, and from the other ships with the heavy oil, then that was a different problem.

EM: Well the ROE, before there were any sinkings, was that for environmental concerns, or what?

SA: Here's what was explained to me. I had pushed back on Schwarzkopf before we had this meeting with Cheney and Powell in Riyadh. He said, "I want you to get up in front of them, and I want you to tell them specifically that any ship that is going to come into this AOR is going to be sunk. No matter where you find them, if they are headed for there with what we think is ammunition or fuel, I want you to declare that you will sink them wherever they may be."

I said, "General, the Law of the Sea prevails here. If a guy's coming out of Yemen and headed around the corner to come up into the Gulf, I don't want to sink him either in the Red Sea, as he gets under way, or out in the Indian Ocean. I would be happy to sink him once he tries to get through the Strait as he comes into the Gulf."

He said, "No, I want you to be able to make the declaration that any vessel that we determine to be a threat, you will sink wherever and whenever."

EM: Was that conference recorded?

SA: Probably. So his and my conversation with Schwarzkopf was recorded. I'm pretty sure that it was the conference where Powell and Cheney were sitting right there in the front row, right beside Schwarzkopf, and I was talking to him. So I said, "Okay, if that's what you want, if that's what you're going to write in the ROE, I will execute it."

He said, "It will be in the ROE. That's the way I want it to be. I want you to be able to execute."

I said, "I can guarantee you that I can do that." So that's the way that last brief went before we went to war.

Well, in the meantime, I guess Powell, when he got back to Washington, had second thoughts about that, and said, "Look, we don't want to deny Iraq an ability to reconstitute their economy when this is all over. And those large oil tankers that they have under their flag will be important to that reconstitution effort. I don't want them sunk." So that changed the ROE.

EM: So that came from Powell?

SA: That came from Powell. And I didn't get the ROE. [Laughter] So I executed just like I was told.

EM: And after you executed, what was Schwarzkopf's reaction?

SA: He called up and said, "I'm going to court-martial you."

EM: Oh jeez.

SA: I said, "General, why are you going to court-martial me?"

He said, "Because you didn't adhere to the ROE."

I said, "We had this discussion."

He said, "Well, we changed the ROE."

EM: And you didn't get it?

SA: "And why didn't you read it?"

I said, "I don't have a new ROE. I'll go right now, I'll go chase it and see." And sure enough, went down to the comm guys and I said, "There's a message that changed the ROE, and this is the date-time group [identifier] of the message." They said we hadn't received it. They went back. They tracked it. It was still sitting in the key to be transmitted to me. It hadn't. With my circuits it hadn't.

EM: So again, the communications were backlogged?

SA: Backlogged.

EM: Was it just a priority message?

SA: No, it was an OP immediate.

EM: And still it was the backlog, wow.

SA: It was their backlog. So my good friend Jerry Tuttle, who assured me there was so much bandwidth that nothing could clog it up—it got clogged up. So that was the first real interesting exercise.

EM: What other disagreements did you have with Schwarzkopf?

SA: We had the—

EM: The IFF [identification friend or foe] business with the F-14s? Or was that with [Charles] Horner?

SA: That was with Horner.

EM: Okay.

SA: I mentioned this to Schwarzkopf too. He said, well you and Chuck work this thing out. So I went to Chuck, and I said look, you know, this is ridiculous. He says, "Well, I don't want any blue on blue things."

I said, "I don't want any blue on blue stuff either. I guarantee you I don't. But we've got the best long-range missile in the world, and these guys are getting away when we've got shots but your guys can't get to them to get a shot. I think we've got enough coverage out there to make sure that we are not going to get blue on blue."

But he said, "No, I'm not comfortable with that. I'll send [Brigadier General] Buster [C.] Glosson out to the flagship, and you guys work it out."

So Buster comes out to the flagship, and I get everybody together and I say, "Buster, here's our problem. We've got this capability. We don't have the same ability to do IFF like you guys do. That's not the way we were designed. We were designed for open ocean war at sea type of thing, and we've got these long-range shoots that we can do." And I said wondering, obviously when we've got an airplane that's coming out of there and he's trying to get away from your guys and trying to get to Iran as an escape route, I said, "We can shoot him from 100 miles away."

"No, well, I'll tell you what, I'll go back. We will develop a couple of free zones, so if anybody shows up in this free zone, you guys can shoot them. And we will program you guys to have a CAP station so that that can happen."

I said, "That's a great fix. I appreciate it." Never happened. Never happened.

EM: There were aircraft that were fleeing Iraq over to Iran. You had shots at those? The Air Force did too?

SA: Air Force had some shots. We had a lot of shots at them. Could have had a lot of shots at them.

EM: They had nothing comparable to Phoenix in terms of the air-to-air, just Sidewinders and Sparrow?

SA: Sparrows and Sidewinders.

EM: That was it for the Air Force?

SA: Yeah.

EM: What's the range of a Sparrow, like 20 miles?

SA: Maybe.

EM: Nothing comparable to Phoenix in their arsenal.

SA: Nothing anywhere close to it.

EM: Was there any hesitation by the powers-that-be in Washington or Riyadh, about why are these planes going over there? Were they fleeing, or were they going to join up with the Iranians or what?

SA: This is where we got in a big discussion, and I'm not sure whether it was just me and Horner, or whether I got into the discussion with Schwarzkopf too. I think maybe it was Schwarzkopf. Mining was one of the subjects that Schwarzkopf and I had a lot of dialogue about, and I'll come back to that. But basically, when those airplanes flew to Iran, I went back into the headquarters there in Riyadh and said, "I need to know what else went with those airplanes. Did transports go with them? What do we see on the ramp over there? Are there weapons being offloaded on the ramp over there? Is there handling gear for these airplanes being flown in there?"

EM: To set up a base in Iran?

SA: I said, "Basically, I need to know if these guys are going to fly those airplanes out of Iran."

EM: Right.

SA: "Because if they are going to fly airplanes out of Iran, I know where they are coming, because their easiest target is me, and they've got this valley over here that we talked about. They've got this valley and they can come steaming down that valley. We can't see them, and they pop up over the ridgeline, and I've got a five minute window to shoot, and that's not enough time." I said, "For one or two that's enough time, but if I've got a whole bunch of airplanes coming out over that, that's calling it too close. I've got to change the way I've got the fleet configured, and I've got to have more CAP stations that I can put facing Iran, because right now we are just doing the minimum."

EM: So what intel were you getting?

SA: None. So I said, "Who is going to give me the intel?"

"Well, we are not looking there."

So I came back into Washington to the Navy circuits and said, "I need some intel coming directly to me about what's happening with these airplanes over in Iran."

EM: Who did you go to?

SA: I went to DNI [Director of Naval Intelligence].

EM: Tom [Thomas A.] Brooks?

SA: Yeah. And we started getting some stuff from other satellite coverages for other things and basically found out that the transports took their families over, not weapons.

EM: No staging equipment.

SA: Yeah. I thought all I had to do was call Riyadh and I could have had that in an instant. And I was told we are not looking. That's not our interest.

EM: So you had to go to Navy intel to get that?

SA: I came back to Washington to Navy intel, and Navy intel opened up the circuits that then they could feed me.

EM: Two questions here. First of all, what kind of support did you get from Tom Brooks and his company, other than what you just mentioned?

SA: We got the normal stuff that they broadcast out to the fleet, but our real tactical intelligence was all coming out of Riyadh. Now, how much Tom was feeding that piece of the pie I don't know, because they were collecting it from several different sources. I guess Connie probably had a better insight into that than I did.

EM: Well, you had your NavCent and Seventh Fleet intel as well, right?

SA: Yeah.

EM: So you were flying reconnaissance?

SA: Yeah. I was still Seventh Fleet for all intents and purposes while I was there, so I still had that stuff coming to me.

EM: P-3s were operating?

SA: P-3s in the area. The EP-3s that I had in the area—of course to me, a NavCent, although they were Seventh Fleet assets they were working the NavCent problem. Now, the P-3s and EP-3s out in the Pacific theater were still reporting to me, but they weren't as Seventh Fleet. They weren't looking for the same things that we were looking for over here.

EM: The other thing is, what indications did you get from the Iranians during the war? Did they make some threatening moves at various times? Did they try to stay away from making threatening moves?

SA: I tell you what, the Iranians were very well behaved. Now, they did a lot of bombastic dialogue, in fact, we had to dust off several plans for going into their places to knock them down if they didn't behave. And so when they made threats about closing the Strait and all this, we looked at their navy facility, the closest there, and we had all of the targeting done and knew what we could do and how long it would take to do it. And we kept that dusted off on a routine basis. I thought a couple of times that we might get the green light to have to go do that. But in reality, on the day-to-day, they were very well behaved. They did not provoke us in a period of time. Well, the navy units—the remaining navy units, did the same thing that the airplanes did. They tried to come out and skirt right around the corner and go into the port facility up there in Iran. And we caught them en route and sank them all. The Iranians—we were very, very close to their border, their territorial waters. And we were right up against it and they didn't push back. They could have, and they didn't. So they were well behaved, as a rule.

EM: But you said there were a couple of instances where you thought you might have to put those plans into play.

SA: Yeah, well, when they were doing some of their bombastic stuff about we are going to send these gunships out and we've got a couple of submarines here, and we are going to close the Strait, block you guys in here. When that noise level reached pretty fevered pitch, we were pretty much ready to do something about it. But they always backed away from that, and we'd watch them really closely there to see about the submarines and their gunships.

EM: If they had taken that step and done some aggressive activity, were you going to defend in place or were you going to pull ships out?

SA: No, we were going to defend in place.

EM: You had enough strength to deal with it?

SA: Yeah.

EM: All right. The fact that you are Com7Flt and in the Persian Gulf, who was minding the store back in Japan or East Asia?

SA: Well of course we had the amphib guy down in Okinawa. We had a TACRon [tactical air control squadron] guy up in Japan.

EM: Was there one guy, one deputy of yours who was in charge of that?

SA: Yeah. And we had the guy in the Philippines, ComNavPhil, and so the day-to-day business, they were all on the circuits, we were on the Seventh Fleet circuits. We had a little trouble because we had to get to the Seventh Fleet circuits, we had to get a double bounce off the satellite, so we had a time delay in the transmissions that we had to play with. But we were able to do the daily things that we needed to do and answer the daily questions that came up about operations in the Seventh Fleet. Of course there wasn't that much, because we had most of Seventh Fleet out with us.

EM: What if the balloon had gone up with the North Koreans? Who would have been in charge of taking counteraction?

SA: Oh.

EM: Nobody was? So you still had that responsibility?

SA: Yes.

EM: That's kind of dicey.

SA: Yes.

EM: Maybe Third Fleet would have come out?

SA: Probably sent Third Fleet.

EM: Had you thought about that? I mean, that's a little leaving you naked.

SA: Yeah, we thought about it. And CinCPac was thinking about it a lot. Yeah.

EM: Who was CinCPac at that point?

SA: Hunt Hardisty.

EM: So you were all a little bit nervous about that?

SA: Yeah. And one of the things that I had to do right away, while I was anxious to get released once this thing was all over, was to get back to Seventh Fleet AOR. It was obvious that some of the relationships there were a little nervous from the fact that they were naked.

EM: The Japanese and South Koreans?

SA: South Koreans.

EM: Yeah.

SA: Even some of the folks down in Singapore, Indonesia and places like this, they were a little nervous. Two things that I wanted to accomplish as soon as I got out of there: one, I wanted to have a presence in the Muslim nations over in the Seventh Fleet AOR down in Indonesia and Malaysia; two, to show them a friendly face, that the thing in the Gulf had nothing to do with a crusade.

EM: Oh, right.

SA: That this was reaction to somebody that did harm to a friend. We would have taken the same action for you as we did there. And the other was to sort of shore up the Japanese and the South Koreans that, yeah, we are still here and we are always going to be here, and we haven't forgotten you, and that was just a short-term fix for a problem that we are going to solve by changing our structure so this doesn't happen again, I hope.

EM: Although—

SA: It took a while.

EM: Yeah, although we are still in a big way in the Middle East and still have a lot of responsibilities out there.

SA: We now have somebody there.

EM: Right, Fifth Fleet is there.

SA: So, Fifth Fleet. That was one of my two primary things that I wanted to do when I came out. We rapidly went around through the area. Now we brought the Japanese minesweepers down, and the Japanese minesweepers were on the way to go out after I came out. So we made a big fanfare having a passing at sea of Seventh Fleet coming home and waving to their minesweepers that were about to help clean up the mess. And that got a lot of good press out there. Japanese Navy was very happy to have that kind of recognition for their forces.

EM: That was their first operational activity since World War II, I think.

SA: Their first.

EM: That's far from the Pacific.

SA: Yeah.

EM: They paid for the war too. A big chunk of money came from Japan for operations.

SA: The other thing was all of our fuel out there was provided by Saudi Arabia. All that fuel came right out of their bunkers.

EM: You mentioned the mining problem.

SA: Yeah, well, you know—

EM: You want to go into that, Schwarzkopf's and your view?

SA: I wanted my patrol assets farther up in the Gulf. My EP-3s and P-3s, I wanted them farther up in the Gulf. I would protect them with the CAP stations, because I wanted to know more about what was going on the surface of the water up there. We knew that there were some mines out there because we had seen some floating mines, and we weren't sure whether they were all old leftovers from before and had just broken their moorings, or were there some new mines. And we were absolutely convinced, because we had a couple of snippets of views from the radars that said they were dropping mines out there.

About that time, Schwarzkopf backed us out. He said, "No, you can't fly anything north of—" I forget what the line was.

And I said, "Well, that doesn't work for me. I've got assets that I can protect that need to be up there to see exactly what's going on, because if we have to go up there and don't know where those mines are, we are going to be in a mess." I said, "If we can find out where their fields are and how they are laying them out, and if we get some good photos or something of what kind of mines they are using," or whatever, I said, "it's going to make it a whole lot easier for me to accomplish the goals that you want me to accomplish."

"No, you damn Navy guys. You always want to start the war." [Laughter]

I said, "Well, mining is an act of war, yes General, but my objective is not to start the war. My objective is to be able to execute our war plan when and if necessary. And yes, if I see a guy laying a mine, I would like for you to be able to tell me that I can sink him. That would be the best of all worlds."

EM: So there were no hostilities going on at the time, right?

SA: No. "All you Navy guys want to do is start a war."

I said, no, no, no. We went around and around. I said, "Look, say you're out in the desert in your tent, and this guy comes around and lays a daggum landmine on the front flap of your tent, and you see him do it. What are you going to do about it? Are you going to shoot him, or are you going to let him lay that mine there, so that you have to figure out how are you going to cut your way out of the back of the tent?"

"Ah, that's not a fair comparison. This is totally different."

I said, "It's the very same thing. That's the exit to my tent up there, and I want to go out there and do it."

EM: But you weren't even talking about taking out those mines. You were just talking about surveillance and what's going on.

SA: I gave him two things. I said, "First of all, I would prefer to be able to sink the guy if I saw him carrying mines and one go over the side. As soon as I saw one go over the side I would like to sink him. But if you can't give me that, give me the ability to at least see what he's doing and where he is putting them, so then I can make my plans for how I get through there when and if we need to do that. It just seems to me to make a lot of sense."

"Nope, nope, you are not going to able to do that."

EM: I guess he was very concerned about having everything in place ready to go at the time he wanted to kick it off.

SA: Yeah. He didn't want anything to disturb that.

EM: Did you get a reaction from him once *Tripoli* [LPH-10] and the cruiser [*Princeton,* CG-59] hit the mines?*

SA: No.

---

* On 18 February 1991 off Kuwait, the amphibious assault ship *Tripoli* and guided missile cruiser *Princeton* struck Iraqi-planted mines that caused significant damage to both ships and injured sailors, but successful damage control efforts prevented the ships from sinking.

EM: No. You didn't say I told you so?

SA: No. No.

EM: So what intel did you have once you went into the minefields? Not enough, obviously.

SA: You remember that we had a couple of frigates up there going around, close into the coast. In fact *Nicholas* [FFG-47] was the first one, the first captives of the war were [captured by the] Navy.

EM: On that island, right?

SA: That all sort of all got forgotten until later on there was an investigation against the skipper on there, that maybe he had done something wrong. In that process, we had found some copies of the mine plan. Now, despite what Schwarzkopf didn't want me to do, we still did as best we could from afar. And from afar, we had described the minefields that we thought were there. When we compared that estimate to the information on documents we captured from Iraqis, it was a close match.

EM: This all happened before *Princeton* and *Tripoli* hit their mines, though?

SA: Yes.

EM: But my impression was they didn't know that they were in minefields.

SA: We were still offset. We didn't have the right distance. We had the right configuration, but they were off by a couple of miles. If we would have had the precise locations, then we would have, but—we were close. But close was only—

EM: Good in horseshoes and hand grenades?

SA: That's right. So luckily we escaped what could have been a lot worse.

EM: There were no deaths; three injuries, I think. Going back to the multinational embargo patrol, things had been pretty well set by the time you got out there, but did you have it down to a routine? How to inspect the ships? How to stop, search, board, and the rest of it; was that pretty well set when you got there?

SA: Yes.

EM: Did you get any interference from Washington about any particular ship that needed to be stopped, do you recall?

SA: We had a lot of interest in—remember the ship that was headed into the Gulf that was going to go up to a port in Iraq that had a bunch of activists on it?

EM: Oh, right. The *Ibn Khaldoon?*

SA: The *Ibn Khaldoon.*

EM: Yes, right, that's it.

SA: So there was lots of interest about how we were going to get that thing stopped and how we were going to get it stopped without having a big incident, killing women and children. So there was lots of dialogue about how we were going to handle that. In the end we were able to get the ship stopped and take it into Oman, and that issue went away.

EM: No one was hurt?

SA: No one was hurt. A couple of ladies yelled at our guys, and a couple of the guys yelled at them to shut up, stuff like that. It all turned out to be okay. The more interesting thing that happened right before the war kicked off was the overrun of our embassy in Somalia when we went in. The ambassador had been screaming. Now this is a case, this is something that ought to be talked about again in today's environment of Benghazi [terrorist attack on the U.S. consulate in Benghazi, Libya, in 2012]. Because here we had an ambassador [to Somalia] screaming bloody murder in there that all the other embassies had been emptied. "They are all here on my turf. The guys are coming over the wall. They are going to kill us all. Can't somebody come and save us? Who is going to do it?"

And he had been warning of this for about a week that this is how this thing was going to develop. We were watching the coverage and, I'm talking to Riyadh about it and I said, you know—

EM: Wasn't Somalia in your AOR?

SA: Yes. Somalia had originally been in the CinCPac AOR, but now it was in the CinCCent [Commander in Chief, Central Command] AOR.

EM: Oh, right.

SA: So I'm waiting for somebody to say something about what are we going to do. The Marines down there on the flagship that was still out in the IO at the time, because they had come back from that exercise and started laying plans, saying we can get in there and get them if we can be detached to go. But we need tanker support. We've got to in-flight refuel those helos on the way, and that has to be C-130s.

In any case, Riyadh couldn't seem to—they couldn't put it all together so I said, "We can do this. I'm going to have to break a couple of rules relative to how far away we launch helicopters, but I think the issue is in extremis and we should take the risk."

And the Marines assure me that they now have some Marine C-130s that can come down into the area and refuel them en route, and so we can get there and get back.

"Okay, but you know the ground war is going to take off. I don't want these guys stuck out there. What we will do, we will go in and pick up as many people as we can and get out of there."

EM: Yeah.

SA: We are not going to send a raiding party or anything else. We are going to go in and do an extraction. So we launched those guys a long way out. They got refueled. They took some extra tanks with them, fuel tanks, and got them in there. The interesting thing was we were planning it. I had been to the embassy in Somalia in our RDJTF days. It's amazing how this all keeps rolling around.

EM: Yep.

SA: What the guy was describing to me didn't sound like what I had seen.

EM: The embassy?

SA: The embassy, and lucky enough, I think the Marines ended up telling me it was a lance corporal on the flagship down there who happened to see the plan or something, said that's not the embassy. I was an embassy guard there, and here's where the new embassy is.

EM: The old lance corporal picked up on that, that's good.

SA: So that reaffirmed what I was saying—jeez, what they are describing? I don't remember all that that the guy did. So luckily they had this young kid down there who had been there too and was there where he could observe what was going on. So we went to the right place, or we would have been downtown at the wrong embassy. They probably would have had people being dragged through the streets a lot earlier than we did have.

EM: We made that mistake in Sarajevo—not Sarajevo, but Belgrade. Remember we bombed the Chinese embassy and we thought they were somewhere else?

SA: It's amazing, it's just amazing in this day and age that we still have those episodes.

EM: Human error.

SA: Yep.

EM: So you pulled out all the evacuees in one fell swoop, or did you have to make more trips?

SA: I think we got them all out at one time. We had a baby born on one of the ships on the way to Oman. We offloaded them in Oman. Never heard back from them. I sent a little note to the parents saying someday give your baby—I forget if it was a son or a daughter—this little note of where he came into the world. I was very proud to have him born on our ship.

EM: Maybe they named the baby after the ship?

SA: Yes, maybe.

EM: Going back, this is before the war now, but relating to joint things, why was the Navy so reluctant to play in joint activities before the Gulf War? You had Goldwater-Nichols in '86 and then the war in '90–'91. Not taking part in exercises in Florida, like command post exercises and things of that sort. Was there a Navy reluctance still to get into CentCom in a big way, or not?

SA: Yes. Yeah. One of the things you remember is that John Lehman was always very much anti-Goldwater-Nichols. And so Navy leadership got their mindset saying the

SecNav, who was really a good SecNav and a bright guy and understood things, was not much in favor of all of this, and maybe with good reason. Maybe we would be putting our assets off on do-little things when we should be thinking big, grander, strategic things. Too much dialogue just about CentCom, we've got Russia. We've got China. We've got Korea. We've got all this and right now CentCom seems to be not one of our highest priorities. All these little exercises that have that kind of focus are probably not things that are of much value to us.

I had pretty much been in that same bathwater myself, until I got into the middle of that RDJTF thing and then saw how it rose to be a CinC, and saw what was going on and what the ramifications were of things really going wrong in that AOR. But again, from a Navy standpoint it's a gulf and a lot of sand all around it. [Laughter] That's not of much of interest to us, and I think we just didn't believe it was in the grand business of our concept of operations that it was that big a thing to be involved in.

EM: It's interesting to see the intersection of different eras too, because we were still in the mindset of the Cold War.

SA: Yes.

EM: Even though the Cold War was over in '91.

SA: Right.

EM: Or by then. We hadn't really made the mental transition to a new era of littoral warfare.

SA: Yeah. And, when I was Vice Chief and we were finally starting to talk about littoral warfare and how important it was going to be for the Navy in the future, basically what I found out there was that the FFG-7s [*Oliver Hazard Perry*–class frigates] were the only thing that I had that could really play close, especially close in. But we had lots of things that could play relatively close in. And then I started reflecting about where we had been since World War II, and even in a lot of World War II the fleet had already been doing littoral warfare for a long time.

EM: Yeah.

SA: We just didn't think of that term. We thought we came racing across the ocean blue and did our thing and steamed back into the ocean blue and off we went.

EM: Right.

SA: We didn't think that really we spent all of our action time in the littorals.

EM: Well, this is not my interview, but that always struck me, that we were always planning for the big fight with the Soviet Union, which never happened, and yet we fought Korea, Vietnam, and a million other things that were sort of off the agenda.

SA: Yeah.

EM: They would pop up.

SA: Yep. My thing was, when I came back—and we will talk about this during my Vice Chief time—we really had to start thinking joint, because now there was no other option: we were going to have to do things jointly. And basically, when it came down to the simple things like numbering of our staffs, an Army staff guy didn't understand who the three [staff officer responsible for operations] was in the Navy compared to who the three was in their staff. So we had to do something. Because again, going back to the subject we talked about earlier, Navy guys think in three dimensions all the time. No matter what their warfare specialty is—surface, submarine, or air—they don't go to sea without thinking about, is there a submarine down there? Is there an airplane over there? Is there another ship over there? And where's my next gas station, and where's my next load of groceries? We are thinking in those dimensions all the time, and our other services don't necessarily. They think we are here. We've got to go there.

EM: Well, the Army in particular.

SA: The Army in particular. So we have an advantage in that mentally, our minds are tuned to those three arenas automatically without even trying to think about it. We have that advantage and we don't take advantage of it.

EM: Okay, you often hear that air power won the war; many wars, but in particular the Gulf War. I think folks down at Maxwell, the Air War College, would say that air power was the predominate factor for our victory in the Gulf War. Is that accurate?

SA: From the standpoint of ending it so fast, yes. We had such unfettered access to the targets after the first night, second night, that basically we didn't give them time to think. But the fact was they had to have another worry in their mind other than, what are we going to do about these airplanes. Every time we stick our neck out, somebody is going to drop a bomb on us. The real factor was there were a whole lot of people with a whole lot of armor sitting on the sidelines, just chomping at the bit to come racing across the desert to wipe [the Iraqis] off the face of the earth. Basically, I think the fact is that the air was a great enabler [for the Americans and allies], but the real issue is that there was enough force available to do anything that we wanted to do if we had to do it. If all they thought was going to happen was airplanes, they would have figured out a way to deal with it.

EM: So you think without the ground component, air alone wouldn't have been able to drive them out of Kuwait?

SA: No. Well, air may have driven them out of Kuwait, but not necessarily so. I think the fact is that if it was important to stay in Kuwait, they would have found a way to stay. If it was just air. Because they would just keep digging and hang in there. But the fact was that they didn't have enough breathing room to get that act together before they knew somebody was going to come racing across the desert and chase them out. So, you know, air as we've seen in Vietnam and we are seeing today—and we've seen in Kosovo and other places over the years that air can do certain things—but if you want to change the geography of the landscape, you've got to have ground people to do it. Because you can eventually hide from air if it's important enough; you can do it. You just have to change a lot of things. You can't be hiding tanks with hot engines in a bunker in the desert. [Laughter] That ain't going to work.

EM: What weaknesses did you find in the naval forces set up? What aspects of your operations could have been better? What equipment?

SA: Well of course, the number one ticket always is mine warfare. We still don't have that business figured out. We are not good at it. We don't pay enough attention to it. We don't think we can afford to do it the right way. We made all of these agreements with our allies that we would let them do that business, and then when it came time to execute it, we wanted to be in charge of it and not listen to them as much as we probably should have. So we ended up with our brand new minesweepers out there, and they were nothing compared—

EM: Not up to the job?

SA: They were not up to the job. And so the stuff that the Brits brought down and all was much better than what we had. Mine warfare was a problem, and it cost us a couple of ships there that were off the line for a while because of them.

EM: What about control of the air?

SA: Control of the air, we found that even though we had exercised a lot with AWACS over the years, we really didn't—the smart people that were in the antiair business. The Aegis [antiaircraft defense system] guys and the E-2 guys still didn't know where the boundaries were between them and what AWACS was covering. And the data links that were distributing the information to the various components weren't compatible at all times, and so what we had—for example the time that we had the sweep [air attack] that came down out of Iraq and came right down the coastline of Saudi Arabia, which we thought was going to turn into one of the oilfields and could have turned out to the fleet and been on top of us in a millisecond.

Now, Aegis had seen him. The F-14s had seen him. The AWACS had seen him. E-2s had seen him. But they [Iraqi planes] were on a boundary, and so there was a time delay in the targeting information so that the targets didn't fuse, the whole system didn't fuse on that as the target. So basically, AWACS with the F-15s from the Saudis, they were on the same sheet of music, no time delays and no boundary, and so the Saudis were able to get there with the first shot. Now, the first shot should have been several minutes ahead of that by the F-14s, but they weren't all in the same link, and so the power of the joint exercises, the more you do those things, eventually you are going to find those things out before—we found it out the hard way.

And we should have been able to find that out. We'd been out there long enough that we should have known that we had a boundary issue. I think Tom [Thomas F.] Marfiak on the cruiser, I think he had indications that this had been going on for a while but hadn't had the time or hadn't been able to pinpoint it. But I think he had a suspicion that there was something going on there that they didn't know. So again, that's the power of exercises, when you can do that, and in a timeframe that is not so severely compressed with yes or no, shoot or not shoot, but pretend you shot and pretend you don't shoot, then go back and reconstruct it. Somebody would have found the fact that, yes, everybody had that contact, but not everybody was synced up time-wise in a link where everyone was confident that they were looking at the right thing.

EM: Did the Iraqis discover that weakness, or it was just happenstance that they came down?

SA: I think it was happenstance, but you never know.

EM: They didn't get that much further, but you don't know.

SA: You just wonder. Did they know something that we didn't know?

EM: You took corrective action afterward to change that with the Air Force?

SA: It took us a while to figure out what happened. I think that part of the game was over by the time we finally figured it all out.

EM: Yeah. Did you put different CAPs closer to Saudi Arabia then?

SA: We layered our CAPs. We had the F-14s out farther, and then we still had some F-18 CAPs. You know we started that effort out there with the Canadian F-18s augmenting the Marine F/A-18s initially, until we got the F-14 carriers in with the F-14s. So we made some changes relative to CAP stations. Hopefully that wouldn't happen again, but I was convinced that we would have got the first shot at those guys, and I was really surprised that we didn't.

EM: They only did that once, right?

SA: Once. They had been probing us. We used to laugh about these exercises that they would run, where they would launch a lot of airplanes up there and they would orbit around, and then all of a sudden they would pretend like they weren't doing anything, and all of a sudden they would turn and race down as fast as they could and then at the last minute, turn around and run back. So we had been watching these probes for a long time, and it gave us a little bit of false security that jeez, we've been tracking these guys all of the time. We know where they are and if they really continue, we are going to be more than ready for them. Then when they did it for real, we weren't.

EM: You mentioned the Canadians. But if you want to speak a little bit about the allied support we had, allied naval support—who were the good players and the ones who really made a difference?

SA: The Brits of course made a big difference. The Canadians didn't have that much out there, but they were very easy to work with, and their commander and I got along really well. He ended up being at the embassy in D.C. after we came back from the war. The Brits contributed not only with the Royal Air Force and the Navy, their Sea Skua missiles off of their helos were outstanding, just absolutely outstanding. The Brits, when I went to the flagship and met up with Chris and—

EM: Chris who?

SA: Commodore Chris [Christopher] Craig.

EM: British commander?

SA: A British commander, two-star. I was amazed. I went aboard his flagship, and all the mirrors had tapes on them. All of the silverware and presento stuff was all stashed.

EM: They were ready for war.

SA: All the silver that they had on the ship that they had shipped back to England in boxes, preserved, so if they came back the silver would be there. I mean, they were on a wartime footing. They took it serious.

EM: Weren't your ships in the same boat?

SA: No, they weren't. No they weren't, but I came back and, "I've just had a revelation here. We had better pretend like we are really at war." [Laughter] We've got to think about this a little bit harder. This is not a peacetime operation. We had the minesweepers. We had German minesweepers and Brits and—

EM: Afterward, you had a whole lot.

SA: Yeah. Had a whole bunch afterward.

EM: Speaking of the Brits, at one point there was somewhat of a testy exchange in regard to sweeping the mines, and the British commodore, he wasn't quite sure the American plan for clearing the minefields was sensible. Do you recall that?

SA: Yes, yes. I went up to the flagship once I heard that there was a flap going on. I went up and sat down with him, and we talked our way through that. The Brits wanted a couple of things. First of all, they wanted an EA-6B [electronic countermeasures plane] dedicated to that force.

EM: For minesweeping? An EA-6B?

SA: Yeah.

EM: In what capacity?

SA: To jam any radars that might pop up on the coast with a guided missile.

EM: When they went into sweep? Okay.

SA: To sweep.

EM: Right.

SA: EA-6B became everybody's favorite friend out there. So he wanted air cover. He also wanted better knowledge of the minefield, you know, "Why can't we find better knowledge of this field? I'm a little afraid that we are going to start in the middle and work our way out, and I'm not sure that's our best effort."

So there were some geometrics about where should we start the thing to make sure we are starting in clear water, and the process of doing the sweeping itself. What kind of patterns and stuff should we use. So they had valid concerns. Their mine force was a much more professional mine force than ours. These guys had thought about this a lot more than our guys had. And that was the burning issue: "Why are we under the thumb of somebody that doesn't know how to do this as well as we do?" So it was a little bit of a manhood issue and rightfully so, I think. But we were able to finally resolve it, and they were happy as long as we met some of their demands.

EM: You didn't put the British commodore in charge of the minesweeping effort?

SA: No. I think we still had the guy on *Okinawa* [LPH-3], I mean *Tripoli,* the guy up on *Tripoli* I think was still in charge, the commodore there. Yeah, I think that's the way it worked out.

EM: What did you have the Canadians do?

SA: When we had the big oil spill, the Canadians had some airplanes, as well as the Coast Guard, that were able to do some surveillance up there that could map the flow of the oil. Canadians also, I'm trying to think what else they were doing. Of course their CF-18s were fully integrated with the air. They ended up being closer to the Marines than they were to the Air Force, which I think sort of surprised them.

EM: Didn't they run the "Piccadilly Circus"? Wasn't that the area where you had all the support ships gathered?

SA: Yeah, we had a protective ring around the logistics guys where they returned. The Canadians were good folks to work with, as always. I'm a little worried what's happening know up in Canada.

EM: Change of government?

SA: The change of government.

EM: What about the Australians? What did they do with you?

SA: The Australians were great. They did a lot of logistics with us. They also were able to—later on, because of the connections we made, we were able to trade Budweiser for Foster's on beer days, which was a very big plus on that side of the business. The Australians were very, very willing to get right in the middle of things. They were not bashful about whatever you wanted them to do, they were there to do it. They were a delight to work with. I spent some time afterward in Australia. My first visit to Australia after years of thinking that as a WestPac sailor I was going to get to Australia. I finally made it. [Laughter] Treated very nicely by them, and it was just like working with our guys, our ships.

EM: Did you have any trouble with any of our allies, including—we had Arabs, Saudis, and Kuwaitis?

SA: I had some problems with the Saudi Navy guy, and he couldn't come up on any of our circuits. We could only work with him in HF [high frequency], and that was very hard. We weren't up on HF that much. Didn't want to be up on HF that much, and so being able to share some of the data that he wanted, the only way we could get it into an HF format was to manually transmit; you know, do a manual transmit, decode it instead of sending digits, do an analog, basically, exchange. And so he was very frustrated that he couldn't get as much timely information as he wanted for his forces. I said, "If you can get the comm we can try to do better, but what I will do—" I tried to give him a special report each day that he could use to get a better understanding of what had transpired.

He was never totally happy and I wasn't totally happy, but we both walked away saying, well, we are both trying to do the right thing. And I gave him assurances that after the war we would do our part to help him get his forces into the right configuration so we could work better together. And I'm hoping that did carry forth. He was a very aggressive guy. He was, and the Saudis tend to be two-sided. Some of them keep their cards very close to their vest, and they will be aggressive on a very particular point, and then they will be very passive on other things. This guy was always in the attack mode, all the time. He was really working the problem hard. It was okay.

EM: Did you have any particular problems you recall with any of the allies?

SA: The first problem I walked into almost—shoot, I don't remember how many days it was after I took over, but Hank had scheduled a meeting on the French flagship, and it was time for the head of the WEU [Western European Union] to change hands from the Brits to the French. The Brits play in the WEU to protect their own interest. They don't really promote the WEU. And the French play in the WEU because they want it to be theirs.

EM: Does it still exist?

SA: I don't know.

EM: You never hear much about it.

SA: I think it's still there. So the first conference that I was going to attend, and meet all of the other people from the other navies out there supporting us, was going to be on the French flagship. And the topic of discussion was the change of leadership of the WEU. [Laughter]

EM: Didn't really relate to the Gulf War?

SA: Yeah. I'm saying, oh, dear me. We were able to sort of get along, and again, he was hampered by communications. And so my inroad to him initially was, "Have you ever met Admiral LeFebvre?" Oh yeah, back several years ago. I said, "Yes, I had him aboard my flagship when I was in the Indian Ocean on *Ranger* and let him go flying in an F-14.

We had a very good relationship, and I hope that we can continue to build the same while we are out here together." So I had a little entrée to say the French aren't unknown to me even though I'm a Pacific Coast sailor. I have had interface with your Indian Ocean guy, and we had a very cordial relationship, so I said I hope we can do the same. If there are things that I can provide you that you don't have, I will be happy to do that. Again, he was sort of a HF frequency guy, and of course I said part of our problem on being able to meet all the HF commitments out here was basically, my Navy has done away with HF.

EM: Because it could be easily read?

SA: Well, because it could be easily read and we were transitioning to all the new stuff, which was all UHF/VHF [ultra high frequency/very high frequency], and basically we'd started to lose all of our HF radios and people who knew how to work HF radios. So I said, "We are at a little disadvantage on being able to meet all the commitments for HF, but if there are things that I can do in the UHF world where you need another transmitter, or something like this that we can work with, we will try to do that." So the WEU sort of laid under the surface, and they didn't try to push it too hard, and I certainly didn't try to encourage it. Again, he was willing to do almost anything that I wanted him to do. He gave me what his druthers were, what they could do. Biggest problem we had with all of the people out there was national command authority ROE. They had to go back to their national command authority for certain ROE things, like if I was under attack, they couldn't help me unless they got approval. It wasn't an automatic thing. They would have to wait until they got an approval. And so that sort of dictated where we could put people in certain stations. Those people had those kinds of restrictions—you had to go back to national command authority, or basically just could not defend another unit.

EM: What navies could operate in automatic defense?

SA: The Canadians, the Brits.

EM: Australians?

SA: Australians. We were all on the same page. Now, you know we had some South American navies. We had ships from Argentina.

EM: Argentina, yeah.

SA: Their CNO came out and said, "I can't do much for you, but can you do things for me?" I said yes, what do you need? He said, "I can do logistics and I can sort of do that corral thing [logistics area]; can you give me credit for being a part of the program?" I said absolutely. I said, "We need everybody out here, and you can rest assured that I will give you messages that you can play back to your leadership that shows that you are here and very active in doing the right things."

EM: If you could give a proportion to the operational or the political benefit from our allied contributions, what was more important? The operational capability they brought or was it the fact they were all in this with us?

SA: The latter.

EM: The latter. Okay.

SA: Yeah. Although, for example, I would not have wanted to do without the Lynx helos on the British ships. I mean they had that figured out right.

EM: Lynx and Sea Skua.

SA: And the antiair missiles on the French ship. We had one missile fired at the minesweepers, and they were very clever about it. They waited for an opening with our shotgun ships [destroyers and cruisers] protecting ships our battleships. What we did was, we had them one in and one out with the battle ship, so that there was always a guy—

EM: Close to shore?

SA: Close to shore facing the threat. What they ended up doing was, as we had the turnover guy out here turning and this guy coming back this way, they shot the missile.

EM: Did they think the Brit wasn't as capable of shooting down?

SA: I guess so.

EM: Yeah.

SA: The Brit got the shot; I mean he was right on the money. The Brits, as small as they now are, they are very professional navy as they always have been. I got a better appreciation for that during that evolution out there that, yeah, if you go to war, you would like to have those guys with you, because they can still teach us things and we ought to pay attention.

EM: Can you speak a little bit about the inshore types of operations. You had *Nicholas* up there, you had the Brits with their Sea Skuas and Lynx helicopters. We had MH-60s. Taking the islands, remember the small islands where we got the first prisoners, and SEALs of course were involved in that too. Were you happy with that evolution?

SA: Yeah, but you know of course for our FFGs, we ended up putting the Army helos on them because again, we didn't have the—

EM: Armed helos?

SA: Armed helos. We are still dealing with that issue. I mean we finally got helos armed, and the attack guys even today are going ballistic because they don't know why they should have a daggum armed helo around.

EM: You know I asked you that question. We [Naval History and Heritage Command, then called the Naval Historical Center] interviewed you back in '92 or '93, I think, and we are still working trying to get armed helos. So that's still an issue?

SA: Yeah. Although they've got them now. We've got armed helos, but the battle group guys don't like it very much. They don't know why we are wasting time with armed helos. The in-close guys, I was impressed with the FFG-7s—at least the guys that I had out there had a mine-hunting sonar so that they could paint the mines fairly well. God bless them, the skippers were telling me that, yeah, they could back their way out of a minefield if they had to. I said not with me aboard. [Laughter]

In the budget wars that we were fighting back in the Pentagon—a lot of that was on the margin, so I don't know whether we have done anything to improve that. I assume that this module that they are trying to put on the LCS [littoral combat ship] is coming aboard with an integral mine-hunting capability, but I'm not sure about that. I bet it's a 50–50 toss of the coin.

EM: What do you say, do you want to call it a day?

SA: Okay.

EM: We will pick up tomorrow. There are more questions I have on the Gulf business tomorrow. We probably won't get into your time as VCNO. In fact, I've got to do some more research on that period, but just so I can start doing that research, what items do you think I should really focus on? What were your hot button items when you were VCNO?

[End of 27 October 2015 interview]

Interview Number 6 with Adm. Stanley R. Arthur, USN (Ret.)

Date: 28 October 2015

EM: Admiral, we were talking about the Gulf War yesterday.

SA: Yes.

EM: I've got a number of additional questions. I did some reading of my book last night.

SA: Okay. I've got one correction. I was thinking what I said about the ATO, and I think I said 24-hour planning cycle. It was really a 72-hour planning cycle.

EM: Oh, right. Okay. Yeah, that's right.

SA: I wanted to make sure. I woke up this morning and I said I think I said 24 when it's really 72.

EM: That's a long time.

SA: Yes.

EM: For emergent targets. One of the questions I had was the TLAM [Tomahawk land attack missile] shoots after the second of February 1991. Schwarzkopf called them off? Told you to stop the TLAMs?

SA: Yes.

EM: Why was that?

SA: We had the press coverage and then the shootdown of a couple of the TLAMs as they came down the highway, and so they said okay, that's enough of that. Let's forgo doing any more.

EM: Were supplies running low of TLAMs?

SA: No, we still had plenty of TLAMs. I think they thought that the code had been broken on how to deal with the Tomahawks, so they had sort of lost their effectiveness, in their mind. Part of the lesson learned from our standpoint was, I don't think anybody had looked ahead, far enough ahead on the planning cycle about CinCPac and CinCLant relative to the fact that you've got to start changing the routes. You can't use the same routes every day.

EM: Isn't that the same with any kind of military evolution?

SA: You just can't say use the same route and then divert off of it for the final runs. And that was pretty much what they were doing. Planning table was, they had their marks back there with the way they did it. I don't know if you had more questions about TLAMs.

EM: I do, yeah.

SA: We will probably get to it later.

EM: Well, another question I have is the TLAMs, were they accepted readily into the ATO? The planning process?

SA: No, oh no. It was a hard fight to get them in there. That was not in the ballgame. And Hank had worked this issue hard before I got there, and then I had to continue the labors and we got the TLAMs in. And quite honestly, I was still very nervous about—I knew they could arrive at the target; I was really nervous about how effective they would be on the targets.

EM: What was their reluctance? Was it Air Force guys?

SA: Air Force, yeah. They had their air-launch stuff that they wanted to bring in, and of course their systems didn't work that well, as far as we could tell—the B-52 stuff that they brought.

EM: The air-launch cruise missiles?

SA: Right. But the TLAMs really did a good job, and especially on that first night when basically the F-117s [stealth fighters] didn't get many into the target area because they couldn't see them. There was overcast, and so the TLAMs really, that first night, delivered most of the punch.

EM: The F-117s didn't have enough clear weather to hit their targets?

SA: They were laser-designated targets. I forget what the final number is. I think Marvin Pokrant in his book has it, or I've seen it someplace, where how many of the 117s were unable to get to their targets relative to releasing on them because of the weather. So the TLAM was very effective. We probably overhyped it at the start, but when we did a very rigorous—against the wishes of my favorite Navy, I had ordered, after I left the theater, I had CNA [Center for Naval Analyses] come out, and I said, "I want a complete reconstruction of every Tomahawk; can you do that?"

They said, "Yes, you will have to work some magic on some of the clearances and stuff, but we can do it." They did, and the Tomahawk ended up being much more effective than we thought from the warhead standpoint, and from their ability to reach the target. We found out the root cause of the ones that missed the target, and it goes back to the planning tables, back at the CinCs where they do the planning.

EM: The warheads were powerful enough? I had heard that they weren't.

SA: They needed to be better, but the fact that the missiles you were using for the headquarters buildings and stuff like that, were the right weapons for the primary targets that night. They came in vertically and punched right through the roofs. Now, if we had a much bigger warhead they could've done a lot more damage.

EM: Yeah. Was that done later?

SA: I'm not sure where we are with the Tomahawk. They were clear up to Block 4 when I last knew of them, and yes there's more punch in the warhead.

EM: What did your fellow aviators think about cruise missiles? Did they think it was a diversion from a true attack with aircraft? A complement?

SA: I think it was a complement. I don't think anyone was worried that they were going take away the mission, but they were a nice complement. The beauty of it is you could launch them anytime, as long as you had the planning done. So if you wanted to keep up the pressure during periods where you wanted to take a little break, you could use the Tomahawks to keep their heads down. The thing that we learned—we did a practice before we did the first launch. We pretended; had the force pretty much in the same alignment we would use on that first night, and then we launched, did simulated launches of the Tomahawk and took a look at where the routes were relative to where we had our ships positioned and everything.

We decided well, that was good. Everything seemed to work all right. Then on that first night when we launched them, we almost had the first one shot down by one of the British Type 21s because the Tomahawk was such a new thing and still pretty much unknown to even our allies. We hadn't divulged everything about when we were going to launch them. So the real launch occurred, and they [British] weren't aware. The first thing, they ran in with their automatic weapons systems in full engage mode all the time instead of having the engagement being manually controlled. So the first thing that happened was one of the Type 21s, his weapon system locked up on the first Tomahawk going, and luckily the operator down there in the war room of the destroyer—

EM: Pushed cancel.

SA: Yes. He recognized, oh that's something going that way. It had to be from those guys, and he put it into safe mode.

EM: Can you just describe to me your feelings that first night, when everything—all the preparations you've helped put together, you are ready to go. The war is about to begin. Of course, we didn't know how it was going take place. Where were you? You were on *LaSalle?*

SA: I was on *Blue Ridge.*

EM: What was your feeling?

SA: I was pretty apprehensive. I thought we going to kick off something here, and there's going to be a lot of things coming our way shortly.

EM: Air and—

SA: Air and missiles. I thought, you know, they would probably find some way to get some missiles down our way. Everyone was worried about gas attacks, and so we all had our gas masks ready. And of course the first thing that happened was, once we started the launch of Tomahawks and everything was a go, the first thing we had—I forget which carrier it was—the guys were talking through their gas masks, and they thought there was a Scud missile inbound, and so we thought, well, here it comes as advertised. Finally we got it sorted out that there was nothing incoming. But yeah, everybody was pretty tense, and we certainly expected to see something coming back our way very quickly, but that didn't materialize.

EM: So you didn't anticipate it being a quick victory?

SA: No. No. I thought once we decided to go, we would be at it for quite a while, because everything was based on the air campaign for the first stage of it, and then eventually the CinC would make the call for the ground to go, and so I thought it would take a lot longer to get everything ready to go for the ground campaign. It worked well.

EM: When did you start to think this might not be as tough as you were anticipating—fearing, if you will?

SA: About three days in, I think. Yeah. They are not reacting. Now, are they holding something back? But all the indications were that they were overwhelmed and figuring out how to consolidate their positions.

EM: The air opposition wasn't significant?

SA: No. It was not significant. We still thought that they had the ability to give us some air opposition, but it wasn't materializing.

EM: Yeah.

SA: I forget how long that was. Finally they went east, and that part of the threat was gone.

EM: When did you take off the MOPP gear [Mission Oriented Protective Posture against chemical agents], the chemical and the gas masks?

SA: We all had it. I kept mine under my bed. And I stopped wearing it I think about the second day, but I kept it close by.

EM: What was your reaction when they shot one—I guess Silkworm, that hit in Dhahran and killed the Army reservists?

SA: Yeah, I think that was a Scud.

EM: Oh, Scud, okay.

SA: Yeah, I think that was when a Scud came down. That's why I wanted the flagship at sea and not alongside the pier in Bahrain. I was afraid of the same thing that happened in Dhahran. And I thought that they were finally getting to the point where they were picking out, able to pick out the targets they wanted to hit with the Scuds, and of course, they missed by just meters. If they'd hit one of the ammo ships that we were unloading up there in Dhahran, it would have been a whole different story.

EM: It landed close aboard?

SA: Obviously that's where they were aiming and it ended up close, but not enough to do any damage to the ships. But that was one of my main concerns about moving stuff in up there, until we really were sure that they didn't have the ability to launch any more Scuds.

EM: What defenses did you have against the Scuds?

SA: We didn't really have that. If the Scuds came out toward the fleet, we had our weapons systems on the ship that could deal with those, but ashore there wasn't that much. I mean the Army had their batteries—

EM: Patriots?

SA: Patriot batteries and the Patriots did engage.

EM: So it came over land then?

SA: Yes.

EM: Okay. All right, let's see. The battleships, could you tell the story of Schwarzkopf's reaction to the battleships' fires and the press?

SA: Yeah. We had the two battleships [*Missouri* and *Wisconsin*] out there, and my theory on the battleships was, we've got them here and the whole world knows that our amphibious tactics from World War II were we always were prepping the battlefield with battleships before we would launch the amphibs. And I said, these guys are going to believe what they have seen in the movies—we keep the battleships back until it's time that we really want to make them believe we are going to come ashore, and then we will get them into a position to fire.

Well, we had the mines that we had to deal with, so if we were going to go way far north, we had to figure a way to get through the minefields. So part of the deal with the FFGs going up in the early part was to see if we could really find our way, find out something about the minefields, something more than we knew at that time. And also to start probing, and so it would look like these guys were the first wave coming up and sort of taking a look, and we had one of the battleships come in as close as they could with the Saudis.

The Saudis were on land alongside the Marines, and there were some targets up north of them in Kuwait that we thought we could hit with the battleship, and so we started. They had a small harbor up there, and they had lots of troops, and we were tracking the food trucks that came down for the evening feedings.

EM: The UAVs were tracking these trucks?

SA: Yeah. So we finally broke the code of how to get the battleships in the ATO, because the projectiles went up to 35–40,000 feet and back down. So we finally found a way that we could account for the battleships in the ATO and they could tell us when we could shoot.

EM: So the Air Force had to accommodate TLAMs and battleships?

SA: Yeah.

EM: Not used to that.

SA: That's right. That was uncharted territory. So we were plinking along and we were knocking off some little harbor craft and destroying some of the buildings, and we think we probably interrupted the evening prayers and food. Of course the newspaper, the press was getting ahold of the battleships, you know, finally the battleships in there. So Schwarzkopf was not very happy that period of time, he says, "You've just got to knock off those battleships."

I say, "I want to shoot every round out of here that we've got, because they are never going to be used again, so why don't we just—as long as we can find targets, why don't—"

"No, no, no, you are not going to do that."

EM: Why did he want to stop?

SA: The information I had was he got a call from Powell in Washington that says knock off all of the press on the battleships.

EM: Getting too much press?

SA: Too much press. [Laughter]

EM: All right. Although the use of the UAVs for targeting, for spotting, I'm not aware that that had been done before.

SA: It hadn't. This is one of the things going back to John Lehman, who early on, as we were talking about in my earlier days in OpNav, that was sort of his idea. "Hey guys, there's a way to make the battleship more effective." Of course he was still holding on to the battleships for a 600-ship Navy, and why not try these things out, which was an unknown concept for most. But the guys on the battleship had practiced out at sea before the thing kicked off, and they had these big red balloons that they would set adrift, and then they would steam off someplace and send the UAV up and have the UAV track them and feed back targeting data. And they would plunk out and see if they—with the 5-inchers, to see—"Yeah, we can, yeah, we can spot you using these things," and so they were feeling pretty comfortable that they could work it.

EM: With the apparent success of the UAVs in adjusting battleships' fires, that's 1991, and yet we really don't get a sense that they made a big deal until what, 2003, with the invasion of Iraq? I mean, there seems to be a lag period there. Were they not readily accepted as, hey, this is a new mode of warfare we've got to grasp and go with it?

SA: It wasn't accepted. There were a lot of things we did out there during that period of time that needed some real follow-up, but sad to say, a lot of the lessons learned just went unheeded.

EM: Faded away?

SA: I'm not sure, and I hold myself to blame for some of this. I thought people were paying attention. I thought that we wrote a very extensive lessons-learned package from there and pointed the finger at several things that we needed to do. The only thing that we really took to heart was the Tomahawk. And I have to tell you, luckily I came back to be Vice Chief, because by the time I got back to the building as Vice Chief, the CNA folks were well on their way to solidifying what they needed to do for the last piece of the effectiveness of the Tomahawk, but the opposition from everybody, almost everybody in the building was, "We don't want to give you that information if it was from the intel guys. We don't want to give you that information because we were using a lot of overhead sensors too."

We needed the information. But "we don't want to give it to you."

The systems command was, hey, we know all we need to know about Tomahawk. We are not interested. Everyone was afraid that the report was going to come back and say that the Tomahawks were not effective. They didn't want to hear that, so they just said everyone was clamped down. So when I came back as Vice Chief and the CNA folks came to me and said, "We are not getting the access that we need," I said okay and finally just sort of rammed it down everybody's throats. I said, "Here you are either going to give us the information or I'm going to raise a stink that you don't want."

EM: The first reports of the TLAMs talked about 85 percent effectiveness. I remember that term being used a lot. And then, I think it was Barry Posen up at MIT [Massachusetts Institute of Technology] came out with a study, and he said first of all, you couldn't determine that every TLAM that hit was just hit by TLAMs at the target. It could have been hit by TacAir [tactical aircraft] as well. His figures were lower than the 85 percent. Was he just a wolf in the wilderness, or did his conclusions gain traction?

SA: There were some who believed that, yeah, he may be right. The fact is that in the process, it was very easy to tell what was a Tomahawk hit and what wasn't. Very different.

EM: Craters?

SA: I'm not sure how much I can say on this, but a Tomahawk hit for most of the targets that we were shooting at, not all of them, but for the preponderance of targets that were the primary Tomahawk, it would take somebody about 15 seconds to tell you—look at this hole and compare it to this hole. This hole is a Tomahawk. This hole is a guided, you know, whatever. And from that moment on, you could look at the field of where the target was, and even if it missed the target, it was out in the field next to the building instead of the building. You knew that that was a Tomahawk, and that the other one was something else.

EM: Iron bombs.

SA: Yeah. It's just absolutely crystal clear.

EM: So from your experience in the Gulf War from that point on, were you a champion of TLAMs?

SA: Yes. Yes. They had a place. They had a role to play. You couldn't run a whole campaign with Tomahawks. The tradeoff in costs and all was not right, but if you had a specific important target that was suitable for that warhead and you needed it before you even sent somebody else over the beach, Tomahawk was the way to go. We had to look at what were the vulnerabilities of Tomahawk to being shot down. Part of that we learned, of course, don't send them down the same road all at the same time. And the other thing was they probably needed some more stealth technology in the system. That's where JASSM [joint air-to-surface standoff missile] came in, which the Navy never wanted to have anything to do with JASSM.

EM: What's JASSM?

SA: It's a joint air-to-surface missile. Joint program, Navy doesn't want it. Air Force loves it. I just happened to have them in my afterlife. That's where my part of life at Lockheed Martin was. We were doing missile work. We built the thing. But basically it's an air-launch system. I had proposed it as a follow-on to Tomahawk, because it's got some really neat features that correct some of the deficiencies we had with Tomahawk. So it's the next generation, and the Air Force has it and they've now gone to a long-range JASSM. I've been reading in the paper.

EM: The Navy is opposed to it?

SA: The Navy has just never been interested. They are satisfied to stay with Tomahawk, and there have been a lot of improvements with Tomahawk along the way.

EM: Backtracking a little bit to UAVs, at the time, you and fellow aviators, were you starting to think maybe the UAVs could take some of the role of TacAir, you know, piloted aircraft, or was that too far in the future at that point?

SA: No, I think a lot of us were thinking these unmanned had some attributes that we needed to look at, and after I came back we started looking. When I was the Vice Chief, there was lots of programs starting to move in that direction. So in the JROC [Joint Requirements Oversight Council] we spent a lot of time talking about these things. The hard part of the unmanned world is, one, the main selling point you hear from the people who have made up their minds that we ought to do away with all manned aircraft and do everything, is that now we don't—we are not risking any lives; the pilot is out of the equation. We don't have to train him. But the problem is the pilot in a normal aircraft is the backup system. All of our aircraft have got autopilots, and you can lock them up [the planes] on inertial and they will fly themselves in auto throttles.

EM: Even land on a carrier?

SA: Can land on a carrier. But the pilot is still the backup system. So either you build a system that has just the systems that already exist with no backup—and can you make them enough so that they are expendable? So now they truly are expendable, because more often than not they are not going to come back, because there is going to have to be some intervention by the pilot or a backup system to make sure. So can you make them cheap enough that you can pretty much put them in the class of expendables, or are they going to be so exotic that they need this additional backup system? Pretty soon you get to the point that, well, if I'm carrying all of this extraordinarily expensive, say surveillance gear, yeah, I don't want that to be expendable. I want to make sure that I can always get this back. Okay. That's no cheaper than a pilot.

EM: Although we haven't expended a lot of UAVs doing surveillance and firing Hellfires and various things in Afghanistan and Iraq. I mean, they are coming back.

SA: Yeah, they are coming back, and those are part of a pretty exotic system. Those are not cheap.

EM: What if we got into a fight with a nation that would have the capabilities to take them out? I guess that would be a different story.

SA: So that was one of the arguments—let's not get ourselves too wrapped up in the cost. In other words, if you really want to do this mission, don't think you can do it on the cheap with a UAV. You've got to spend the money so that it doesn't become an expendable system. Don't make the argument that the UAV is so much cheaper than a manned aircraft because it may or may not be, but it's not in the order of magnitude that people like to think. So you get through that argument, and then you start arguing about what are the right chores for this thing to do? Is it going to be an interceptor that goes out and intercepts incoming raids and does dogfights and stuff like this?

EM: Hmm, no?

SA: The fighter guys still want to do that, of course. Is it going to replace all of the attack airplanes? Well, can the UAV do close air support? Really heavy close air support? Um, I don't know. Do we know how to work 10 or 15 airplanes or UAVs in the same airspace at the same time and the same target? So my take was, the easiest role for them to assume and for us to get a lot of experience with them was recce. Use them for our reconnaissance and the long-range reconnaissance. Put the sensor packages on there that you want.

EM: You worked this problem when you were Vice Chief?

SA: We went through these arguments, yes. So that was where I wanted to start, was with a good long-range reconnaissance package for what the Navy would need to do their part of the warfare, and start working them into the ships. And as we did that, that was a mission that they could do well. We already knew that, because we had lots of experience with that particular use of UAVs. And then, as we grow in our knowledge, then back it down into the other warfare aspects, you know, ASW, strike, antiair. Because we will learn that way. You get so involved with the arguments. People want to go one direction all of a sudden, and I'm a believer in sort of take the easiest, what appears to be the easiest use, and one that's already had a lot of data on what kind of sensor packages and all you need.

EM: Evolutionary process?

SA: Do the evolution.

EM: All right. If you could talk about the effectiveness of the Navy's frontline aircraft during the Gulf War. The A-7, the A-6, F/A-18.

SA: Yeah. It became very apparent to me early on that the A-6 now was the A-1 from Vietnam. The A-6 was now that version for the Gulf War. That was not very popular among my A-6 brethren.

EM: Why did you think that?

SA: Well, the A-6s still felt more comfortable down on the ground. The Iraqis had a pretty impressive antiair defense system, and of course we dismantled the sensors and stuff pretty early on. But the fact is that if you wanted to go trailing in on the ground, there was somebody out there that was going to be able to shoot you, and I think right at the end of the war we saw the same thing with the Harriers. In the last few days the Harriers got engaged up in Kuwait and the "road of death." We lost a couple of Harriers up there. Again they were low, so my take was that we needed to get the A-6 replaced as soon as we possibly could. And that's again why the things like what we were trying to do with the A-12 and the A-X program were essential to replace that capability.

EM: Were the A-6s able to hit targets with, you know, PGMs?

SA: Yes.

EM: They had laser designators?

SA: By this time they had laser designators. But if you put them in the same airspace as an F-15 from the Air Force with their Strike Eagles, which had basically the same or better capability, it was no contest. The Strike Eagle was a much more effective program. So when it came time to knock down all of the oil wells after he [Saddam Hussein] set the oil well fields on fire, the Strike Eagles just knocked off the heads of the shutoff valves. It required Red [Paul N. Adair], the super guy to put out the oil well fires.

EM: Was that Red played by John Wayne?

SA: Yeah.

EM: Red Adair?

SA: Yeah. Basically you just put a big explosion right over the thing, and it would rob it of all of the oxygen with enough time for it to just go out. And so theoretically that would have been a great thing for the A-6s to be doing, but by that time the Strike Eagles were doing it.

EM: They were doing that to various oil wells?

SA: After he lit those oil wells fields, that's how we were knocking out those heads.

EM: Interesting.

SA: Yeah. The F/A-18 did well. The multi-mission role showed up right away. We had the incident of a guy, you know, shooting at a MiG and flicking the switch and going into dive bomb, bombing runs. That aspect, the dual engine of the F/A-18 showed that they were more survivable than we thought from that standpoint. Again, they take a lot of fuel to get very far. The A-7, again,. it performed, it did well. But if you had your choice between an A-7 and an F/A-18, you would probably take the F/A-18. We just wished the F/A-18 had some more legs to it, more staying time.

EM: On the F/A-18, were you involved as Vice Chief with the upgrades to the Super Hornet?

SA: Yes. Started that evolution back when I was OP-50—we were working toward the upgrade to a Super Hornet.

EM: To give it greater legs, more distance?

SA: More distance and more capability as far as ordnance and stuff. And that was a good move, I think.

EM: So as far back as when you were OP-50 you were involved in the upgrade to the F/A-18 E and F variants?

SA: We were starting, yeah. And again, I think what we learned there in the Gulf War made the E/F version a must to bring that fleet into today's environment. I haven't operated with them so I don't really know. I know when I was Vice Chief they wanted me to come out to Lemoore to speak at a gathering, so they sent an F/A-18 two-seater back to Washington and I flew out and then back. The guy that was flying me around was an F-14 guy who was getting to transition the F-14 guys into the E/F, and I was asking him his thoughts. He was very unhappy; he loved his Tomcat. All the Tomcat guys loved the Tomcat.

EM: Yeah.

SA: Earlier on, I'd checked out in the F-14 just to fly it a couple of times to see why they loved it so much. As this guy was flying, he says, "This F-18 is a pretty nice flying machine, it really is." But the thing that they were sorry to give up was the Phoenix missile. Of course, they never shot the Phoenix missile at anything, so—and that is a great loss. I mean, as you look at the scenario that people are talking about with China and all, you need those long-range shots.

EM: It's not still in the inventory?

SA: There's nothing now that can carry it. The F-14 was the only thing that could carry it.

EM: That seems shortsighted.

SA: I think I'm right on that, but I don't think the F can carry the Phoenix.

EM: There couldn't be an equipment modification to do that?

SA: I think the size, I think the size of it.

EM: Oh.

SA: We probably need to check that out. That's my interpretation. So we've still got some work to do. We really have to think ranges, still. It used to be you come out of World War II and Korea the word was, speed is life. And I remember Mel Paisley, who was a P-47 guy. He said, "You guys have got to erase that lingo from your memory bank.

Speed is not necessarily life. In some cases it is, in some cases it's not." He said, "Us P-47s guys didn't have much speed, but we could take a hit pretty good."

And we sort of learned that in Vietnam, that speed wasn't everything, because of the Viggies [Vigilantes], the RA-5Cs. Faster than a speeding bullet, but we lost them, and the Recce F-8s the same way. So being able to out-range the guy or see him first, all those things are key ingredients to what you need to build to the future. That equation changes with every evolution of the airplane and the weapons systems.

EM: Right. How effective were your carriers in the suppression of enemy air defenses? SEAD [suppression of enemy air defenses]?

SA: Good, because we had the HARM and they were very effective. We had several incidents. We had evolved that system so that it had tremendous maneuverability, and so you had to be careful that you had it locked up. That it was locked up on the right target, because left to its own devices it could sometimes see something back behind it and go. So you had to be careful.

EM: That happened, didn't it, during the Gulf War?

SA: Yeah. And people didn't believe it, you know. The first couple of incidents were that we had ships hit fragments, shrapnel, and they all believed that it was from a shore-based Iraqi system. Of course I was reading all of the message traffic, and we had the ship hit, and he was certain he was hit by something from the shore battery. Then I'm looking at another message about aircraft firing a HARM at an AAA radar on the beach, and I'm looking at the geometry and I'm thinking, there's something—

EM: No coincidences.

SA: I don't think it's a coincidence. So I was very anxious for them to get somebody up there and make sure they had somebody expert take a look at the shrapnel on the ship, and sure enough we found serial numbers, and we knew that it was HARM. I think we also had something with the Air Force that we might have clipped, but luckily there was no major catastrophe. But it showed that you really had to be careful with HARM. Part of that lesson I took, much to the chagrin of my VP guys, who wanted to get involved in using Harpoon, especially if the Iraqis stuck their nose out and wanted to take their ships over to Iran like they finally did—VP guys wanted to be ready to be able to shoot Harpoon at them, and I backed them out of that. They weren't happy, but I was too afraid that with such a target-rich environment, we would have a Harpoon wandering through the fleet.

EM: So what was the antidote to that so that you could use HARMs without fear of them coming against you?

SA: There was no such thing as all of a sudden you see a radar signal pop up on the horizon and you shoot a HARM at it. You really had to be aware of what that envelope was, so if that radar signal was here and you were firing it from over our units or any kind of system that we might have that could come up on the air, you had to make sure that they were out of the envelope. So it was much more the old "shoot from the hip" that we had done with Shrike in our Vietnam days. As soon as we saw a radar signal close to the target area, we pickled that off. With a HARM, you had to understand that if you had units nearby—

EM: So you had to clear everybody out before you could fire in that direction?

SA: Not necessarily clear them all out, but you may have to close with the target a little bit closer than you would want. In other words, the HARM had such a radius, had such a maneuverability radius that although you had it locked, you had the signal and it was seeing the signal; if it saw something similar to that out here where it had room, it could maneuver enough almost to the point where it could almost do a 180.

EM: Seems like a real limitation to the weapon.

SA: Well, it was. There, we really had to say, guys, you've got to really pay attention. You just can't shoot these things like old Shrike guys. This thing has so much more maneuverability that you've got to really make sure—

EM: Was anything changed, or does it still have that problem?

SA: I don't know. I think there were some changes so that you could restrain the missile from picking up other objects. Once you had it, target locked on not to be looking—

EM: Well, it's still in the inventory.

SA: Yeah. It's a very capable system.

EM: All right. Here's a more abstract question. Do you think that Schwarzkopf understood naval warfare or navies in general?

SA: No. [Laughter] He didn't, and he had a bias against the Navy, and the bias came from the standpoint—and I don't want this to reflect badly on him, because I greatly admired the guy. I thoroughly enjoyed working for him, and he had his thing; Stormin' Norman was an accurate phrase. But you just had to toughen up a little bit and everything was okay.

From the time that I was NavCent I always said the Navy ought to be in the rotation for that job, because that job had a lot of water AOR associated with it. The Navy had been in the region ever since the end of World War II with Commander Mid-East Force, and so there was no reason that the rotation for that CinC should be just Army and Marine. The Navy guy should get in there. Well, finally somebody, I guess, and I was not aware of it at the time, that it had got to this point. But when Norm had been nominated or in the process of being nominated to be the CinC, I guess at the last minute the Navy had made a major play with Hank Mustin for that role. And he [Schwarzkopf] didn't forget that. [Laughter] He made it clear when I got there. He said, "You know, the Navy really tried to get me out of this job."

I said, "I was not a part of that. But I believe a Navy guy should eventually—"

EM: Well one has, right? Admiral [William J.] Fallon? He was CentCom or CinCCent.

SA: Yeah.

EM: Yeah okay. But I think he is the only Navy guy, a couple of Marines.

SA: So he had that little bit of a bias. But he didn't understand all of the capabilities that we could bring. If NavCent had been sitting closer to him, he would have been more amenable to some of the stuff that we were doing.

EM: Did he appreciate the number of ships that were in the Gulf area?

SA: The first thing he argued about was, "Why have I got all of these damn carriers out here?" [Laughter]

I said, "Well, I can tell you part of this because when I was back in D.C. as OP-04, one of the very first questions we got when Desert Shield kicked off was how many carriers were available to get out there. We had a big study, and I had a little part to play in it, but most of it was done through the 06 organization." And I said at one point in time, the number was 10. If absolutely critical, we could have brought 10 carriers. Eight was a number that was also on the table, and six was a number that—if this was an extended conflict, we could start with six, and we could probably hold on to that number by some rotation and stuff. I said, "The number of carriers that are out here didn't happen by the Navy saying 'I want to send six carriers.' The Navy sent six carriers because the Navy was asked to send six carriers."

So, I said, "I don't know what the arguments were from your standpoint or anything else. All I know is that number was there when I left the building, and now that I'm out here, that's pretty much what I've got. We have already rotated a couple of ships back, and we've got the new guys here or the last guy was on his way. So I can't get into that argument with you, because I don't know why, what the number six was, but I know that the Navy was asked, and here's how it fell out."

And he said, "Well, can I use them?"

I said, "They are here and we can use them. I don't plan on just putting guys out there and letting them steam around and do nothing, so I want to work to get them integrated into the plan. The plan is right now we will work from the Red Sea and from the Persian Gulf, which gives us a two-axis approach for the carrier air, and if in fact we get into a position where I need more air, the closest range to hit most of the targets in Iraq is from the Persian Gulf." So, I said, 'if that becomes necessary, I will rotate carriers; I will take them from the Red Sea and bring them to the Persian Gulf."

EM: Which you did, right?

SA: Which I did. I said, "The fact is I've got a lot of flexibility. I can move them east or west."

EM: Why did you bring them around? It was kind of late in the game, wasn't it?

SA: It was late in the game, but the big push was on—the land thing was kicking off, and so with the possibility of needing more close air support or more battlefield oriented air support, I could do more from the Persian Gulf. "So I'm going to bring the other guy over so we can generate more sorties for the targets that I think are going to be accessible."

EM: Could you talk about the amphibious option from Schwarzkopf's standpoint. Was he against it from the beginning, or did he come to that point? Describe his evolutionary thinking.

SA: He didn't think there was ever going to be a need for an amphibious operation, I'm pretty sure; and he didn't want one to be done. But he recognized that the geometry of the thing was basically, he was going to be bringing everything up from the south right up through Kuwait, except for that big wide sweep piece of the pie that he had oriented, which I couldn't influence much with an amphibious operation. So one of the things that he wanted was, "How do I fix all of those [Iraqi] troops and guns that are all aligned on the coast? How do I keep them in place so that they don't turn and collapse in on the eastern flank?" And so from that standpoint, having an amphibious option and making it look real was very important to him.

That's one of the reasons I had that amphibious group out that went to Somalia and brought those people back right before the war started. The reason they were out there was to gain some press on their amphibious exercise down there with the Omanis, down on their coast. It was to say okay, they are down practicing, and we had some practice stuff here in the Gulf that we did with the amphibs. The issue kept coming up, and that's part of the reason Jack Sheehan now was with me and from the Marines regarding an amphibious landing.

We finally decided that we were going to have a final, "Okay, what are we really going to do? What is the CinC going to do?" So I hosted a meeting on the flagship, because I always wanted to get Schwarzkopf out to the flagship. "You've got to come out to the flagship."

"Yeah, I don't want to go to sea." He was afraid of getting seasick. [Laughter] Anyhow, he flies into Bahrain, and I have the helo Blackbeard go in, pick him up, and come back. First thing is he's grousing about is that daggum hydraulic fluid leaking all over the place in the helo. I said, "General, that has been the nature of that helo since the day it was built. We just learned to live with it and pour more hydraulic fluid in—" [Laughter] He wasn't happy about that. Got his uniform a little splattered. [Laughter] But anyhow, we sat down to have this conference, and of course Walt Boomer was there and everybody.

EM: Sheehan was there too?

SA: Yeah. And so he said, "Well, what's the story?"

I said, "Here's the outlay. Here's how we would do it if you needed one. Walt and I are pretty much in agreement with the fact that I should be prepared to conduct an operation in case he gets in trouble on the route up, that I can come in and reinforce him.

The alternative to that is if that doesn't happen, then I've got a lot of assets here that can be his reserve that I can now flow down south, and then bring them back behind him and put them ashore down here, and let them come up if he needs more reserves. So those are sort of our options." I said, "Now, if we execute the option up here where we come straight in because he's been pinned down in here, there's a couple of things that give us grave concern. One is this—not propane, but basically a gas—"

EM: Natural gas?

SA: "Natural gas plant down here. It has to go." I said, "It has to go before I put people on the ground. And the landing area is in view of a whole long row of apartment buildings, and we already know that most of those apartments now have got guns sticking out the windows and stuff like this, so the battleships are going to take those down right away."

And so he says, "Okay, I'm not about to save Kuwait by destroying Kuwait."

EM: À la Vietnam?

SA: Yeah. And, he says, "Yes, I understand that you've got to keep that option open, and I know what you are planning there is being able to provide him more reserves. That all sounds fine." He says, "So here's what I want you to do. I want you to do everything that you would normally do," and I talked about the minefields, and we talked at length about how long it would take to clear the minefields. He wanted to keep the option open to the very last minute.

And I said, "If somebody is in trouble, I've got"—I forget the number of days specifically that I told him, but it was, you know, five, or I think it may have been 10 days—I said, "I've got this amount of days that it is going to take me, at best, to get lanes cleared to get the ships in to do this. It's not an overnight decision. So I've got to have ample warning to do the preparations to make this happen."

Schwarzkopf said, "Oh, no, that's not right."

I said, "Well, wait a minute, that doesn't mean that we have to commit it, but I do have to commit the minesweepers. I do have to commit the battleships, and here's my plan. I'll get two lanes swept. Because we don't know as much as we should know about these minefields that I need to get started at least on this date. I need to start sweeping them, and then once I've got them swept so I've got the lanes," I said, "what I plan to do is, I will put a battleship in the north lane, and I will have him protected by surface ships for any antiship stuff from the beach, and I will use him to shoot at designated targets in the Kuwait area. This looks like the preparation for an amphibious landing." Jack Sheehan had also proposed an excursion up there on Bubiyan Island.

EM: Bubiyan or Faylaka?

SA: Faylaka. "And we've got some good targets, or we think we have some good targets up there, and we will do a raid in there."

EM: That came from Sheehan?

SA: His group had helped plan that with [Harry W.] Jenkins [Jr.] and the other amphib guy. So I said, "We've got that, and that raid will also be a piece of making sure that they think we are going to come." And so he finally ended up buying off on that. And that's how we ended up in the final. The third piece of that, or whatever number it is by now, was he wanted to send Sam [Samuel A.] Donaldson [Jr.] out to the flagship.

EM: The reporter?

SA: Yeah. Did you ever see that interview?

EM: I don't think I have.

SA: Oh, it's a great interview. He sent Sam out to the ship and said, "I want you to convince him that we are going to do an amphibious landing." [Laughter] I said yes sir.

So Sam comes out, and it was a fun interview. Anyhow, we had the charts out pretty much as I've laid out here. We had all those charts and took him up on the flag bridge and down in the ops center so it was, you know, big picture stuff, and he could see these arrows and the ships and Kuwait and all this stuff going on. We talked about amphibious landings and how dangerous they were and how many people you lose, and I agreed with him. He finally said, "Would you really do an amphibious landing?"

I basically told him, "Sam, we are here to do that," and "no, I'm not going to do it just to have an amphibious landing." I said, "If we have an amphibious landing, it's going to be because it's necessary to have an amphibious landing. I can tell you that we are ready to execute."

EM: So what did he report back?

SA: He reported back pretty much that.

EM: It's going to happen.

SA: I think Sam thinks he was had.

EM: Did he say that to you?

SA: I heard later on that he was not happy that we had sort of done the reverse on him.

EM: Well, I'm sure you were very upset at that.

SA: Yeah. But it was a fun interview. We strolled around the ship. I'll tell you, just as a sidelight, you see these people, and you don't really understand how good they are at their business. Until I saw the piece on TV and then reflected back on what we were doing as each thing took place. He had the ability—from wherever he was standing and talking to you, he had his own cameraman and you could see him moving him around. When I saw how it all turned out, he had the ability to see what the guy was seeing through the camera. He knew exactly where his face ought to be. How his face ought to be turned. I mean it was uncanny, and I didn't pick up on that until the fact that I had been there with him as we were doing it, and then I watched it on TV and then remembered how he had sort of walked and maneuvered himself. It was amazing.

EM: A real professional.

SA: Yeah, he had the ability to see exactly what the guy was seeing. It was very impressive. And he comes across as not so pleasant sometimes, but he was very, very nice. He was very fun to work with.

EM: Okay. Jumping ahead a bit, but do you think the ground war was stopped too soon?

SA: No. No. Of course there has been a lot of talk about that over the years. Schwarzkopf queried all of his commanders. He said, "Hey guys, what do you think?" You know we had the Highway of Death thing going on. He said, "I'm going to be asked a question—should we press on. We have accomplished the mission. We've accomplished what we said we were going to do. We pushed the Iraqis out of Kuwait," and he said, "How much further do we go?"

My reply to him was basically, we've done our job. "I don't think it's in our best interest to go any further." I forget how I phrased it, but my concern was that we didn't want to turn this thing into a war against Muslims, which I was afraid that we were going to do. But we had had pretty clear guidance in the UN sanctions and everything else that we wanted the Iraqis out of Kuwait. Nobody had said anything about a regime change or anything else in Kuwait. Nobody said anything about destroying the whole army or anything like that. So my vote was we've done what we came here to do. Let's call it quits.

EM: Well, hindsight is 20–20, but the fact that the Republican Guard, a lot of it, got out and then subsequently beat up on the Shiites and the Kurds. Was that the right decision that you made at the time? Knowing what happened afterward.

SA: Well, knowing what happened afterward, yeah, a couple of things happened that I was very surprised that we allowed. In the negotiations that then transpired, which I didn't have any part of, Schwarzkopf sort of handled that with the guys back in Washington, and the rest of us—I don't think any of the component commanders were much involved in it either. Schwarzkopf allowed them the use of their helos. That was a mistake. We should have imposed some sort of, "Here is how you've got to behave or else," but you know, Saddam asked for and was granted the use of them within days. When the thing was all over, he was off and running, consolidating his positions.

So yes, from that standpoint one can make a case that we should have maybe done something different. My problem is I'm not sure that we—I'm pretty sure that at the time we didn't, and I'm pretty sure even today we don't—we still don't have enough people that understand that culture. We don't understand the Kurds and the Shiites and, you know, it's all sort of a mixture that nobody really understands where the—how do you even out the revenge? I mean basically it's all tribal, it's all tribal warfare, and I'm not sure that we understand how to get into the middle of the campfire and say, "Okay guys, you've got to figure out a way where you are satisfied here and you are satisfied there and you are satisfied here. And, for the common good, you've got to have a central way, a central government that will serve the needs of all three." I don't think we know how to do that.

EM: Was there consideration of having the Safwan conference not at Safwan but on board *Missouri?* I have seen that. Were you consulted about that being on the ship?

SA: No. I would've played very hard for that.

EM: It was probably settled in Washington.

SA: Yeah. [Break]

EM: All right, we are back on the air. Okay, I wrote a few notes last night. Let's see. The Iraqi missile boat, were they Komars?

SA: Komars, yeah.

EM: Did you have a plan to take them out while they were still at Umm Qasr before they fled to Iran?

SA: Yes.

EM: It hadn't happened yet, before they fled? Were you in the process of doing it?

SA: Wasn't one of the priorities on the ATO. I'm not sure why I couldn't get them in there. We did a lot of damage. We were allowed to go and damage some of the facilities there, not all. I'm trying to think—I think we got involved a little bit in an argument about were there still some merchant ships close by these guys, and was there a possibility that they could be hit?

EM: Foreign country merchant ships?

SA: Yeah, and that sort of sticks in my mind as one of the reasons I couldn't get them until they came out.

EM: When they came out, they became an emergent threat that you could deal with quickly?

SA: Yeah.

EM: Did you call Schwarzkopf and say can I do this, or you just did it?

SA: We told him. I didn't speak to him directly. But we got clearance out of Riyadh. Zap Zlatoper had a plan all along for that possibility, so he right away executed and was able to get most of them.

EM: I would have thought it was a big concern, because years earlier the fact that the Cubans and the Russians, everybody had these Komars with missiles. Wouldn't that be a real threat to your ships?

SA: You bet. Again, you know, threats to my ships were not a big concern.

EM: In Riyadh?

SA: Yeah. You guys always want to make things bigger than they are. No, we just want to protect ourselves. That is the one area where your earlier question about did Schwarzkopf really understand things from the Navy standpoint—it didn't rise to the same scale. If he had some super-duper kind of tank that was sitting in the weeds waiting to come roaring down the road, he would have been much more concerned than about Komars coming down into the Gulf.

EM: All right. What do you think of Saddam Hussein as a war leader?

SA: He was not. He led his country by fear and intimidation. My take is that he thought possessing certain capabilities was all it took to utilize them. I think he had quite an array of equipment that could have been much more effectively used, and so that it wasn't comes back to his abilities to form a team that could execute. I think we gave him way too much credit because of the known capabilities of the equipment that he had. I don't think we understood that the ability to utilize that equipment in the manner in which it was designed was—I don't think we understood they weren't as good at being able to execute as they were able to accumulate. So I think he held the country together by what most dictators do, through fear, and make sure you cut off a few heads and hands along the way and everybody will behave.

He had a big array, got a special group of troops, the Republican Guard that everybody else can form behind. The other thing, the big mistake that he made, I think, is he played into our hands in that he let us dictate the time. I was amazed the whole time I was there. I said, we are working on a timeline, very slowly developing timeline, but we are working this timeline, and he's doing nothing to disturb it. And I just kept thinking, he's got to attack someplace in the middle of this, he's got to do something big that makes us all of a sudden jump back and say, whoa.

EM: Well, there was the one instance. The surge at Khafji and they came down. What was your reaction to that?*

SA: I mean it was done and over with so quick, and the Marines did a great job of containing that. Thank goodness that they did. But what that did was, it really solidified what Schwarzkopf always worried about, the Saudis on one side of him and the Egyptians on the other side of him, and I think that episode really bridged the—okay, we are all in here together, and we can all work together, and I think it eased a little bit some of his concerns about, you know, am I just in the middle of two guys that are going to run the other way?

EM: It impressed him that they were going to stand firm?

SA: I think so. We haven't talked that much about it, but I think he was much more at ease after that than he was before.

EM: The logistics support you got for your naval forces, was it timely? Was it what you needed—the equipment you needed, the supplies you needed?

---

* The Battle of Khafji, which lasted from 29 January to 1 February 1991, was the first major ground action of the Persian Gulf War. Allied coalition ground troops and aircraft repulsed most of the Iraqi Army's attacks, even though several units temporarily occupied the town. The aborted attack cost Saddam Hussein's forces almost 700 casualties.

SA: There were at least a couple of emails from me back to my good logistics friends. One that stays in my mind, and I know some of the Supply Corps guys have got it pinned on their walls in certain places—if you ever run into Ed [Edward M.] Straw, he can probably pull it out of his pocket. Now we are at the point we are thinking about what's coming next; this thing's going to wind down, and we are going to end up winning, and we are going to have decorations and awards. So, you know, "How many air medals do we have to present?" Well, jeez, there's just a handful out here available. I said, "Well, let's send in an order for some and get some out here, just so when the time comes to start making some recognition, let's take a look."

EM: He was the N4 on your staff?

SA: Let me, commander, my names catch me.

EM: Who is this fellow you are talking about?

SA: I'm talking, oh Ed Straw was my supply guy back on the N4 staff, so he ended up with this message. Actually, the message went to DLA [Defense Logistics Agency], and he eventually was DLA by the time I got back. But instead of getting, let's say 1,000 air medals or something like that, I got cots.

EM: Cots? [Laughter] How did that happen? Somebody pressed the wrong key?

SA: Somebody pressed the wrong key, so I sent back a sort of a snippy message about, you know, I ordered air medals and I get cots, who is in charge of that type of thing. It caused a lot of [laughter], a lot of giggles, and a lot of red faces too. But in general, we had everything that we wanted.

We had some issues early on with, I guess it was Sidewinders. The Air Force had some. We had some. One or the other of us either wanted a different variant, or I forget now what the issue was, and we had to scramble around a little bit and share the wealth there for a week or so. And this was before the war really started, we were sorting through stuff. But in general I had everything I needed plus. And so the Air Force started running short of bombs, and I had extra bombs that I could get to them.

EM: Did you have enough PGMs?

SA: I had enough PGMs. I didn't get into a position where I had to ration them. The only supply issue that continued to give us grief along the way—and basically that was my discussion on the refueling. The JP5 against the JP4. That was just an issue between, really, me and the Air Force, because there was plenty of fuel out there. It was just that getting the right fuel to the right guy was a little bit harder than it should have been. But all the preplanned, the prepositioned stuff that we had at Diego Garcia and all of that, I don't think anybody could have expected more from that. It came off, we learned a little bit about what it takes. We had everything that was designed to accept that prepositioned stuff. You won't find many ports in the world where you had access to as much alongside space and equipment to move that stuff. That long range hauling for the Army, for their ability to take it back, far back in the desert, was done mostly by contract trucking. So no, from my standpoint we were well supported.

EM: Can you speak to the support we got from our Arab allies—the Saudis, the Qataris, the GCC countries? The remains of the Kuwait armed forces? How did they figure into your operations?

SA: Well, I visited the UAE [United Arab Emirates]. Of course from my standpoint, it was port facilities that I was most interested in, and they were able to handle that. Oman was very gracious to us as well. So the things that I needed from them were pretty much intact as far as they were there. They let us use them. It was pretty much the access that we wanted. The Bahrainis went way overboard for everybody, turning their airfield, the airfield down south—basically Royal [N.] Moore [Jr.] and his Marines owned that place and lots of equipment there. They allowed me, because of our Mid-East Force days, I had a spot at the pier for both flagships out there whenever I wanted to put them there, and other ship visits. They had port security, provided port security for me.

EM: What about combat operations? Were there any boats or units up there?

SA: The Saudis had some ships, some small ships that they gave, put into the basket and UAE, no, but they let us use their facilities. All of the Gulf allies had some command facilities that they let us use. Some let us use some of their beaches for amphibious workups, so we got used to the tide shifts and what wave action in the Gulf was like. So they weren't reluctant to let us use the stuff. They were very receptive. If we had a request to use some of their military spaces for different things, they would let us do that.

So it was much different, I think, than what I had seen during my CentCom days and RDJTF days. Then we were really new out in the region with Mid-East Force, and they were just used to looking at that one guy for anything military that had to do with the U.S. They weren't near as guarded in their reception to me. They were on the side of, is there anything here that we can do for you.

EM: What did you do to put out information to the sailors and the Marines that you are in a culture here with a different religious base and some other cultural differences from what we are used to?

SA: Well—

EM: Like alcohol.

SA: Well, Schwarzkopf took care of that. He had—what was it—Field Order One, or whatever he called it, outlined all of that, and it was preached to everybody that came into the region, "Okay, here's the do's, here's the don'ts." And he had outlined it very clearly, and so all we had to do was sort of re-publicize his directive. Now, I had the interesting piece of being in Bahrain for the most part, you know, if I was ashore. Bahrain worked on different schedules, so alcohol was permitted. I mean they had nightclubs and all this, and that's of course where the Saudis came to play. And so there was a little more free rein in Bahrain, and I just had to be careful with our sailors if we went to other ports and they went into other parts, Jubail and Dubai and all those, they had to behave, although there was alcohol there. We had to constrain it to certain locations.

My biggest concern was the Cunard liner that they brought in and put at the pier, and we had all of these guys that had been out in the wilderness—this is before the shooting started, all these guys were out in the wilderness and had been there for months on end and nothing had happened yet. Schwarzkopf wanted an R and R [rest and recreation] facility for them, and so they sent in this cruise liner from Cunard Lines, and I was going to be the big daddy to make sure that nothing went wrong, you know.

The first thing that I did was, as soon as the ship moored I arranged for a meeting with the captain. I go over saying here's my concern, and he was an old Royal Navy guy. He said, "Look," he said, "we are in this business of entertainment, recreation. We know how to do this. You are welcome to come over anytime you want to." In fact, he asked me for dinner that night, I think. But he said, "Here's how we do it. You guys are going to be responsible at the end of the pier when these guys get off their buses. There's going to be a place there where they've got to check—if they are carrying any weapons or bringing any weapons or anything else, they are going to be left there, and so then once they are through there, they belong to me."

EM: He said this?

SA: Yeah. He says, "They are under my responsibility, and I can guarantee you nothing is going to go wrong." [Laughter]

I say, "Look, these guys have been out there in the sand and the bushes for months on end."

He says, "I know."

EM: What did he have on hand to assure their entertainment?

SA: Oh, he had all sorts of stuff. He had the dancing girls. He had everything. I gained a new appreciation for how much time and effort they spent working to make sure that they knew how to control—make sure they had people that could see trouble and get in the middle of it and get it done. Basically, he was right.

EM: Had no trouble?

SA: He was right. They did a wonderful job.

EM: Probably just a fraction of the forces that were there had the opportunity to go to that?

SA: I forget how many they handled at a time. I think it may have been close to 1,000 at a time. The troops had a wonderful time. They just loved it.

EM: Well, they needed a break from the desert. They were sitting out there for a long time.

SA: Oh, yeah. Yeah.

EM: Okay. Back to the aircraft. The F-14s used TARPS [Tactical Area Reconnaissance Pod System], right, for reconnaissance, tactical reconnaissance?

SA: Yeah, and we had very limited TARPS capability out there.

EM: We didn't have the Vigilantes any longer, or the RF-8s. So the TARPS, F-14 TARPS handled all of the tactical reconnaissance—was that enough?

SA: No. We wanted more TARPS, but we just didn't have it at that time.

EM: How were they used? What type of tactical information?

SA: I think most of the photos of the Highway of Death were TARPS photos, as I recall. So we got a lot of the oilfield stuff from the TARPS guys. I'm trying to remember now, because I remember we agonized over the number of TARPS pods that we had. It was not what we thought we should have.

EM: So what targets were not covered that you could have if you'd had the equipment?

SA: I don't think there was anything that we really missed, and I think it was just the fact that we might have had to wait a little bit for the next sortie, or something to get the TARPS bird in the package. I think we would have been hard pressed if we had gone more, had ranged farther up into the country, if we did want to push Saddam clear back into Baghdad. Then I think we would have been a little starved for information.

EM: How did the heat and dust in the Persian Gulf area affect materials, maintenance?

SA: Well, we had to do a lot of preemptive stuff. We put extra screens on our intakes and more mesh and stuff, so that we could collect the dust before it got into our systems. So we did okay. But it took some pre-planning to get all of the intakes covered and make sure that we did the PMS on the equipment a little bit more frequently than we would normally do. So in the long run, it was the preparation took some time and effort, but we really weren't impacted that much.

EM: How was your operational readiness, aircraft in particular?

SA: Our operational readiness on our airplanes, we were very high. I'm trying to think if we had any real showstoppers that popped up, and right now off the top of my head I don't remember any.

EM: All of the ships were air-conditioned, right?

SA: Yes.

EM: How was sailor morale during this long waiting period for things to start?

SA: You know, they did well. I was amazed. Everybody knew what they were there for, that this was important. There was that nervous factor of, gee, how big is this thing really going to be. Then it got to be a little bit of, well, we've been here quite a while. When's this thing going to happen? They had to be talked to, and the battle group guys did well. We had some entertainment come out into the theater at Christmastime and make its way around the ships. And we kept taking the ships into port, so from the sailors' standpoint, we had it a lot better than the Army and the Air Force guys, because the Air Force guys were in Riyadh a lot and on some of those remote bases where they were. They were in some good places, but they were also in some remote places.

But from the sailor standpoint, the biggest thing that we never solved, and I've had more arguments with the Supply Corps on this, was mail. We had, of course, ships from the Pacific Fleet and ships from the Atlantic Fleet. Fleet Post Office Atlantic and Fleet Post Office Pacific.

EM: The Pacific ships couldn't get the mail quickly?

SA: It ended up none of the ships got their mail quickly. All of the mail came into Bahrain, and then we would distribute it out to the ships. So Bahrain became the big mail post office of the universe. And sometimes on some unique spare parts, this same thing applied. If a unit was, say a Pacific Coast unit, his mail had to be routed through Fleet Post Office San Francisco, and Fleet Post Office San Francisco would send it out to who knows where, Guam probably. And then Guam would—they would come that way, whereas all we really had to do was a one-day mail transfer from anybody in the United States to Fleet Post Office New York, or whatever it was, and come the short way. Could not solve the problem.

EM: They got their mail late or it didn't arrive at all?

SA: We had stacks of mail that would come, and it would take us forever to get it sorted. Some of it got into a bounce thing, where it would bounce back and forth a couple of times before it got there. It was the most frustrating—I spent a day in the post office there in Bahrain trying to figure it out, because I thought it was our problem. For a while I thought we just had a problem in Bahrain trying to sort it. We did have some problems. But our mail was just atrocious. It was just absolutely atrocious, and I was I fussed about it the whole time I was there, and I fussed about it when I came back.

EM: Did your sailors give you comments about it?

SA: Oh yeah. Yeah. Yeah. But with everything else, I thought jiminy Christmas, you know, if I needed 20 more tanks I could probably dial them up in a minute. If I wanted a letter from the States in one week I couldn't get it.

EM: Did that problem go away after the war with the decommissioning—not decommissioning, but all the troops went home and sailors?

SA: Yeah, and I don't know.

EM: You didn't have to address that when you were Vice Chief?

SA: No I didn't, although I asked a lot of questions when I was Vice Chief when I got back about what went wrong, and why couldn't you sort it out and send it to me the shortest way.

EM: Were you pleased with the support you got from the American people for your war effort?

SA: Oh, yeah. Yeah. They had that parade in Washington and then the parade up in New York. I didn't want to come back. I was ordered back. I said no, no.

EM: By whom, Schwarzkopf?

SA: The CNO. "You've got to come back and be with the rest of the gang." I didn't want to do that. But the American people were just absolutely super. Sent a lot of nice letters and stuff out to us. Got thousands of letters from schools and kids. The normal thing was, when you got a package from a school, or anyplace where there was a group of people gathered, an organization, you sent them an autographed picture of the flagship, a nice letter. And so I got this letter from this school in Maryland, and their teacher was a classmate of mine from high school days. I wrote them all a nice letter, and I said, "I see you have a wonderful teacher, and I'm sure she makes sure that you do everything right, and you should do as the teacher asks. But I just have to remind you that if she is getting too tough, just ask her about her days playing in the band on the bus going to the football games." [Laughter]

EM: So she was also in the *Osky Wow?*

SA: Yeah. Years later—I don't know, it must have been four or five years later, after the war, I'm back and she's married to a Navy guy. She comes up to me and says, "We send you this nice letter and all you do is send back a nasty note and a picture of yourself." [Laughter]

EM: How effective were Naval reservists in your operation?

SA: That was a real pleasure. They had so many different tasks that they performed out there, and they really did them well. Of course, we had a reserve staff that had been associated with the CinCPac or Seventh Fleet staff for years, and they came out and we were pretty well integrated. By the time I got to the Gulf to take over from Hank, our staff was twice the size that it would normally be, and all of that came from the reserves, and you couldn't tell them apart. They were integrated right in with the thing. Most of the port security stuff was the reservists, and Seabees for some of that, and everywhere you went, the field hospitals reservists were absolutely superb.

EM: The two hospital ships, were they ready to go?

SA: Hospital ships, they came. We had some interesting dialogue. I was a little chagrined to find out, I think it was the second one that came through the Suez Canal decided to have a formal dance on the flight deck as they were transiting the Suez Canal.

EM: Not well appreciated?

SA: I said this is not the way you come to the war zone.

EM: Especially through a Muslim country.

SA: Yeah. Then Cheney, one of the trips that he was out there, he wanted to go visit one of the hospital ships. He had a deputy, was it [Harvey LeRoy "Lee"] Atwater? He played again in the political arena in Washington for a while. Anyhow, one of the reasons I found out after we got there was Cheney had been approached by somebody. This couple must have had some political clout. On that hospital ship we had a doctor and a nurse who were married, and in a combat zone, you know, back from the Sullivans days [five brothers lost at sea on one ship in World War II], we had restrictions on family members being together. And from the Navy standpoint, cohabitation, you have cohabitation—it's not right to have a couple out here, a married couple on board.

So we were in the process of going to separate them once we found out about them. So Cheney is up there talking to the assembled group. He's out there to pat them on the back, and so his aide comes up to me as says, "You've got this married couple on board and he doesn't want them split up."

I said, "Well, by Navy regs that's what I'm going to have to do."

And he said, "Well, you're not to do it." And so I said okay, I understand. So then I had to come back to Washington, the CNO's office and say hey, you know, this is above my pay grade out here. Who is going to make this call? Do I split them up as per our regs, or are you going to tell me to ignore it?

"Ignore it."

EM: Nothing further happened with that?

SA: No.

EM: How did he get wind of that?

SA: I don't know.

EM: One of the two complain to him?

SA: Like I say, I think they must have had somebody that they could talk to in the political circuit, because it was a surprise when I called back into the CNO's office. But that was the only thing. They weren't as ready to do business as I thought they would be when they arrived, but they got into shape pretty quick. They had a tremendous amount of capability. What we had to do was spend a little more time with them on their flight deck. How to work their flight deck. We spent some time with them on getting that up to speed, but, boy, they had tremendous facilities. The people were well motivated. Luckily we didn't have to use them that much.

EM: The hunt for Scuds, how did that impact on your naval operations?

SA: Well, the old Scud hunt. It was just one of those things that—now the focus of the sorties was, somebody find a way to go find Scuds, and that was not a very successful operation.

EM: Did it prevent those aircraft from looking at other targets that you had on the list?

SA: No, not really. It's sort of hard to tell with the targeting and the ATO process in Riyadh, but for the most part we were pretty much on schedule doing the things that we had normally been doing. It was very frustrating to Schwarzkopf. I don't know how much Chuck Horner, what his reflections have been on the war, but it seemed to me like most of us thought this was not going to be anything that was going to turn out well. We weren't going to go out there and discover a bunch of Scuds laying around. They were pretty well—

EM: They were moving them around.

SA: They were moving them around.

EM: Hiding under bridges.

SA: We weren't going to see them until they shot them. But the Scuds into Israel was very disturbing. Someone could launch a chemical Scud into a high population area or even into our areas. That was going to be a real problem.

EM: Going back to the preparation for chemical warfare, what did you do on your ships with the personnel for that eventuality?

SA: We had them all briefed. We had a prophylactic pill that we had to take for X number of days. I think we made everybody a little—not feeling good, so we didn't take them very long. [Laughter] But, we went through all of the MOPP gear and how you responded. We went through all of the alarms and what you were supposed to do to not have to be shut down.

EM: Did all of the ships have a washdown capability?

SA: All of ours did. I'm not sure about the other guys.

I'm not sure whether we exercised it out there. I'm trying to remember. No, I sort of remember our washdown drills coming out of overhaul; they were always exciting. I can't remember whether we tried one of those out there or not, but probably not.

EM: What actions did you take to prevent collateral damage and civilian casualties in the strike operations?

SA: Ever since Vietnam we had been into the thing that we knew. The aircrews were well briefed, and if you saw something that was not right, you weren't cleared to drop even though it was an intended target. But mostly it was pretty clear cut. I mean, we didn't have a whole lot. Right toward the end as they were running out of Kuwait, some of that had to be pretty well managed because you had so much population sitting there. Right after it was over we went into Kuwait. The airfield was strewn with these metal baggage containers that they slide in and out of the airliners; they were all over the field. I don't know how they got there, whether the Iraqis threw them around to just make it hard for people to move, or whether we did some of that on the way through. I'm not sure.

We spent an awful lot of time, from my standpoint—because we had most of the targets in Faylaka and Bubiyan islands, there are some ancient ruins which now the ISIS [Islamic State in Iraq and Syria] guys are hell bent on destroying. There were some Mesopotamian ancient ruins that were on the no-hit list, and there were some allegations as the war went on that we were hitting them and that my guys on a raid had done something. One of the very first things I did as soon as I got a clearance was, I went into Bubiyan Island to see if we had hit any of the heritage units or heritage monuments there.

EM: On Bubiyan Island or Faylaka?

SA: Bubiyan first and Faylaka later. First thing we found was, we had a brand new BMP [amphibious vehicle], one of these Russian personnel carriers sitting there. It was brand new, almost spotless. I had my intel officer with me. I forget how many guys I had, three or four guys with me, a couple of Seabees and this one kid and another young sailor. Anyhow, we walked up to this thing and looked around to see if it's got ordnance or anything, and then we climb around in it and no keys, and so we are scratching our heads and saying, boy, if we could get this thing running it would be a good way to take a tour around the island here. This kid says, not a problem. He had this northern New York accent. [Laughter]

EM: I do it all the time back home. [Laughter]

SA: One minute later, *vroom, vroom, vroom,* we were off and running.

EM: Wonder what his youth was spent doing? [Laughter] That's funny. So you toured the island in the BMP?

SA: Yeah. We hadn't hit anything. But it was a great concern, as it should have been.

EM: The Al Firdos bunker business. Now that was a TLAM, right, that penetrated the top level and went down and exploded into the civilians that were below.*

SA: Yeah.

EM: What resulted from that? Did Schwarzkopf contact you about that?

SA: No, because that was a designated target, and again this is part of the ebb and flow of the ATO where you've got this 72-hour cycle. I don't know if anybody prior to the hit had discovered that it had been converted into a bomb shelter or not. I don't recall how that investigation turned out. But it was one of those casualties of war. It was the assignment. We didn't miss the target. We hit the right target.

EM: You didn't know that there were civilians there?

---

* On the night of 13–14 February 1991, Air Force aircraft destroyed Baghdad's Al Firdos bunker, a command center, with 2,000-pound bombs that killed not only military personnel but also civilians who, unbeknownst to allied planners, had sheltered there.

SA: We didn't know. That is still the disconnect between the intel world and the operational world. Who knew that was a bomb shelter? Could anyone have known it? I suspect that somebody knew, somebody in the system knew that that was a bomb shelter. I'm not sure.

EM: But it didn't get to you?

SA: It didn't get to the planning. It was a headquarters building basically.

EM: Commodore Chris Craig, the British guy in charge, you worked closely with him, I'm sure. In postwar writings he complained that there were too many changes of command among the American naval commanders, like in the middle of an operation one guy is replaced by another guy in sort of a routine way. Was that a valid criticism?

SA: I think that was probably true. I think part of that came about in the renumbering that I did when I first got there—that I might have influenced that a bit. The fact that I came out there was—that took special dispensation from Schwarzkopf, because he had made an ironclad rule that none of the commanders were going to leave the theater until he said so. And he wasn't going to say so until the thing was over, and then, you know, the relief from Mauz came about, and he finally agreed with the CNO, apparently, for me to come out and relieve Hank Mauz.

So that was an aberration. And then somewhere along the line, I won't name who it was, but I had one of my flag officers out there that I had to send home for a humanitarian reason, and I had to go in and talk to Schwarzkopf about it before I did it. His first reaction was, you're not going to do it. Then I explained the circumstances, and he says okay, yeah, you can do that. That affected some. Chris may have seen that, I'm not sure.

EM: I think he was talking about the surface operations.

SA: Yeah. Part of that was—what we did was we moved guys around, and we would put people in a certain area for a certain period of time, and we would rotate them down to another mission. So another guy, another—say the cruiser skipper was designated to be a CTG [commander task group], so-and-so, so-and-so, then we would rotate him out and the new skipper would now take over.

EM: Why was that done?

SA: We did that as just a little bit as change of pace. Once we got into the routine, we moved guys around a little bit so that they had a different—

EM: Experience?

SA: Less stress. But, yeah, I think we probably could have eased that a little bit. Of course—and I really like Chris, and I'm surprised that he didn't talk to me a little more about that, but I would agree with him. I had trouble keeping track of some people sometimes. [Laughter]

EM: Well, you had a lot of people to keep track of.

SA: We had a lot of people. Every now and then we—well, like the amphibs, we would rotate them in and out, so there was some friction there between with the two groups there a little bit, that we had to resolve.

EM: All right. A few questions about the immediate postwar period. [Raynor A. K.] "RAK" Taylor, Rear Admiral Taylor, you put him in charge of port clearance, clearing the harbors and the minefields as well, right?

SA: Yeah.

EM: Okay. How well did he do in that job?

SA: Well, if you've ever met RAK, he's got—

EM: I have met him. We interviewed him.

SA: He's got an energy level that, [laughter] you know, you just sort of point him, and off he goes. He's a very, very bright guy, very bright guy. So yeah, he did a good job. He took charge and he was off and running. I went up—when he was finally clearing out the port there in Kuwait City, I went up there and spent a day with him and the crew that was doing that, and they were doing great stuff.

EM: Well, he was your successor as ComNavCent, right?

SA: Yeah.

EM: Why didn't he get a third star after his experience there, and it went well?

SA: You know, I don't know.

EM: Yeah. He was an upper half or lower?

SA: I think he was an upper half. Yeah.

EM: He didn't go beyond. In fact he retired shortly after the war.

SA: And that was the mode of the Commander Mid-East Force. The arrangement was, they came out there and it was a good two-star billet, and then they all went home afterward. And then they finally broke the mold later on with Tony Les and all the three-stars. I don't know. For me he was a the right guy at the right time. He was a quick learner. He had a lot of smarts. He had a lot of good ideas, and I thought he performed extraordinarily well.

EM: He was not a proponent of aerial minesweeping. He didn't think the helicopter sweeping was an effective way to do mine clearance. Did you share that view?

SA: My view of mines, helo minesweeping is—and of course one of the other things we had to do for the Brits and the rest of the mine force was, they wanted to see the helos sweeping first; they wanted the first passes by the helo. So quite honestly, most of that sweeping was a placebo, because we still hadn't perfected the art of dragging those things. I don't know how effective they were. I don't think they were effective from the standpoint of the other people in the mine warfare community thinking, okay, the helos have made the first passes, and so it's okay for us to start our work. If you thought you were going to do clearing of all the mines with the helo minesweeping—I don't think that would have ever happened.

EM: You weren't confident they were picking up?

SA: No. I was not a great big fan of helos either. I'll tell you my heart is out to those guys, though. You talk about dangerous flying. That is dangerous flying.

EM: Because you are tied to a sled?

SA: You are tied to that sled, and, yeah, you've got explosive bolt cutters and stuff back there to cut you free, but by the time you do something, it's too late. Most of the time you are right at the torque limits on your gear boxes and for the rotors and all, and so you are working the helo at absolutely right at the limit of what it can take, and you are doing it in superhot weather.

EM: But it didn't go away. We still have aerial minesweeping.

SA: We still have it.

EM: Did it come up when you were Vice Chief?

SA: Not that I recall. I can't remember a time in my life when we weren't talking about minesweeping at one point in time. I don't think any big issue came up during my time.

EM: Well, the folks that do it must be confident that it is effective; the helicopter minesweeping squadrons, what is it? HM-14 and HM-15.

SA: Like anybody, you buy into your mission and you do it. Yes, they had some successes, but I was never confident, and I know mine guys were, and I know RAK was not that big a fan. It takes a lot of hardware to support them, but out there we executed the way it was supposed to be done. I'm still not a big believer that the money spent in aerial minesweeping is well spent. But then you start looking around at how they can cover a lot of water in a hurry compared to surface craft.

EM: When you came back to the States, you were in both New York and Washington for the parades?

SA: Yeah.

EM: And you made your first visit to the Vietnam Wall when you were in D.C. Did you do that for a reason, or just as an opportunity to go down there?

SA: I did it for a reason. Yeah. I had never been there. I didn't think I wanted to go. I get a little choked up on this.

EM: I do too.

SA: I was afraid to go. So then I'm back, and we had a pretty successful ending to a war, and I thought it was time. It was really—it was very moving. A lot of friends on that wall, some very close friends, a real close friend particularly. I was glad I went. I don't know if I could go back again.

EM: Really.

SA: But yeah.

EM: It's very moving, you know, in its simplicity.

SA: Yeah, it is. I just decided that it was time to go face the Wall, and it was worth it, and I'm glad I did. I think in reality it's probably the best—when it was being built, I wasn't sure about it. But I think it's probably—spiritually, it is the most moving of the memorials.

EM: It hits you. All of the names, and they look back at you and you see your reflection.

SA: Well I'll tell you, the guy that had that vision had it right. I didn't think so.

EM: She, Mya Lin.

SA: Yeah. I didn't think so at first, but I did, especially after I had seen it.

EM: Were you involved—you left the Gulf in April of '91, right? Did you go out to your Seventh Fleet duties, or did you go right to Washington to take on VCNO?

SA: No, I went to Seventh Fleet duties, so I started the grand tour out there. I went to Indonesia. I went to Malaysia. Went to Australia. Went to the French island out in the Pacific—

EM: Oh, Tahiti or Bora Bora?

SA: No.

EM: Réunion?

SA: I did go out to La Réunion. I went out there as Seventh Fleet, because the French gave me their Legion of Honor. They won't do that unless you are on French soil.

EM: So those three Americans that recently got it that were on French soil saving French lives.

SA: You had to be at an embassy or something like that. So it was. I went to the Seychelles. I went to Ceylon, Sri Lanka, as part of that trip. The other island was New Caledonia. A beautiful spot.

EM: When you went out to your Pacific Fleet duties, were you still involved in things like [Operation] Provide Comfort, you know, giving the supplies to the Kurds up there in the north, or the continuation of the Maritime Patrol—redeployment of our forces from the Gulf?

SA: No, most of the redeployment was done by the time I left. In other words, we had finally got everybody back home that was going to go home, and they set the new stage for what was going to be the presence in the area. That was all done by the time I left. Basically, that was it. Once I left I didn't have anything that I really had to think about going that way, other than, I guess, the routine movement of ships through the AOR that were going to be out there.

I spent my time trying to go back to all of the places that Seventh Fleet had in the area of responsibility, and also to some of the places that had been supportive of me during the Gulf War. I went to places like Kenya and Mozambique. In Kenya, we had had access in there for many years from my days at PacFlt during that Iranian hostage extraction that was going on in the Pacific Fleet and Indian Ocean in those days. We needed to place some munitions and special munitions and have VP access, and so and we had used them during the Gulf War for some access. So I went to places like that, wanting to thank the people and just to show that their efforts on our behalf over the years had been much appreciated. That's why I just sort of went around the big swing. Went to India.

EM: Did you get to Australia?

SA: Australia and—that island northeast of Australia.

EM: Papeete?

SA: No. But the French were good so that was a good place to stop, and La Réunion. I think that award came through while I was still Seventh Fleet, so I went down there. That was a great, great time. Great people there. Then Malaysia, Singapore, Indonesia.

EM: With the flagship, right?

SA: Yeah. Well some of it was flagship, and some of it was my P-3. India was the P-3, and that time—and that was not necessarily associated with the Gulf War, but it was associated with the fact that back in Washington they decided that they wanted to start being more proactive in getting onto a better posture with India. They said, "Could you go to India and take a P-3, an operational P-3?" And I think there was a move to sell the Indians P-3s at that time, too. So I did that.

I was able to take Jennie Lou on all of these trips that we went on. So we did the tour of India. First thing, because of what the Washington guys wanted, they wanted me to really demonstrate the capabilities of the P-3Cs to India. So when I got there I told the attaché, "I'm prepared to fly Indian aviators that are in the ASW business and take them for a ride in the P-3 and let them see our equipment and all."

They said, "Oh yeah, that would be a great idea." Then they came back with "but who is going to be demonstrating the equipment?"

I said, "Well, the aircrew on the airplane."

They said, "But your operators are all enlisted, aren't they?" I said yes. "Well, the Indians don't want their operators who are officers being lectured by [laughter] enlisted operators."

EM: Oh, jeez.

SA: I said, "Well, you know, that is sort of what I've got. All I can say is that they will be very respectful that they are talking to officers, and they will describe their gear and the capabilities of the gear that they can, within the limits of clearances. But they won't be trying to talk down to them and say here's how you have to this."

"Okay, we think it might work, but there is a possibility that this could come off the rails a little bit."

I said, "Well, let's give it a try." So that part worked, and then the quid pro quo for that was the Indians were going to let me fly their simulator for their Harriers to get their guys ready to fly aboard their carriers. I had never flown a Harrier or Harrier simulator. So I was a little suspect of what I would find, because years before when I was in the NavCent job and went to Somalia, I was given an opportunity. I think we talked about this. I was going to fly a MiG-21 and got into their trainer. So I'm thinking, well, maybe I will see the same thing here.

But, oh man alive, was I surprised. We went through double doors. You went through one door, and you had to get rid of your shoes and you had to tuck things in, and second door was basically, now you were in a clean space with the simulator. It was pristine, and everything worked just as advertised and the instructors were extraordinarily good and the equipment was very nice. So I was really impressed.

EM: It's not an easy aircraft to fly?

SA: It isn't. No, it isn't. So it was a challenge for me to fly it, but it was a very good simulator. A lot of realism to it.

EM: Well, the previous story though, you didn't have any trouble with the enlisted guys explaining their equipment?

SA: No. It all came off without anybody getting their feelings hurt, and the Indians treated us so well. We went down to Goa which is, you know, a Portuguese settlement really, and got to see—what's the saint that is buried there that did all of the exploring—came through and went all the way up to Japan, and then was headed back from Japan.

EM: Portuguese?

SA: Yes, Portuguese saint.

EM: I don't know.

SA: Anyhow, his body is preserved and laying in a glass coffin in a big cathedral down there in Goa, and so I went to pay our respects there. I ate some of the best cashews you will ever eat in your life.

EM: Your big issue, even before the Gulf War, was that ComNavCent needs to be on an equal plane with some of the other component commanders. That it should be a three-star billet. Admiral Kelso absolutely opposed that, and it took what, a year and a half before he finally relented. Why was his opposition so strong? Didn't the Gulf War experience show that you needed someone of that clout?

SA: It did. I don't know, and we will talk a little bit about this when I'm Vice Chief, but part of his problem was we didn't have enough three-stars, and he thought he had better use for his three-stars than in that particular spot. Again, remembering that the big daddy of Com Mid-East Force was Admiral Crowe. He was sort of the godfather of what Mid-East Force was to become, and now you moved down the wiring diagram from Crowe to Watkins to Kelso, all submariners, all out of the same school, and Frank knew the pressures on three-stars, and of course there was a new thirst for three-stars with the Joint

Staff now building itself bigger. Again, it's way out there and there is not much going on. Yeah, we did have this big excursion here, but maybe it will turn back to what it used to be, Sleepy Hollow with two destroyers and a flagship. So I think he just wasn't prepared to think that hard about it.

EM: The fact that after the war—supposedly the war was over—but then you had Operation Provide Comfort. You had Southern Watch. You had the continuation of the embargo patrol. Did that perhaps influence him?

SA: I think, and as you know, when I came back to Washington or shortly thereafter, we reorganized the OpNav staff and a lot of the three-stars left the building. The three-star submariner, the three-star aviator, the three-star surface guy were all gone.

EM: Right.

SA: Now we had to fix.

EM: So you had a surfeit of three-stars?

SA: We now had some three-stars free that could go to operational billets. I will tell how that story happened once we are talking about the Vice Chief job.

EM: All right. This is sort of the wrap-up time here. Should we have gone in and deposed Saddam instead of just freeing Kuwait?

SA: In retrospect, from what we know now, yes. From what we knew then, no. I thought that the job that everybody had signed off on, the UN sanctions and everything else, all said here's what we want to do. We want to eject the Iraqis from Kuwait and return Kuwait back to its original state. We did that. We probably went a little bit further than they wanted with that big wide sweep and getting as much of the Republic Guard that they wanted, but that was part of the plan on how to get these guys to move as fast they could out of there. So at that time, the answer was no. Today, yes. Lots of provisions to that yes, because then we would have owned Iraq. Were we prepared to put in place a regime that could run Iraq? Later events show that we still don't know really how to do that very well, because we did it later on.

EM: You are saying things that would say maybe it was not a good idea to invade Iraq in 2003?

SA: I think yes; go ahead and try it. Now, would we have been as successful as our first task was? No. We would have inherited a lot of problems, and I'm not sure we would have dealt with them very well. We certainly could have done it much easier than the second time we tried to do it, because they were really beat. They'd had the same problem. They had been in there for months on end, and none of them were going home. They were all there and locked in place and nothing was happening, and then all of a sudden, in a few days really, the whole world sort of crumbled in. So they were running for the hills, and they would have kept running when they had seen the rest of the stuff coming.

EM: Yeah.

SA: So yeah, we could have got Saddam. We could have started a regime change, but again, I don't know. I don't think we would have known. Schwarzkopf knew that region well. He knew those people well. I will tell you a little story offline that I just found out last week that I didn't know about Schwarzkopf and his experience in that area.

EM: You don't want to speak online?

SA: I don't think I want to because I don't know how far I can go. Well, I'll tell you the basics of it. The basics of it were that although he was really an Arabic specialist—I mean he knew the region, he knew the language, he had lived there, he knew a lot—he never displayed all of that knowledge in all the time he was out there with the Arab counterparts.

EM: Why?

SA: I don't know. But—and I knew that he knew all of this—I didn't know that he never even greeted them with the traditional Arabic greetings.

EM: Maybe he thought he was representing the United States and he had to keep it to that?

SA: I think that was it. But that was something that I didn't know, because I didn't know specifically that he never addressed them in their own traditional way.

EM: Are you learning this from someone who was close to him?

SA: Yes. But what the reasoning was—

EM: Well, I mean, who knows, but that's interesting.

SA: Sometimes I would have thought that he would have, but he didn't.

EM: Interesting to ponder the reasons for that.

SA: Yeah.

EM: What actions did the Navy take to profit from the Desert Shield, Desert Storm experience? It's a broad question, but did the Navy profit from the experience?

SA: Oh, we profited, certainly. And I think the main point that I tried to drive when I came back as Vice Chief was, okay, we the Navy, who parochially understand jointness as well anybody, we don't know the language. We didn't know the language. We didn't have enough people experienced at doing joint tours who then came back and did other things. We didn't know the numbering system of the other staffs. They didn't know our numbering systems. We just had this schism of knowledge that should be shared back and forth, and so the thing that the Navy had to do was, the Navy had to get off of its butt and stop being completely—"Okay, this Navy staff looks exactly like this, and this is the way we do business, and we all know who the personnel guy is and we all know who the logistics guy is and we know what their designators are." But we don't really know who the other guys are who we marry up with on the other staffs.

And so we need to learn the language, and we need to get people in the right billets so that they can influence the good things that the Navy can bring to the table so that the other guys have some appreciation of the capabilities. Too many of our counterparts believe that we live on the ships. We are all self-contained. We don't want to talk to anybody. [Laughter] We don't want to do anything. We just want to show up, and if you ask us to do something we will do the best we can, and then we disappear again. We need to get into positions that show that we can do a lot better than you've seen us do. If you let us do things a little bit differently, or maybe we can learn to do it your way better.

So, jointness—we had to start looking at jointness as a big J instead of the little j. I think that was the big thing that I brought back out, and I think a lot of people understood that too, once they—you know, having an AWACS out for a fleet exercise out in the middle of the open ocean someplace is a lot different than having an AWACS over land controlling airplanes from the sea and from the land bases all at the same time.

And the water-land interface issue that our E-2s have—and the same issue that the AWACS has on the reverse side of it—is, we've got to learn how to dissipate those discrepancies and blend together, so we don't have this blind spot that so many things happen in. [Break]

EM: I interrupted your thought.

SA: Yeah, I was saying that moving on to weapons, we needed to have more precision weapons. We needed to continue the evolution of where the technology was going to take us with weapons. I found that we were outclassed in precision weapons by the Air Force out there, and needlessly.

EM: It wasn't just that you didn't have the numbers that you needed, it was the types?

SA: Types. In other words, do you have a delivery system to deliver the system with the accuracy that you have to have, and it's an interface between the airplane and the weapon to make sure that happens. That's why the F-15 Strike Eagle was a better platform than the A-6 with all of its gear. So there is some more work to do in precision weapons. There is more to do with airplanes. Again, the fact that we had designed an air-to-air missile system that had extraordinary range and kill capability that basically, in a land war, was not going to be accepted because of the identification process to label the guy as an enemy.

So we had to solve those issues. If we had to have two separate types of identification systems for identifying a foe on our airplanes, then that's the way we had to go. If there was another way to do that with the E-2 and the airplane, then let's figure out so that it matches up. We now know we are going to have to fight the war in a joint manner, so let's make sure that we have the same ability to shoot as our Air Force compatriot does. If you've got two airplanes out there and one's Navy and one is Air Force and one can shoot and one can't in a joint arena, then we've made a big mistake. The charge against the Navy that you only do things your way and nobody else's way is valid in that respect. And there was a good reason for it, because we were always thinking about, we are out in—

EM: In the middle of the ocean?

SA: In the middle of the ocean, and the bad guys are coming after us, and if they are coming after us, there's only one person out there that's bad, and we are going to shoot him. Now we've got to think of it in a different perspective, and so we've got to make sure that our equipment allows us to think in a different way and act in a different way. So hopefully we brought all of those ideas back. I'm not assured that we have. In fact, certainly during my time as Vice Chief we made some strides. We didn't make the giant strides that I had hoped for.

EM: Well, I think you've answered most of my questions. Oh, George Bush made the comment that the Gulf War banished the ghosts of Vietnam. Do you think that is a valid conclusion?

SA: I think it depends on the generation that you are talking to. My generation—Chuck Horner, Walt Boomer and that crowd—I think it did. I think we were all at the age where we had to lead people in Vietnam, and most of us saw it right up until the end. We finished this task [Gulf War], and it was a task that we were told to do, and it was done. It was complete, at that point in time at least. We thought we had accomplished exactly what we were told to do, and we did it with very few losses, and we did it with enough precision that Kuwait was still able to recover. I think from our standpoint—and the other part of it was, we had a lot of voice in how things were going to be done.

EM: The military?

SA: Yeah. We had a lot of input to the process, and although we had a few tweaks from Washington back at us as we moved along our path, for the most part the targeting was done in theater. We operated in theater. There was no breakfast club or whatever, the Tuesday meeting back in Washington telling us what we were going to do next week. We were sort of the masters of our own fates to a much greater degree than we were as leaders during the Vietnam War. So from that standpoint, yes, the Bush comment was valid for us.

Now, all of the other stuff associated, the bad things associated with Vietnam, the race relations, drugs, and all that, basically, again, we did really well. Sadly, I don't think—to some degree society today is not as well settled as we were when we came out of the Gulf War. I think within the service, within the people in the service, it may well be settled, and maybe it should be better. But I think from the outside looking in, the outside world would say there are a lot of bad people that came back out of the Vietnam War, and we don't know much about the—people don't really know much about the Gulf War anymore. I'm surprised about how short the memory; I guess I shouldn't be surprised, because I've gotten older much faster than I thought. [Laughter] You know, and people's minds today, I don't think they—it was a war. I don't know whether they think we are any better than the guys that came back from Vietnam or not.

EM: What do you think of the postwar period, including the continuation of the embargo patrol? We struck Baghdad a number of times with TLAMs and air strikes, and then we eventually went to war again in 2003. What do you think about the state of the postwar American policy toward the region?

SA: The state of the policy is in poor shape, I think.

EM: Did we take the right actions after the war, when Saddam was doing his various things?

SA: I'm mixed on that. On some days I will say yes, we did the right thing. Then on other days I will say no. Again, I'm becoming more reluctant to get involved in what's going on over there, although deep in my heart I know somewhere along the way it's got to get resolved.

EM: Yeah.

SA: Again, I just wish we knew, we had people who knew more about the culture that we are dealing with, and how deeply rooted their beliefs are in what happens. Until we've got a cadre of real statesmen who can sort this out and say, look, if we remove this irritant from the mix, how do we influence the next century? Who do we support? How do we support and keep all of these other entities quiescent to the point where they are willing to wait and see if this will work?

EM: Is that something we can do?

SA: I don't think so. So once you say that, that says keep your finger out of the pie. You can't fix it; just get out of it. Stay out of it. I think we have some really tough times ahead. We've let the genie out of the bottle, and there are other people now who want to take over our role who will face the same problems, except we will pay the price for it.

EM: Well, thoughts to ponder. Do you have any last words about the Gulf War?

SA: No. I guess the one thing that I will always remember from the Gulf War is the people. We asked them to do a lot of things. We asked them, as a word of faith to take it on faith that we are going to do okay. We are going to get out of here, and things were going to be okay. We didn't all know that, but we believed we could. But the people just performed extraordinarily well. Everywhere you turned, there was somebody doing something special. I guess it's always that way, but it's one of the things that from a military guy, you know, people talk about—you get to meet so-and-so up here at the top, and the real joy, the real joy of being in the military was that as long as you wore a uniform, there was an 18-year-old that was no more than two arm lengths away from you, and you could watch them and see them do their thing. Sometimes they disappointed, sometimes they didn't, but when the chips were down, they always—

EM: The rank had less utility at those times.

SA: They really impressed you. Much earlier you asked about the message I sent out before we kicked off the air campaign. Here it is:

> The next few days will set the course of the world order for decades to come. If we must fight, we will. If we can restore freedom to Kuwait without a fight, we will. For you and I, the choice is not ours to make.

If the call to fight is sounded I am confident in your abilities, your courage and stamina. We are joined by professional, talented allies with impressive skills, courage and dedication. We are not alone.

Now is the time to ensure that the final checks are made. Reach inside and put the all-too-normal fear to bed and replace it with calm, cool, collected thought. War is won by those who can think best and make the fewest mistakes.

There's a marvelous line from the book *The Bridges at Toko-Ri:* "Where did we get such men and women?" [Arthur added "and women."] Where indeed? I'm proud of each of you. Good hunting and God speed.

[End of 28 October 2015 interview]

Interview Number 7 with Adm. Stanley R. Arthur, USN (Ret.)
Date: 19 January 2016

EM: Okay, we are on the air. This interview is being conducted with Admiral Stanley R. Arthur by Edward J. Marolda of the Naval Institute on the 19th of January 2016. Good morning, Admiral.

SA: Good morning. How are you, Ed?

EM: I'm very well, thank you. Okay, so we have gotten through your early distinguished career, and now we are on to the Vice Chief of Naval Operations job.

SA: Yes, sir.

EM: Second in command of the Navy. Were you surprised to be nominated for this position?

SA: I was. I was. Of course, after the Gulf episode out there, I was back on the flagship. We'd been back to Japan for a while, and then we were out and about the area and I got a call from BuPers there and they said, "Hey, by the way, you are nominated to be Vice Chief."

I said, "Oh, that's interesting. When is this going to happen?" Of course they were involved with the Tailhook flap back there at the time, and I said, "Oh, you finally found the only aviator that hadn't been to Tailhook. [Laughter]*"

EM: Because you are out of the country.

---

* The annual meeting in September 1991 of the Tailhook Association, a group that looked at issues related to the Navy's aviation community, degenerated into raucous parties and lewd behavior fueled by alcohol in certain areas of the hosting Las Vegas hotel. Female officers reported that inebriated male aviators had groped and verbally abused them as they made their way through the hotel. Lieutenant Paula Coughlin, a helicopter pilot and an admiral's aide, reported the egregious behavior to her superiors. Dissatisfied with the Navy's investigation of the matter, she went public, and the activities at the Tailhook convention soon became heated national news.

SA: Well, I had never been to one either. So, Bob [Robert] Erskine was the flag matters guy that did all of the corresponding back and forth between the building and us guys out in the field. He says, well, you know, there are some problems, but we think this nom [nomination] will go through pretty quick. In the meantime, Tim Wright had already been detailed out to relieve me, and so he came aboard the ship for a while, and I think we went around a couple of months floating in and out, and still nothing. So I said, "You might as well go back up to Japan, get settled. Get to meet everybody up there and spend some time before you relieve me," because we were eventually going to go to India. I had been to India by airplane. I had taken the P-3 to India and did some first talks with the Indian navy. The push was to get a little bit closer to India and find out what they were up to.

EM: So did you go down with *Blue Ridge?*

SA: This was on *Blue Ridge*. So *Blue Ridge* was then going to go back to India. When they first told me I was coming back to be Vice Chief, I didn't think I would be able to make that trip [to India]. Then it looked like, well, maybe I would. So we were on the way, and finally got a call that said, "No, you are going to be approved." I said okay.

EM: Why do you think there was that delay?

SA: I think they were having the problem there on the Hill. Everybody was being looked at very closely relative to had they been to Tailhook and all of this stuff. So then they said okay, it's a go. So I called Tim up in Japan. He flew down and got back aboard the flagship. We were down by Singapore. So on the third of July, there at sea, we had a change of command and I helo'd into Singapore and caught a flight up to Hong Kong. Hong Kong to San Francisco. San Francisco to D.C.

EM: Was that sort of a routine step, from numbered fleet command to Vice Chief, or do others go on to PacFlt or, you know, LantFlt?

SA: Yeah. Usually I guess it's sort of halfway back and forth. Usually the fleet guys, Seventh Fleet and Sixth Fleet guys, have traditionally either gone on to CinCPacFlt or Vice Chief. I think Hunt Hardisty was one that did that. There was a Sixth Fleet guy who was in the running to be Vice Chief. I'm not sure at the same time as I was, but then it fell through. So it wasn't uncommon for a fleet guy to go to Vice Chief, no.

EM: I don't want to make light of it, but who was your supporter in Washington, or did you have someone who was pulling for you for that job?

SA: I think the original guy that was pulling for me for that job was probably Larry Garrett, the Secretary of the Navy, but, by that time he had left. He left the job by the time I got back there.

EM: What was your connection to him?

SA: We had been in the building together when he was Secretary, when I was OP-04, during that period of time. So we had gotten to know each other pretty well. I think he was fairly comfortable with me. I have a feeling—we never talked about it, but I have a feeling that he was probably the one promoting me. We had met in Australia after the war. I had the flagship down there, and he was on a tour, and we had dinner together down there. Our discussions at that time, he was very upset about Tailhook, and we had talked at some length about what was going on back there because I was just getting fragments. I didn't realize how bad it was.

We were talking about it and I said I was afraid that something would happen at Tailhook that year, and I said—in fact I sent a personal message back to Dick Dunleavy, who was OP-05 at the time, the aviation guy on the OpNav staff and said, hey Dick, I'm concerned. I said, "The aviators, although they did a fantastic job out here, they feel like they are not appreciated, that the Air Force has gotten most of the credit. That's just the way the PR business goes, but they have nothing to hang their heads about. They participated at and above the level that their strength was out here." In other words, if you wanted to measure the number of guys that were out there flying airplanes and all of this stuff, they did more than their share, and they did it very well. "Now, we in naval aviation found some things that we need to do much better. We need to get much more involved in precision weapons," and stuff like this, "and that's the job you guys back in Washington have got to take care of. But make sure when you are talking to them out there that you don't let this sort of creep in that they were pushed aside."

EM: But the problem at Tailhook wasn't so much they were dissatisfied with their role in the war. I mean, did you get an inkling that they wanted to blow off some steam?

SA: Well, I figured being a little underappreciated would result in blowing off some steam as well. We'd had inklings before. When I was OP-50 and Ed Martin was OP-05, there was a Tailhook—I forget which year this is now—but Ed went out. Chuck McGrail, who was one of the other deputies for Admiral Martin, he had the 51 shop and I had the 50 shop. Both he and I stayed back in Washington to be the keepers of the flame back there while Tailhook was going on. Ed came back from that Tailhook and said, "Boy, I am really discouraged about the behavior of some of the young guys out there. We've got to do something about that," and we all sat around and talked about it and said, "One of things that has to be done is you've got to talk to the type commanders and say somebody's got to be in charge. Somebody's got to look at this," and I think Ed did. I think he had personal conversations with both AirPac and AirLant, saying, "Hey guys, you need to take a harder look at what's going on at Tailhook."

EM: Do you recall who they were?

SA: Let's see. No, I can't off the top of my head.

EM: I can find out.

SA: I think probably Bob [Robert F.] Dunn was AirLant somewhere along that timeframe.

EM: Yeah.

SA: Vice Admiral [Crawford A.] Easterling was out at AirPac for a while. He may have been out there at that time, I'm not sure.

EM: All right.

SA: But it was known. In fact, the reason I never went to a Tailhook was the very first Tailhook was in San Diego; actually it was down in Rosarito Beach, down in Mexico about the time that I was a JG. At that time the guys who were forming up Tailhook, the Tailhook thing, were mostly the Miramar guys. The fighter guys up at Miramar, and so we S-2 guys down in North Island weren't much involved. But what happened on that first one is, the guys got down there in Rosarito Beach, partied a little too hard. A couple of them were scheduled to go on deployment on the *Bonhomme Richard,* as I recall, and didn't make ship's movement. So there was this laugh and giggling about big partying guys miss movement and they are in hack, and Jennie Lou said, boy, you'd better never, ever go to a Tailhook if that's the way they are going to behave.

EM: Wise advice. Sage advice.

SA: So over the years, Tailhook wasn't something that I wanted to do.

EM: Well, other than partying, Tailhook also involved speeches and addresses by people in the Navy.

SA: As it grew; the first one was nothing but a party.

EM: Right.

SA: Then as time went on, they decided, hey, since we've got this attraction that draws in people, we can have some of the old guys that can tell us the history of naval aviation, and the guys from Washington can tell us what's the future. We can have some professional aspects to the game. And as it grew, it grew to be a professional symposium with some pretty hard partying on the side. So Larry and I had talked about that. He says, "You know, I don't know what's going to happen with all of this stuff. But we've got a problem on our hands back in Washington with this right now."

EM: All right. Well, going back to the nomination process. You were finally called forward for the hearings, right? You had to go through hearings?

SA: No, I didn't go through a hearing, as a matter of fact.

EM: Really?

SA: No. I had the change of command on the third of July off Singapore and Saturday was the fourth. I was in the air flying. I had flown from out there and got back into Washington, I think it was either Saturday evening or Sunday morning, and went to work Monday morning as the Vice Chief.

EM: I thought it would be required at that level?

SA: Yeah, you would have thought so, but somehow or other I escaped that. [Laughter]

EM: Lucky you. All right. In a general sense, what were your responsibilities as Vice Chief of Naval Operations?

SA: Well, the responsibilities as I understood them were, one, I'm the guy that's supposed to make the staff work like it should, make sure that the staff is providing the information up to the CNO that he needs to do his job, and that he shouldn't have to worry about how it's getting done. It should always be there for him in the format that he needs. So I sort of looked at is as just like a squadron XO job. I was the XO to the squadron skipper, and I was to make sure that everything worked right. There was a lot of representation that I had to do that the CNO couldn't; he couldn't meet all of the obligations relative to representation of the Navy at many, many, many things. So there was a representation aspect to it that really was a lot more demanding than I had realized.

EM: You mean giving awards and attending ceremonial functions?

SA: Giving awards, attending ceremonial things, especially social things in the Washington scene, at embassies and stuff like this. So basically, I found out that you could be out five nights a week and attend two events every night if you wanted to.

EM: What did Jennie Lou think about that schedule?

SA: Some of it she enjoyed, some of it she didn't enjoy. But it was pretty hectic, and basically what I finally did with the staff was say look, we've got to figure out a way to keep track of the functions and things, and make sure that I'm not doing double duty, you know, that I have really not been there before. That I don't need to be there again this soon, and try to get it down to maybe three nights a week that I could be out and about. Many times a lot of those were sort of stag events, and so that gave Jennie Lou some relief. I tried to keep it to no more than three night events a week, if possible. That was sometimes pretty hard to do. The schedule inside the building was, everybody and their brother had meetings, and they had to have some representation by the Navy. And then I found out that as time went on—I think it was Joe Prueher who relieved me and took a step in the right direction by tightening up office administration. What we had before that was a shop that was sort of dedicated to the mechanics of doing the paperwork and all of this other stuff.

EM: What OP would that be?

SA: That was 09B. So in there we usually had a brand new one-star. Sometimes we had a very senior captain. All of the other services had a direct deputy to their Vice Chiefs who was a three-star. So basically they turned over all of this outside representation stuff and a lot of the meetings inside the building to that individual, which I didn't have the luxury of doing.

EM: Joe Prueher instituted that?

SA: I think it was on Joe's watch that they finally upgraded the 09B job to a three-star billet, and I think that is still in place today, which is a very good move.

EM: Naval History came under 09B, so I recall that transition.

SA: Yeah. So I think that would have been a great relief. The job was to make sure that everything was working, and that meant attending a lot of meetings, and especially during that time, where the Joint Staff was really taking over the power structure. So there were so many meetings associated with the JROC. It eventually got formed, and so we had that event to cover.

EM: Traditionally the XO of a unit or the Vice Chief, they are considered to be the hard asses of the equation.

SA: Yes. Yeah.

EM: Did you see your role that way, that you had to come down hard sometimes?

SA: Well, you had—

EM: Did you relieve Admiral Kelso of that responsibility?

SA: Yeah. If somebody got in trouble—and of course I inherited Tailhook, so I had to get into that business. So, yeah, sometimes you had to call people in and tell them things they didn't want to hear. Sometimes you had to do things for, let's say cosmetic reasons, like as we tried to work our way through Tailhook; and I'm sure we will talk some more about Tailhook. Eventually what I had to do is, I had to take every flag officer that had been to Tailhook.

You know, a lot of people wanted everyone fired, and I wasn't having any part of that. But I realized that somewhere along the way, there had to be a definitive declaration of where we were headed and what was acceptable and what wasn't acceptable. So working through the process, it eventually came out that I ended up talking to every flag officer who was there, and for most of them, I presented them a letter; a letter that they had and I had, and it wasn't going to see the light of day unless something else happened under their watch. It said look, this is not acceptable, what happened out there.

EM: So it was like a first warning?

SA: So it was sort of like a first warning, and I had one young aviator who had worked for me as a lieutenant in BuPers, he said, "I guess this ends my career."

And I said, "No it doesn't, and don't think that. Just understand that we are closing the door on this event now, and we are making sure that everybody understands where we are headed, and as long as nothing else occurs that you would have had the opportunity to interject yourself into and stop, then it's all going to go away." This individual eventually became CNO. [Laughter]

EM: Who was that?

SA: Jay Johnson.

EM: Oh. All right.

SA: So, you know, yeah, you had to sit some people down every now and then, which is not a very pleasant task.

EM: Right. All right, when you took over the job, what did you think your greatest challenges would be?

SA: I knew that Tailhook was going to be a great challenge. I had not realized how badly things had gotten back in Washington. There was a lot of finger pointing at people, but I didn't realize how severe the rift was between the Secretariat and SecDef in the rest of the building and the Navy staff. You know, there were allegations that the investigation had been whitewashed and people had lied and all of this other stuff, and it turned some of that was true. I had not realized that things had gotten to that point. So as I had to get involved in that, I knew this was going to be something that was going to hang around for a while and we are going to have to make our way through it.

The other thing was that it was very obvious to me from the Gulf War experience, and it had been building up ever since the RDJTF days, my basic feeling was that a naval officer was more joint in his thinking than any of the other services. That in every case where the Navy was involved, we always thought about the air side of the house. We always thought about the at-sea part of the house, both above and under the sea part. And we always had to think about what we were going to do with our Marines. How we were going to support them? Could we support them? Where did they want to go? What did they need?

So our basic instinct was to think of those three things every time we were engaged in anything. But we didn't speak the language. The language had been dictated by the Army and the Air Force. They spoke the language. They didn't think like we did. We—you know, this is very parochial, but I was convinced that a naval officer who had grown up in the operational side of the house was more joint in his thinking than his counterpart in either the Air Force or the Army. It's just that we couldn't speak the language, and we were organized completely different than the other services. The numbers on the staff were not the same, and so before, when I found out that I was going to come back to be Vice Chief, I sat there on the *Blue Ridge* and took out a piece of paper one day and I started creating what I thought the OpNav staff should look like. This thinking sort of started when I was OP-04 when I got involved in getting everything ready to go to Desert Shield/Desert Storm, because we were having the language difficulties then.

EM: When you say language, clearly you think naval officers thought jointly, but they didn't express jointly, joint jargon.

SA: Yeah, they couldn't put it in the same verbiage as the Air Force or Army.

EM: Well then Secretary Lehman, he really downgraded, degraded Goldwater-Nichols.

SA: Yes.

EM: He didn't think that was appropriate, that the Navy, even though the Navy saw itself as always joint, didn't need to be put into that structure. I mean, is that in the same ballpark?

SA: In the same ballpark, yeah. Yeah. So I had drawn up a wiring diagram, and I floated it out to my guys on the staff out there and said, "What do you think? Does this look about right?" And they said yeah, yeah, if we could do that, that would be good. So that was in my hip pocket when I came back to Washington. That was something that I wanted to get to at some point in time.

It just so happened that one day, Frank Kelso walked down and he says, "I want to share this piece of paper with you and get your thoughts." And he gave me a piece of paper that had a wiring diagram that was almost identical to the one that I had.

EM: Well, that's fortuitous.

SA: He says, "This is something with Paul David Miller, Admiral Miller." He and Admiral Kelso were very close, and Paul David was really a brilliant mind, and Paul

David had sent this to him and said, hey boss, I think it's time that you think about doing something up in Washington.

And so he gave me the piece of paper, and I said, "Just a second. I've got a piece of paper I will share with you too," and so I went and found my little wiring diagram, and I said, "I sort of had this in my hip pocket that I drew up out there before I came back, and you will see that if you want to execute the piece of paper that you gave me, just tell me when. We are on the same wavelength, and we need to get some of the three-stars in the building out in the forces, to where we have more three-stars out there so they can be the Joint Task Force Commander when these operations come up and not be like we were at Desert Storm." You know, we had a one-star out there, and then we had to drag Hank Mauz off of Seventh Fleet and bring him down to be the component commander. We should have already had that done. So we got started renumbering the OpNav staff and getting the three-star submarine, surface, and aviator jobs out of the building, consolidating them.

EM: Because you used to have DCNO air, surface, and subs, right?

SA: Yeah. All three-stars.

EM: So those went away?

SA: They went away.

EM: And these were Admiral Kelso and your ideas about the changes that were needed?

SA: Yeah.

EM: Have they lasted?

SA: Yes. They now have the three-star aviator out at AirPac with his deputy as a two-star at AirLant and the submarine guys and the surface guys all the same way. And that was pretty traumatic for some people because they always thought of themselves as, we've got the big daddy of surface guys and submarine guys and air guys with their three-star job right there at the right hand of the CNO. But I think it has worked out pretty well.

EM: So he thought this would be a better interface with the other services out in the operational areas?

SA: Well, I thought part of that would do that, but not as much; the biggest point of getting those three jobs out of the building was more from the standpoint of, we left the job in place for a two-star who then worked under N8 directly. So that took out a little bit of the parochialism. The parochialism had to be fought down at a lower level than at a higher level.

EM: N8 is what? Resources?

SA: Resources, yeah. But it freed up some three-star billets that we could put in the operational game out there, where they could be tapped to be the Joint Force Commander for a standup operation. So it gave us a lot more visibility in the joint world with a three-star.

EM: Is that why it took a while to get, because your recommendation when you were ComNavCent was that this job ought to be at a higher level.

SA: Yeah.

EM: And it took what, four or five years before that came into place. Wasn't Admiral Kelso reluctant at first to go that route?

SA: He was reluctant out there, yes. He was still not convinced that he needed a fleet commander out there. But eventually it was going to happen one way or the other, especially as things started to unravel out there.

EM: Did you get a reaction from the Joint Staff or from the other services to this?

SA: Yeah, I think they were all pleased to see that they now could understand the OpNav staff a little bit better. That the OpNav staff looked much more like their Air Force staff or Army staff in the building. So they knew who the guy was back in Washington that was on the Navy side of the house, they could look at their guy in their own service and say, who do I need to talk to in the Navy? It's got to be a guy that has sort of the same number as this guy that's on my staff, and we fixed that. They didn't have this confusion of having to ask some captain out there, who do I talk to on the OpNav staff? I know the number, because it's like our number.

EM: So the residual Tailhook problem, talking about reorganizing the staff. Were there other challenges that you saw when you came out? How about on the operational side, like weapons?

SA: Well, the big lesson learned from Desert Storm was that we had to get into precision-guided weapons much more than we had. The thing that really disturbed me when I got out there for the effort was that the Navy ended up being excluded from some target sets because they didn't have the precision-guided weapons that they should have had. From my OP-50 days I knew that we had worked on them, and I was a precision-guided guy because of my last squadron when I had VA-164.

We—in the '72 campaign out there in Vietnam, I had the only designators for laser-guided bombs. They were hand-held designators, and we put them in two-seat A-4s. We could go out and bomb with the Air Force on targets that were cleared for LGBs, for laser-guided bombs. First of all we could deliver them, but we couldn't designate the targets because we didn't have a designator. China Lake built us two boxes that we took out there and got the Marines to give us two combat-configured TA-4s, two-seaters, and then we put guys in the backseat that could use the designator and we could self-designate then. We could go out on a flight and have two bombers and one designator, and we could do it just like the Air Force did.

EM: So as VCNO then, did you put a push on certain weapons systems?

SA: I did. I pushed on, you know, let's get the laser-guided stuff so that we can self-designate. Let's go to the next generation, whatever that might be. We'd used a lot of Tomahawks out there. I had chartered a CNA staff to come out and revisit the whole Tomahawk thing for the war. We started out right after the war, and Bob [Robert J.] Murray was CNA at the time. He sent out a team and they went to work. What I wanted them to do was reconstruct every Tomahawk that we fired and find out exactly where it ended up as best as they possibly could. In the process of doing that, I ran into monumental disagreements with the OpNav staff. They didn't want it done. They didn't want anyone to know what, whether the Tomahawk was as good as some people thought it was or whether it was bad or not. They didn't want to know the answer. We don't want to know. We don't.

EM: Does that mean there was a little bit of concern that it might not come out as well as people thought?

SA: Yes, a big concern. The systems guys were against it. We just ran into all sorts of stuff. Fortunately I came back to be Vice Chief, so the first time the team that was putting this together for me at CNA came up and said, hey, we can't get access to this stuff, I said luckily I'm now here and we will fix this. [Laughter] So we did. I got them into all of the classified arenas and they were able to reconstruct every single Tomahawk.

EM: What were their findings?

SA: Basically the findings were good. They are never as good as you are led to believe at the beginning, but they were much better than we expected. And the reason I wanted to do it was, if they didn't perform in the manner that we thought they would, why didn't they? What were to root causes of why they weren't there? We found one right away after we got access to all of the information. One was, as you may know, the Tomahawk planning at that time was done off-site. In other words, the only two planning cells were CinCPac and CinCLant, and they had a designated cell that planned the Tomahawk routes and the accuracy. In other words, you would send back, "I want to hit this point here," and then they would develop the route package and all of the final stuff that you needed to get the pinpoint accuracy. There was an error in—I won't tell you which command—but there was an error on their planning table.

EM: Why not say which command? It's history now.

SA: Yeah, well it was Lant. The Lant planning cell had a flaw in their planning table that the Pac didn't. Not a large inaccuracy, but there was enough of an inaccuracy that for the warhead, it really diminished the returns on what we expected to find. So we found that right away. All along the way, we just found other things that said, hey, these are all fixable items. These don't mean we have to go buy a new weapon system at all. All we need to do is make sure that we account for these different things that we were finding.

So it was very productive, and I think—again, I'm a little prejudiced, just because it was what I had hoped to accomplish—the Tomahawk is a much better system than it was when we first started using it. Much better. It's those kind of things that you can do. It was just serendipity that I ended up coming back to be Vice Chief. Somebody else would have agreed with the rest of the OpNav staff and the systems commands, as I say.

EM: Well, I think that goes to show the wisdom or having someone who was an operator in a key operation involved in all U.S. forces come back to put some of that experience into play.

SA: Yeah. And we do that, and we are not very good at that. You know, we talk about lessons learned and history and all of this stuff.

EM: Going back to the TLAMs though, I recall one civilian analyst, Barry Posen I think was his name, who came out with a report, because earlier you had the figure 85 percent. The TLAMs were fired, they hit 85 percent of their targets, and he said, no, not quite, and he really downgraded their effectiveness. Where did his analysis come out during the process? Do you recall him?

SA: I do vaguely.

EM: He's with MIT I think.

SA: Yeah, and I don't know where he got all of his information. But a lot of the information was gathered by some of the visuals. As you recall, we had the visuals of the Tomahawk running down the highway and by the hotel windows and all that. So from some of that, we knew that there were some misses there that we thought should have been hits, and a lot of that information was out there. And he wasn't far off of the mark.

EM: But your study validated them?

SA: The study validated that, yeah, a lot of the things that we weren't happy with were an error not on the missile side of the house, but on the planning side of the house. So that was an easy fix.

EM: What about the size of the warhead? Was that a concern?

SA: The size of the warhead was always a concern to me, and why initially, in the years leading up to the Gulf War, when I was OP-50 for example, I was not that enthusiastic about Tomahawk as much as a lot of people were, because the early Tomahawks had a Bullpup A warhead in them. I haven't fired the most Bullpups in history, but I've fired a lot, because I ended up—when I was in VX-1 as a project guy, I had the Bullpup integration on P-3s. So I spent a lot of time out at Point Mugu, and I probably fired 50–60 Bullpups out there, and then when I got in an attack squadron I fired some more.

Not for much use; they were—first of all our training for the Bullpup was all wrong. I had trained with the French on their wire-guided missile, and their training was superb. They trained you to hit. Navy trained you to be close. Everything was CEP in the Navy. Everything for the French was hit or miss. There was no CEP. It was you hit.

EM: Circular error?

SA: Probable.

EM: Probable, yeah.

SA: You either hit or you missed, and that's the way they trained you. So I was very disappointed with the Bullpup warhead, the performance of the warhead. I always had that in the back of my mind, that as long as you guys are going to use the daggum Bullpup A warhead, this thing is not going to do what you want it to do.

Of course, time went on. They went to the Bullpup B warhead, the bigger warhead, and that started in the right direction. Then they've done other things since. So I was not that super big on Tomahawk until I got out there, and then we started firing them, and then that's why I wanted that study. First of all, I thought we would find out some things about the warhead that we needed to find out some more about. Surely there were some other things going on with the Bullpup or with the Tomahawk that we would find out, and we did. So on the weapons side of the house, we had to do better. I brought that back as well.

EM: What about aircraft?

SA: Aircraft. You know, we got a lot of criticism from the A-6 community because basically I came back and said the A-6 was the Spad of the Gulf War. We had the Spad in the Vietnam War, while the A-6 is the Spad of the Gulf War.

EM: Yeah.

SA: It is not a survivable platform in the world we live in today, and we've got to do better. We've got to get something better.

EM: So survivability was the key thing?

SA: Yeah.

EM: Yeah, or lack thereof?

SA: Lack thereof. Their survivability was all built on, go in low and fast, and that only works for just a short period of time. First strike, and after that if you are down low you are not going to survive. So the A-6 community was very unhappy with me, and a lot of the guys, they would still be throwing things at me if they could find me, I think. [Laughter] We had—from my OP-50 days, we had already started the replacement. We had all of the arguments about was it going to be a side-by-side like the A-6, or was it going to be fore and aft. Of course, John Lehman wanted the side-by-side version, and most of the aviators wanted the fore and aft, although many of the NFOs liked the side-by-side too. So we had already—

EM: For practical reasons or for sentimental reasons? The NFOs like to think they are the equals of the pilots because they are side-by-side?

SA: Well, I think they feel like, yeah, that their situational awareness is probably a little bit better if they are up there side-by-side than fore and aft. I don't know. You will have to ask Joe, who will probably give you some words on A-6s and the replacement. We started a replacement program, had it gone, but of course it got killed.

EM: The A-12?

SA: The A-12. Then we put all of our eggs in the basket of the F/A-18 and went to the E/F.

EM: What was your take on the F/A-18? Did you think it should be upgraded?

SA: Yeah, the 18 needed to be upgraded, and so the E/F was a step to getting that to fix a lot of the issues with the first generation F/A-18s. The E/F sort of took care of some of those things.

EM: Were you pushing for the E/F?

SA: I pushed for the E/F. More on the E side of the house at the beginning, and then as we recognized that the F-14 was getting long in the tooth and that the F version of the E/F would probably make a reasonably good substitute for the F-14. Certainly not as good in the air-to-air business as the F-14 was, but of course the F-14 was never allowed to use the Phoenix. It never shot a Phoenix in combat. So some of those features that the F-14 had were never used, and we made a bomber out of it and it was a good bomber, but you could make another good bomber cheaper than using it. So the E/F was good.

I challenged McDonnell Douglas that the F was going to be a viable part of the program. That's what I pushed on them. I had a couple of meetings there in my office with people at McDonnell Douglas that I knew well and said, "Look you guys, the EA-6 has been a wonderful platform, but it's got some of the same problems as the A-6 had. Now, without the A-6, the commonality is going to be a problem." One of the other things that I brought back with me as a to-do as the Vice Chief was to figure out how to get more people out of the equation. In other words, what I used to talk about in the meetings with the people was, look, if we've got a two-seat airplane today, we should figure out a way to make it a one-seat airplane. If we've got a four-seat airplane today, EA-6 being an example, we ought to make it a two-seat airplane. We've got to try to get more people out of the equation, because we spend more money on people than we spend on our systems.

EM: Yeah.

SA: And if we keep going the way we are, I said, we are not going to be able to afford the platform. We will afford the people, but we won't have the platforms for them.

EM: Was that putting too much on the individual who has got to handle all of the tasks?

SA: Well, figure out a way. You know, if it's too much, then maybe we take three people instead of four. But the entering argument is cut it in half.

EM: Yep.

SA: Then work it from there. Now we have the G, the two-seat replacement. I don't know how effective it is. I just don't know what's going on out there now. But those were the things that I came back with, that we had to be able to play better in the joint arena. We had to have better weapons. We needed to really think about how many people we needed to do for a job and start figuring out how we were going to afford the systems that we were going to have to buy.

EM: All right. Off to your VCNO days. Who were your principal subordinates that you recall?

SA: Well, the guy on the Navy staff that I had probably the most dealings with was the N8, who was [William A.] Owens to begin with, and then turned into Joe [Thomas Joseph] Lopez after Bill Owens. [Leighton W.] Snuffy Smith was the operator, the plans and policy guy. Steve [Stephen F.] Loftus was the N4, the logistics guy who had relieved me, and then after Steve was a surface guy. Bill Hancock lives right up the road here in the Villages. Jim [James] Hinkle was my EA [executive assistant]. Pete [Peter H.] Daly worked with Jim Hinkle in the office.* My aides, I had two aides. I had a Marine aide and a Navy aide. So they were great people.

* Vice Admiral Peter H. Daly, U.S. Navy (Ret.), became CEO of the U.S. Naval Institute in 2011.

EM: Did they go on to greater things?

SA: My first Navy aide, Corey Tisthammer, he got out. He was a helo pilot. He's had a very successful career with, I think it is General Mills. He's up in Minnesota. He got married while he was on the staff. His replacement went on to be—is right now naval attaché to Spain. My Marines, first guy went off and got a battalion after he left, and then his wife was a Navy nurse and I think he retired as a colonel. Then my last Marine aide, Darren Kearns, he stayed in and got medically disqualified for diabetes. So he ended up retiring I think at the 20-year point. We stay in contact all the time. He's had a very successful career in industry and in the defense industry.

EM: Who was the female—I sat next to her on the plane on one of my trips down here—and she had been on your staff, and I think afterward you exchanged emails. Do you recall her name?

SA: I'm trying to think.

EM: She was a commander.

SA: Oh—

EM: I'm sure I've got it somewhere in my email files.

SA: Yeah, I remember the email and I'm trying to think now. Gosh, I can't.

EM: Well, I will track her name down.

SA: Yeah. But the other thing with the Vice Chiefs job was you really got to see better the continuity that the senior civilian staff brought, what the SES guys brought to the game. When you worked with them, when you had your particular SES that worked with you in your OP-04 job or in the OP-05 arena, from your standpoint they were narrowly focused on that job, but as the Vice Chief you got to see that they had a lot of influence around the building. They had their network, just like you had your network, and were really, really top-notch people. Irv Blickstein, who had been with me in the 80 shop, the 90 shop when I was a programmer, he could do almost anything. He works for RAND Corporation now.

EM: And Nemfakos?

SA: Charlie Nemfakos was a very polarizing element in the staff. Really good.

EM: Why is that? Outspoken?

SA: He was very outspoken. He knew his business better than anybody could possibly know, and he didn't suffer fools lightly. So you had to tread carefully with Charlie, but I tell you, if you needed something done that was in his area of expertise, he could deliver. He was a phenomenal guy. He also works for RAND Corporation now. So RAND is very good about picking out talent and hiring them.

EM: The staff during your time there, were you reasonably satisfied with its performance, or were there certain areas that you said, we've really got to make some changes here with people or with functions?

SA: I think the one that was nearest me was the 09B organization that we really needed to do. We really needed to do more with them, and I didn't get to the point that I should have I think with that organization. I think Joe Prueher did. The staff had a little trouble with the reorganization. There were some ripples in there, but they smoothed out rather rapidly, I think. Joe Lopez was a great influence when he came in and relieved Bill Owens of the job. Both of them did superb jobs. Joe was more of a mechanic than Bill. Bill was a big picture guy, and when he became Vice Chairman that skill came to bear in a hurry at that point. So I was well served with the staff.

EM: No real problems with individuals?

SA: No. No, you know, of course, your memories tend to remember the really good things and the really bad things. I don't remember any really bad things with the staff. There are always some ripples. I found out as Vice Chief that there was never a problem so little that it couldn't make it to the Vice Chief's desk. So you sort of played with a lot of nit and grits, and things that you just didn't think that you would ever have to be dealing with at that level. But on the other hand, the fact that they could get to the Vice Chief's desk meant a lot of people worked very hard to make sure they corrected a lot of things before they had to get there. [Laughter]

And you could see that, you could see that happening. In fact the EA would say, "Oh God, boss, we are going to send you in a paper here in a little bit that's just going to drive you crazy," and the next thing you know, he'd say they just pulled that paper back. "We will wait and see—" [Laughter]

EM: They took care of it. [Laughter] Which I guess was the best outcome, right?

SA: Right.

EM: Didn't have to come to your desk.

SA: That's right.

EM: Could you describe a typical day as VCNO, if there was such a thing?

SA: Yeah. It was complete chaos. There was never a minute that you weren't scheduled for something.

EM: When did you get to the office?

SA: I usually got to the office about 6:15. I would get up there and try to get the day started, and of course people who had to get to you would recognize when there was an opportunity, so a lot of times somebody was waiting for you right there. But you sort of got into the routine. What happened overnight? Was there anything that you had to get on top of right away? What was the next schedule? But you were scheduled—basically you had the morning brief at about 7:30 or 8:00 I guess.

EM: Your staff would come in to brief you?

SA: Well, the staff would come in and brief me a little bit earlier than that, and then we would have the operational brief in the OP center with the CNO about 8:00. Then the schedules of meetings would start. There would be a 9:00 meeting, a 10:00 meeting, and right up through noon, and then I usually ate lunch. I had lunch brought down to the office from the SecNav mess. I would have a little salad and a soup, and that was about it. So I would take about 30 minutes, and a lot of times people would drop in and we would sit there and talk over an issue while doing that.

Then the routine picked right back up at 1300. You'd have meetings every hour on the hour, right up to 5:00 usually. Then you would hope that the 5:00 meeting was going to be short. Usually had an end of the day meeting with the ACMC, the Assistant Commandant of the Marine Corps. I would meet with the Under Secretary, especially during [Richard J.] Danzig's days. We would meet up at his place and spend 30 to 45 minutes going over, okay, what have we got headed your way, and what are the things that we need from you? What do you need from us? What have we done that's disturbed the Secretary's staff? Because there was always a lot of head knocking with that, and so that was the day. I mean, it was just one meeting after another.

EM: Then you would go home?

SA: I would get home about 7:00 usually.

EM: Right, then off to—

SA: Yeah. Then off to evening events. Had a driver. We had a driver, a civilian driver, so a lot of times he would pick up the wife and come to the Pentagon with her, and I would just go out and hop in the car and off we would go. So that was a day in the life of the Vice Chief.

EM: Did you get out of the building a lot, or did your responsibilities pretty much keep you there?

SA: Most of the time in the building, but there were lots of things that happened outside of the building that you would be off and doing. I mean, mostly representation type stuff and award ceremonies, you know, go over to NavSea to present some awards. But no, normally we brought them into the building to do it. It was nice to get out every now and then. Go up the Hill or to BuPers for something. But most of the time you were pretty locked down, unless you were on travel to go out to the hinterlands a little bit.

EM: You mentioned Under Secretary Danzig. How was your relationship with him?

SA: It was fun. The other thing that happened when I first came in as Vice Chief was, we had a lot of turmoil relative to Tailhook and things. Secretary [Larry] Garrett had left. Sean O'Keefe was in as an interim. Then he was the Secretary. Dan Howard was the carryover from Larry Garrett through Sean, and then on the 20th of January, they all left because of the Clinton election in November. So on the 20th of January they all left and we didn't have an Under for several months. Danzig, I think it was almost in the fall before he came aboard. Kelso was Acting Secretary for a while. Then Frank left early, you know. There were lots of things going on relative to—not only were we still fussing around with Tailhook and all those problems, but then we had all of this turnover of the staff with the election, and in typical fashion, the Democrats were always—I will get in lots of trouble for saying this—[Laughter]

EM: Go ahead—[Laughter] What are they going to do to you?

SA: My reflection: in the building, you could tell who was a Republican and who was a Democrat by whether they maintained their schedule. Democrats had no clue about a schedule. If they had an idea it became the most important thing, and in the middle of anything they would stop and wander off this path, and in the meantime things that had to be decided were just forgotten. So first of all, they couldn't make up their mind who they were going to bring in the building to hold down the jobs, nor how to get them approved. So you had to switch your mind. Typically the Republican incumbents of jobs were all—time was money, and they—boy, in and out, in and out, in and out.

EM: Many businessmen.

SA: Yeah. The Democrats tended to be lawyers. So I remember my first meeting with Danzig when he was going to be announced. He wasn't authorized to sit in the job yet, but I invited him down for lunch and we talked about things. I said, "The most important thing on my mind, the thing that I want to transmit to you is, we are not your enemy. I've had it up to here about the military guys not being part of the team. That we are off doing our own thing. Nothing could be further from the truth." I said, "You will never find a group of people who are more convinced that civilian leadership is a must, that we don't want to be in charge of ourselves. We want to perform the mission that we are designed to do, and that is be an element to protect the United States. We are not wanting to own or rule the United States, and so I hope that you will come to the job with the understanding that we are going to be truthful with you. That we are going to do the best to make sure that we do what it is you want us to do. If we have concerns about what you want us to do, we are going to tell you. We don't want you to take it personally. And if you say I hear you but that's not what we are going to do, we know how to say we are going to execute, *boom.* We will take care of it."

EM: What was his reaction to that?

SA: Well, he's such a pleasant guy and a bright guy to be around.

EM: Historian background, I might add.

SA: Yeah. So, he said, "Yeah, I understand it." Now, but he couldn't keep time. I mean, it was just not in his nature to keep time. But he was fun to work with. He had a quick, agile mind. It was fun. He still gave off a little bit of the image of "Well, I'm not sure you guys are—" Another thing was, a lot of times you got the feeling that they didn't think our education was what it should be. Part of the time that was true. [Laughter] But in general we were reasonably well educated, I guess. But we all didn't attend Ivy League schools.

EM: Right.

SA: But no, I found him to be easy to work with. We had lots of differences. I could push back at him and he would let me push.

EM: What were those differences? Do you recall anything?

SA: Oh, no. Most of the differences revolved around alternatives. He would push hard for more alternatives, and basically my pushback was, look, if you've got something specific that we haven't thought about, yes. If you want us to go revisit a bunch of the alternatives that we already discarded before we got to this point in time, it's a waste of time unless you really think that we didn't do the job that we were supposed to do. We didn't just bring you this one solution. Yes, there is one solution here, but before we brought it, there were many solutions that we went through before we got to this point. Now, if you want us to go back and revisit all of these things with you, then why don't you come in at the process at the beginning instead of at the end?

EM: What does the Under Secretary routinely look at? Are they interested in weapons systems or the personnel side of things?

SA: They usually send the weapons system and stuff like that over to the ASNRDA [Assistant Secretary of the Navy for Research, Development, and Acquisition], and that's another discussion we need to talk about, is the RDA. So the weapons procurement side of the house, unless you are Mel Paisley—now, when Mel Paisley was in the building, you know, everything sort of went through him one way or the other. But of course he had that role, and the Under during most of Mel Paisley's time was a budget guy. So those things were back in my OP-50 days. But for the Under, most of things revolved around personnel, personnel-like issues, health and—

EM: So Tailhook was still an issue when Danzig came in?

SA: By the time Danzig got there, let's see. I'm trying to think.

EM: Look at your timeline.

SA: Look at my timeline here. We still had some work to do with Tailhook under Danzig, yes. Most of it was resolved. But when Kelso left, Danzig still wasn't on board. He hadn't taken over the job yet. So [John H.] Dalton, Secretary Dalton, we had to deal with him a lot right during that period of time with him without an Under, so to speak.

EM: All right.

SA: The Tailhook went on a lot longer than it should have. But for Danzig, of course, women in the operating forces and stuff were all hot topics by that time. We were still fine-tuning some of the OpNav realignment. But generally he was sort of involved with the popup issues of the day, you know. Why did this happen? Who was involved with this? How are we going to do this and how are we going to do that?

EM: Now I would like to ask your impression of some of the folks you worked for, some at a distance. For instance, Colin Powell, you had contact with him when you were out in the Gulf.

SA: Yes.

EM: During Desert Storm, and he was Chairman of the Joint Chiefs through October of '93, so a good bit of the time that you were there. Did you have any contact with him?

SA: I did. Met with him shortly after I got to the building. He wanted to see me, and I went down to pay a call on him. We had the usual pleasantries, and basically he said, "Now look, you need to bring the Navy along into the joint arena." [Laughter]

And I said, "General, I know that. I will do it. I will do my best to get us so that we can play a little bit better in the joint arena. I'm fully on board with that." Then we talked a little bit about cars. He liked to restore Volvos, and so we had our pleasantries, and then I would see him quite frequently when I represented the CNO down on the Joint Staff for the meetings there. So we had a very cordial, friendly association.

EM: Professionally speaking, did you have disagreements over the way the service should go? Any other major issues?

SA: During the Gulf War there were a couple of things that I thought, basically, he was not in support of Navy getting credit for certain things.

EM: Which?

SA: Just—for an example, on one occasion we had the battleships out there, and I pretty much told him that if we want to make sure the Iraqis believe we are going to do an amphibious landing, I've got to make the battleships more visible. Because in their minds, in anybody else's minds, the battleships are there for one reason and one reason only. And that is shore bombardment. And shore bombardment is meaning we are getting ready for the Marines. So, I said, "These ships, when this is over, those two battleships are going to go back, and we will never them at sea again. And we've got 16-inch rounds growing out of our ears here, so my philosophy is, if you ever give us a chance to shoot them, we are going to empty the magazines on those things."

Well, that was all well and good until all of a sudden we had the one battleship come up from the south and go inshore. We had it led in through the FFG-7s to bring it in, and we could lob into the southernmost outpost of the Iraqis, and we had drones. We had drones, and they could see the meal trucks coming down in the evening, and they had a little bay there that they had their little coastal craft in. So we got clearance, after we figured out how to get the 16-inch rounds into the ATO—[laughter] because they went so high. So we were doing well. All of a sudden the press got ahold of—boy, the battleships are out there shooting up; boy, the word came from Powell right away. Shut down the battleships.

EM: Too much press?

SA: Too much press. That's my interpretation. The politics of who did what was still alive and well, even under his tenure. But in general, for the most part I was comfortable with him. I thought he treated me well.

EM: Were there any programs or weapons systems that the Navy wanted, but he was reluctant to go with?

SA: No. He didn't get involved in that. I mean, the Vice Chairman really took over, and Dave Jeremiah was Vice Chairman. Dave took over the JROC, and he ran the JROC with a good, steady hand. Dave was—of course he's an old programmer, Navy programmer. He knew how to program. He knew the business really well. So he brought a lot of order and discipline into that new process, the JROC, which was really, you talk about a time consumer. That really ate up the Vice Chief's time.

EM: I want to get into that.

SA: Yeah. So, from that standpoint, Colin stayed very close to the political arena, very little into the weapons.

EM: What is your impression of his role as Chairman of the Joint Chiefs?

SA: I thought he did a good job. I know his concern with Norm. I think he probably got a little too much engaged with Norm's business, but on the other hand—

EM: During the Gulf War, you mean?

SA: Yeah. But in the long run I don't think he overplayed it, especially after the war. He stayed pretty even-handed in the political arena and kept a lot of that stuff away from us guys.

EM: You had the Weinberger Doctrine, which he adhered to. Do you think that was still valid after the war? He had pushed that himself, and my understanding was hit the enemy hard and fast and get it done with.

SA: Yeah. Build it up to the point where you had the overwhelming force, and then go get it and get it done. Yeah, I think he still, he still felt that if we were going to engage, that's where we had to be.

EM: Were there any things that you were less than satisfied about his role as Chairman?

SA: No. No. I was very comfortable with his role and the fact that he let Dave Jeremiah as the Vice Chairman really take over, be the guiding light on how we did business. How we did military business per se, but how we did the business of the building. How do you put together the POM, and how do you get the programs together, and who makes the decisions, and all of that? I thought that he was well served by Dave, and I think Dave was well served by him as flying cover and giving him the authority to do that.

EM: Okay. John [M. "Shalley"] Shalikashvili became Chairman after Colin Powell. He was there from '93 and you left the job in—

SA: '95.

EM: '95, right, so a couple of years with him. What was your contact with him?

SA: Shalley—I had contact with him of course in the joint arena. Shalley, early on in his tour, we ended up on the case of [Michael] Scott Speicher, our aviator that we lost on the first night of the war. We spent a lot of time with him and the SecDef.*

EM: That's pretty low down in the greater scheme of things, the loss of one pilot, but the Chairman of the Joint Chiefs was interested in that issue?

SA: Well, do you know how that issue all—

EM: Well, whether he was captured or whether he was missing or, you know, the categorization of Speicher went on forever.

SA: I guess this is probably a pretty good time to talk about it. We lost Scott that first night, and initially there was a lot of information that we were trying to sort through. Some said he was shot down by an air-to-air missile. Some thought he was shot down by a SAM. It was at night, there was a big fireball, and he was gone. So you know, the inter-service rivalry started right away—was he, yeah, he was shot down by a MiG, and it's the Navy's fault that they didn't have the right F-14 or something. And others, it was a SAM, et cetera.

EM: Was there any thought that it might have been friendly fire?

* Michael Scott Speicher was a Navy pilot shot down in Persian Gulf War. Some information, later refuted, suggested that he might have been a prisoner of the Iraqis. His remains were recovered in 2009.

SA: Yeah, there was a glimmer at one point, but that was pretty well resolved very quickly. That all went to bed, and basically we never heard from him again. We didn't have any indications that there was anything. Then one day when I'm back there as Vice Chief, one of the people in the black world that worked for Snuffy Smith in the Ops area, Ops and Plans, came in and said, "I need to brief the Vice Chief on a sensitive subject." So it was an eyes-only type of brief, and basically we had information that the wreckage of his airplane had been found and we had some photos of it. They showed me the photos, and I said, oops. I said, he got out of the airplane. He did not come to ground in that airplane.

EM: Why? The seat had been shot out?

SA: The canopy was—so I said, unless somebody got him out on the ground and carried the canopy off, if the canopy is where it was when that plane landed, and it landed pretty flat, so the airplane was more intact than we would have thought. And it was out in the wilderness, I mean, it was way out in the wilderness. So we started to go back into the system to find out could we get better photos of the thing. The first photos we had were from a hunter, from one of the other Gulf Coast countries that had gone out hunting in that area. So everyone got together, and we looked at all the relevant information on Speicher to see if we could figure out how to get a better picture of his loss.

And we got some other intel, photographic intel that said, yeah, you know, there is a possibility that he in fact did get out of this airplane. So now what are we going to do about it? Well, that's how we ended up with Shalley and the SecDef and me and one of my guys and a couple of guys from the Joint Staff in a closed-door meeting about what are we going to do about this. How are we going to get in there to find out what's going on? The Special Ops guy on the SecDef staff, who I had never worked with before, we worked independently on what each of us knew. We both had the same information. So this meeting was to sort of figure out where all of the players were on this. He and I ended up being on the side that said, look, this is such a remote area that we can get a team in there, a professional team to take a look at it and tell us, very rapidly, whether they think we've got somebody that survived the crash or didn't survive the crash.

EM: The site is in Iraq?

SA: It's in Iraq.

EM: Yeah.

SA: But way on the western side. So the SecDef and Shalley were on the side that said no. You can't do that. If we get somebody in there and they happen to get caught, we will be embarrassed and it's not worth the risk. We will have to go through the Red Cross, which was an alternative that we had looked at, of course. Basically if you go through the Red Cross, it's a leaky sieve. As soon as you tell anyone in the Red Cross, it's going to be all over the place. "Well, no, we think there are ways that we can protect the information for a while and then look at the exact location," and all of this other stuff. "So no, we are not going to authorize a covert operation to go in there and do that." Well, the guy on the SecDef staff went ballistic and he wrote about it later on.

EM: Do you recall who that was?

SA: I've been trying to think of his name and I can't remember his name. I can find it in a wiring diagram though. Maybe we can research that.

EM: Sure.

SA: But he was really upset, because he was convinced that we could do it. I was convinced that we could do it too, but the decision was made that we were not going to do it that way. This went on then, I think, for almost a year before they finally got the Red Cross team in there. Of course, it had already been staged by Hussein.

EM: But eventually his remains were actually found close by.

SA: Many, many years later, in '09 I think it was.

EM: So maybe the remains had been moved from somewhere else?

SA: Well, I don't know. By that time I was out and I didn't have access to the information, and they were very reluctant to share much with me after I got out.

EM: Were there other major issues that you recall with Shalikashvili?

SA: No. He was not as decisive as Colin Powell was. But again, you had the dynamics. Colin Powell worked with a very decisive SecDef, Cheney. The two of them, *boom,* went *boom,* decisions made. All that was gone with Les [Leslie] Aspin [Jr.] and then [William J.] Perry. It was much softer.

EM: Cautious?

SA: Yeah, very cautious movement, and Shalley—in my mind, Shalley was really a nice guy. I appreciated his demeanor, and he was a very friendly guy. But again, I didn't see the same quick—he was a much more thoughtful guy. He worked the solution down a few more steps then you would have seen with Colin Powell.

EM: Did he appreciate what naval forces could do, Shalikashvili, because he had a really strong Army land-warfare background.

SA: Yeah, yeah. I never really got to see him in a place where he had to make that kind of a choice about Navy. My interface with Army guys, traditionally, is that they have very little appreciation for what the Navy does. It's out of sight, out of mind.

EM: Okay. Let's go on to a major function of yours. I've gathered this just hearing you talk about it before today and today. The JROC, Joint Requirements Oversight Counsel.

SA: Yes.

EM: What do they do?

SA: They are the arbitrating body of what programs each service is going to be able to put into their POM, into their program. So you have to get the JROC's blessing for any major program that's going to come in. We ended up building a staff on the Joint Staff that is bigger than any of the services. You basically have to come and plead to the JROC. All of the Vice Chiefs sat on the JROC, and we all had a deputy within our own organization that was part of the JROC team to bid for a program.

So each year, as you finally put the POM together, you are forced to come to the JROC with the big programs and get the blessing, whether we are going to continue it, or if it's a new program whether it's going to go into the program or not. So it is a big consumer of time and effort. It becomes a bidding war, and the most clever guy there may or may not get his own thing.

EM: How often do they meet?

SA: Well, as the POM period goes, you meet at milestones for programs, and for major programs where you have milestone events. The JROC is involved with those milestone events as well. So part of that is not only do you have to get your systems command and everybody else together and get your act together, and then you have to go brief the JROC on that, and then at the end of the day when the final POM is being put together, you have to come back in and defend the numbers or whatever on those same programs. The working guys are meeting routinely on a weekly basis. The principals, we would meet probably on average at least once a month, sometimes more often.

And a lot of paperwork associated with it, where we would get briefs from our guys and have to clear them to go on and then go down and brief the JROC at the working level. So we spent a lot time, probably one major meeting a month and then several two or three minor meetings where we might not meet as a JROC itself but in preps for JROC. You could lose your shirt down there if you weren't careful, so you always had to be prepared. With Dave Jeremiah—since he had been the Navy programmer at point in his life, he really knew how to do it. Bill Owens, of course, had the same background. Bill came to the job with a much more forward-looking bent.

EM: Owens followed Jeremiah. Two Navy guys. You would think they would rotate the service?

SA: Yeah. They didn't in that particular incident.

EM: Do you know why?

SA: No. I don't know particularly why. Let me see on my timeline if I can figure out why that might have happened. No, I don't know exactly why. But Bill was quite the intellectual. Bill Owens was a very, very bright guy. He had been the N8, so it had helped a lot in the reorganization effort of the N8 organization when Kelso and I had sort of formed and left it to Bill to put all of the pieces together, which is really the job of the guy. So his going to the JROC was a plus for the Navy. It was not well accepted by the other services as much, and Bill didn't do a whole lot to endear himself to the other services, because Bill was convinced that the wave of the future was that there was no place to hide. That the technology would give us perfect—I think these are the words he would use—perfect and profound knowledge of the battlefield in all cases. That basically the matter was all solved by technology, and that all we needed to do was get on with the business of finding the technology.

And to that I sort of said, "Well, wait a minute, there are a lot of systems that we have right now that need to be upgraded, and we need to get to the next generation."

"Well, we really ought to be getting to the generation after that."

"But we can't get there, Bill. We don't have it in hand yet. We don't want to get caught with not being able to get there to begin with."

So, well, "No, you guys just aren't thinking big enough."

So that tension was always there. I will give you a classic example of a meeting, and I think we may have talked about this earlier. We were having so much pushback to

Bill, and Bill was doing a great job within the Secretariat, within the SecDef staff. He would have these breakfast club meetings with all of the principals on the SecDef staff.

EM: SecDef or JCS?

SA: SecDef. So he would have all of these meetings. He had to work with ATL [Under Secretary of Defense for Acquisition, Technology, and Logistics], the logistics guy, and the acquisition guy. Anyhow, they had these sort of closed-door breakfasts, a breakfast club, you know; we weren't involved with it. He would have them all geared up to say okay, this is what we are going to tell those guys to do. And then we would come in and fuss at him and say no, no, that's not going to work. We are not going to do that.

I'm trying to remember whether it was with [John M.] Deutch or [Paul G.] Kaminsky who were the Under Secretaries of Defense for ATL. I'm not sure whether it was John Deutch, because they overlapped it, but I think it was Kaminsky was going to have a breakfast with all of the principals for the JROC, the Vice Chiefs. So we all met earlier down with the Army with Benny [J. H. Binford] Peay [III], who was the Vice Chief of Staff of the Army, myself, and I think it was Walt Boomer. No, it was Rich [Richard D.] Hearney was still there as the Marine Corps, and Tom [Thomas S.] Moorman [Jr.] the Air Force guy. But I don't think Tom was there that day. I think his two-star deputy was.

So we go. We all meet and say what are we going to do? What's this breakfast all about that we are going to? I have a feeling that Bill has got the Under Secretary of Defense all geared up to beat on our heads and tell us to be a little more cooperative with Bill on some of these issues.

And [laughter] so I said, "Well, what are we going to do?"

"Well, it just depends on what he says. We will just have to wait."

And Mike Carns was there—but why was Mike there? That's right, Carns was there instead of Moorman, and for some reason his deputy was also with us. Oh, I think his deputy was working with Bill. So we are at this breakfast, and sure enough, Kaminsky says, "This JROC thing has got to work. We have got to get a lot better on our acquisition process," and, yeah, we all know that. We are on board with that. "You guys have to be a little bit more forthcoming when you come to these meetings. You know, you are sort of entrenched and not being as cooperative as you should be. Admiral here is doing the Lord's work for us, and you guys need to cooperate a little bit better. So what's holding us back?" [Laughter]

And Mike Carns pops up and says, "We don't trust him." [Laughter]

EM: Not much subtlety there.

SA: No, there was none. And it was just classic. The Air Force guy, the two-star that was there with Mike, and I think he was part of Bill's team, but he and I worked together later on at Loral and then Lockheed Martin. He almost choked. He was eating a piece of sausage, and he almost choked on it. We talked about this over the years; I don't know if you've ever met Mike Carns?

EM: No, I haven't.

SA: We went all the way back to the RDJTF days, so Mike and I were good buddies. We lived in tents together and flopped in the sand together, with shooting matches with the Pakistani guys. But anyhow, very, very bright, Mike was. I hope somebody preserves—he takes copious notes. He has a notebook, and I mean everything goes down in it. I worked with him after he retired, and he was the same way. He must have a stack of them this high. But somebody ought to get those, because he's got—every incident that he's ever been involved in is written down someplace. Mike usually was a little more subtle than this, but he just said, "We don't trust him."

EM: What was the reaction to that?

SA: Well, the eyes got about this big.

EM: Kaminsky?

SA: Yeah. Said, "Well now, we've got to work through that. He's got a job to do, and you guys have got a job to do."

EM: What was your relationship with Owens? I mean, you two Navy guys.

SA: We got along well because we had worked together at N8.

EM: But did you have issues?

SA: Oh, I did. I had lots of issues with Bill on some of this profound knowledge, and I said, "Bill, it's just not going to happen." I said, "You know, the world is as we know it has always been: the guy develops radar, the guy develops anti-radar; a guy develops stealth, a guy develops anti-stealth. It's one of these things where you never—but for an instant, do you have the perfect solution and then it's gone. So we've got to understand that we are never going to get to the point where we've got perfect knowledge to go do a one-time event. We've got to be a little more robust than that, and understand that we are not always going to know what it is we are getting into."

Bill was convinced; I think he's still convinced. I see him every now and then, but he spends most of his time in Asia these days, making multimillions. But we got through it. We all got through that.

EM: So it was sort of a compromise between his far vision and your more immediate needs?

SA: Yeah. Yeah.

EM: He was willing to compromise?

SA: Yeah.

EM: Did the other service leaders come to accept that this guy is—you can convince him?

SA: We really formed a strong bond, and unintended by Bill. He was always so much further down the road than we were. It forced us to get our act together so that we would support each other. I can remember one time we were down there on the JROC meetings, and Bill was after them on the V-22. He was going to come in and cut a big part of that program, and my Marine counterpart had just run out of gas, I mean, he'd used every argument that he could use and was just completely frustrated by the thing. Basically, Bill threw down a challenge. I forget exactly how he did it, but he threw down a challenge on the V-22. Rich Hearney didn't reply. I'm thinking—

EM: He was the Marine guy?

SA: Yeah. I said, holy mackerel, he can't let that go. And Rich was really a good guy, and Rich had worked with him when he was N8, so he had a pretty close relationship with Bill from those days too. But I had to step in and make his arguments for the V-22 when he just ran out of gas.

EM: The Navy was supportive of the V-22 program?

SA: We were for the Marines. We weren't as supportive as John Lehman wanted when the program was first built. I was OP-50 when the program was first built. I spent a lot of time down with the Secretariat in those days with Mel Paisley on the V-22. In fact, one of things that—Mel's problems later on were part of what happened around that program. We wanted to up-engine the thing early, because the Air Force, what their requirements for the thing were, and the Marine version, there was a need for a bigger engine. Bigger power plant. There were two candidates on the line at the time for the program. One was at the limit of its heat; in other words, there was no way of running the engine hotter to get more shaft horsepower out of it without a real breakthrough in metallurgy, but the other engine was not near the limit. It had more growth potential. So we were all on the side of the more growth potential. It was the more expensive option and we had briefs after briefs after briefs, and just one night out of the clear blue sky, Mel Paisley said no. This is the engine, that's it. That's gone. Done deal. No more arguments.

Well, now we are back several years later, and we are having this argument again about Navy pretty much, you know—John Lehman wanted us to replace the S-3 with the V-22. The CODs with the V-22, and we weren't having it. We weren't buying into that.

EM: Too expensive?

SA: We didn't think it was going to do the job. We said we would wait on the sidelines for it and keep that option open, and that was enough for John Lehman to say go ahead with the program. And they built these big numbers for the program so that they could get the unit costs down. Had a bunch of Navy planes out there that we weren't ever going to buy. Now we are. Now we are going to have a COD version of the V-22 all these two decades, almost three decades later now, from the time we were first talking about it. So those are the types of things that we did down there as the POM went to bed. When we would take the JROC trip, we would all hop in an Air Force executive jet, usually one of their 135s I guess. Off we would go around the world. We would go to all of the CinCs. CinCEur [Commander in Chief, European Command], CinCLant, CinCPac, you know. These JROC trips, the young kids, the aides, I can't imagine the stories they tell, because we would get on these things and we are flying all night, and then we are getting to a place and they want to entertain us with a dinner or something like that, and then we would have the briefs with all of the big staffs. Then we would hop in this thing and off we would go again. So we would have four-stars lying in the aisles, sleeping in the aisles. [Laughter]

EM: Pretty grueling agenda.

SA: It was a grueling agenda. [Laughter] Some of the guys—I wouldn't take them because I never, ever wanted to take a sleeping pill, but what is the drug that everybody uses?

EM: Dramamine?

SA: It's not Dramamine, it's one of the sleeping things that if you take it, you can't wake up in the middle of it, so it's a four- or six-hour thing, and you will wake up and you're really ready to go. But if you're interrupted there someplace, it throws you completely out of rhythm. Some of the guys would take those things and they would crash, and so you would have one four-star lying in the aisle and another guy sitting near me not sleeping, but then I would have to get up and go to the head, so I would have to crawl over a guy. The aides, they made up T-shirts for us and all sorts of crazy things.

EM: So the object was to hear from the operators out there about their equipment needs?

SA: Yeah. We would go out and say, okay, here's where it is. This is your chance to either say yea or nay, and then we would fly off to the next destination.

EM: Do you recall any briefing where they convinced you, hey, this is something we need to pay more attention to? Anything stand out from these various briefings?

SA: No. We had lots of heat.

EM: Lots of heat? What do you mean?

SA: Well, where their cherished possession got lost in the shuffle.

EM: I don't understand how a different CinC would have a different equipment need. I mean, maybe the Pacific is so wide CinCPac would want planes that could fly farther?

SA: Well, yeah. You would have CinCPac who was much more maritime in thinking than CinCeur, so in CinCPac if the ship count was good, the carrier force was good, the Aegis cruiser upgrades were good, and the Marines had a new amphib on the way,

CinCPac was happy. CinCEur, you know, where's my missile defense? Where are my bombers?

EM: Who got the most play during your time? Was there a favored CinC or favored service that comes to mind?

SA: No. It was pretty even. I think we did a pretty good job of satisfying. Nobody was ever completely satisfied of course, but I think in general it worked out pretty well. There was always this—as always—if I were a CinC I would be pushing back, and they all pushed back. The CinCs were much more "today" guys than the Washington scene is. So there was always that friction of, hey look, I don't care what you are telling me for 10 years down the line. Next year I need this and you're not giving it to me. Why can't I get this? Without this I can't do exactly what it is I want to do and should do, and so you always had that tension. Bill was the big picture guy, so he would sometimes gloss over some of those immediate needs, and that would always get a little bark from the CinCs saying, you know, who are you to tell me.

EM: Were those the only two Vice Chairmen of the Joint Chiefs of Staff, Owens and Jeremiah, or were there others in that position while you were there?

SA: No. Those were the only two when I was there, yeah.

EM: Do you think the JROC establishment is the right way to go?

SA: Yeah. Yeah, we can't get away from it in this joint arena. It was a good place to force guys out of the mode of, hey, look, is a destroyer more important than a tank? Is a new fighter more important than a new bomber? Left to the services' desires, those choices are easy to make. But in the joint arena those choices are hard to make, and so yes. Reluctantly I have to say that yes, in today's environment and in the future, you have to have something like the JROC to bring it all together.

EM: Did your counterparts make an attempt to understand Navy and Marine Corps needs or naval warfare, or did they see it as some alien science?

SA: No. As a part of the process—we called ourselves the Vice Squad because of one of the other great things that happened on these trips. All the Vice Chiefs would bundle up their desk and take it with them, so these long hours sitting on an airplane, we'd be leafing through the business of the staff. I found it very interesting, even though I went through long turmoil of personnel-related issues during this period of time, that my stack of bad guys was much smaller than theirs. I remember the Army guy sitting there with a stack this high of general officer misbehavior things.

EM: Were they sympathetic about your involvement in Tailhook, or did they say, whoa, we've got similar problems?

SA: No. They were very sympathetic with that, yeah. And they were very sympathetic later on, when I got involved with other issues. But what we did was, we socially got together once a quarter. We took trips usually once a quarter, other than just the CinC trips, but we would go to other people's facilities.

EM: So you got to know these fellows pretty closely?

SA: We got to know them really well.

EM: Was that beneficial?

SA: Oh yeah. You could make a deal. You could down and say, this is what I'm going to come forward with, and I need your support, and here's why I think it's in your interest, best interest, if you would support me on this thing. And if you can't I understand, but let me make my case to you. We had the relationship so that you could go down and sort of bare your soul, and we had the relationship that said if you are in trouble, if you need help with anything in your business that you think I may have seen or can give you some help with, just call. I will come down and talk to you. You come down and talk to me.

EM: Do you recall a particular instance where you supported the other guys?

SA: We all had incidents where somebody got in trouble and you would say, have you seen this before? How did you handle it? We had those kinds of dialogues. We had—with Danzig, between Rich Hearney and myself, with our meetings with the Under during his period of time, we had lots. This of course didn't branch out to the other services, but we were able to solve a lot of the heartburn that was going on between the Marine Corps and the Navy on certain issues.

EM: Like what?

SA: Well, there's always, you know, the Marines want more amphibs, and you buy us an amphib before you buy yourself another cruiser. Airplanes, we don't want the E/F. We want the A/B/Cs. We don't think we need the E/F. Don't build your program on betting on us to buy the E/F. So we talked about a lot of those issues in the Danzig/Hearney/Arthur rotation there, so we had a good forum in which to clear the air on some of those issues, although the Under wasn't really that interested.

EM: Who stood out—I mean after all these years, who stands out among your counterparts in the other services? Real effective go-getters?

SA: Well, Mike Carns on the Air Force side. Mike was a dynamic guy. Always well prepared. Smarter than anybody at the table. I think Bill Owens thought he was smarter than Mike Carns, and Bill was awfully, awfully bright.

EM: Yeah.

SA: But Mike was at a different level than almost anybody. He could work in the weeds; he could work in the ether. Very different kind of guy. He was a different kind of guy, but because we had sort of grown up in our first flag jobs, you know, in the RDJTF living in tents out in the middle of the desert, we formed a good bond. Tom Moorman, who followed him, was a space cadet. I mean, he was a true Air Force space guy. His dad I think had been a previous general in the Air Force as well. Tom was a very smart guy as well. Very pleasant to deal with. Didn't have as many sharp edges as Mike. He had softer edges, but he was a good guy to work with.

I went through three ACMCs I think. Started out with Jack Dailey and that was only for a month, but I had known Jack when I was captain of the *Coral Sea;* Jack was the air wing commander for the 1st Marine Air Wing out on the West Coast, and his F-4 squadrons were my fighter squadrons. We got acquainted there so I knew him when we got back, but then Walt Boomer relieved him, and Walt and I worked all the way through the Gulf War. Boomer was a super, super guy, and Rich Hearney who followed him was a good guy to work with. I enjoyed working with Rich.

EM: You didn't have a nemesis among the group? Someone who you always clashed with?

SA: No. In the Vice Chiefs, no. Now, Binnie [J. H. Binford "Binnie"] Peay [III] was the first—well, Denny [Dennis J.] Reimer was first, and Denny was a real gentleman who came along later as Chief of Staff of the Army. Binnie Peay was small, a smaller guy, very active. Binnie could flash in a moment's notice, but not at us. We sometimes had to keep Binnie from getting overloaded with Owens.

EM: So Owens was kind of abrasive?

SA: Well, Bill was not an abrasive guy at all. He's extra-smooth, and so it's all very friendly, and sort of out-in-the ether a little bit, you know. But he had his own ideas.

EM: Some folks didn't trust him. That goes to a personal side?

SA: They didn't trust him because in being overly gentlemanly about everything, he sometimes left people with the impression that well, I'm going to give it some thought, when basically—I knew him a lot better than the rest did, and I knew when Bill had his mind made up, that was going to be it. You weren't going to change it. You might as well figure out a new way to go do business, because once Bill had latched onto something, he was going to latch on to it. So the others couldn't see that quite as quick as I was able to. John [H.] Tilelli [Jr.] was the last of the Army guys that I worked with.

EM: He went to the same school I did. Pennsylvania Military College.

SA: Yeah, do you know him at all? Have you ever met?

EM: No, in fact he was three or four or five years ahead of me.

SA: Yeah. He was sort of a hardnosed guy. But good. I mean we had a lot of fun with him. He flew us all up one night. There was a ball being held up on the *Intrepid* in New York City, and John said, "Why in the hell do we all have to fly up to whatever airport and have them drive us in there? I'm going to land my helo up there. We will just land on the flight deck and we will all be there." So all of us, in our mess dress and with our wives.

EM: Wives too?

SA: We jump in this Army helo.

EM: What, it must have been a big helicopter?

SA: Oh, it was one of the—

EM: 47s? Chinook?

SA: No, it was a passenger version of their H-60s. So we all went up there and did that. But no, I think in today's world you need the JROC. Now, the problem I've got with the JROC and the services now is that the Joint Staff is so big it is overwhelming to any of the service staffs, so they can overwhelm you on a one-on-one basis. So the next decision that has to be made is, are we just going to throw all of the programming into the Joint Staff and have it done in one place, or are we still going to have the services. Now, that's a big—

EM: Current issue.

SA: But it's going to be an issue, yeah.

EM: Is there anything else you would like to say about the JROC?

SA: No. Let's see if I wrote anything down here. No, I think that covers the JROC pretty well.

EM: We touched upon Tailhook, and of course it happened while you were still out in the Pacific, and I guess I've gotten your impression of what it was—what your impression was from afar that things were not right here. Was sexual harassment a problem in the Navy at the time? Did you see that from your position as Com7Flt or ComNavCent?

SA: No. I didn't. In NavCent we had the field hospitals and stuff out there. We had the hospital ships, females, you know, on board.

EM: Were there lot of, or any incidents of sexual harassment or rape?

SA: You know, I can't recall one. I can't recall one. I think back in Japan, I think we had some episodes there, and the Marines had some down in Okinawa with rapes, but not among the service members, I don't think. The thing that caught my eye in Desert

Shield/Desert Storm was the first issue that came to the plate, I think we talked about that earlier, was a husband and wife team on a hospital ship that I didn't think was appropriate, and Cheney and his guys thought it was. I made my protest and protested to the CNO, and they went their way. The funny incident that happened was when one of the hospital ships—I forgot which one, when it came through the Suez Canal, they decided to have a formal dance on the flight deck.

EM: In the Suez Canal?

SA: In the Suez Canal. [Laughter] I said, you can't be serious. They didn't do this. They said, oh yeah. We even got pictures. I said, oh, no, no. Yeah.

EM: Probably a big hit in Cairo?

SA: Yeah. So we had some of those crazy things, but I visited the field hospitals and the females out there. They were working as hard as their counterparts. Everybody seemed to be doing their job. They were good. I remember an old air wing buddy of mine from my days in VA-55, he was in VA-163, his daughter was a flight deck officer for one of the hospital ships, and so she left word one day while the ship was still in Bahrain, and this may have been after the hostilities. I think it was probably after. So she left word that she was over on the hospital ship and would like to make a call on me sometime.

So I got in touch and said, "Yeah, come on over, and if you've got a couple of friends or something that you would like to come, come over and we will have lunch or dinner." I forget. "I will host you for lunch or dinner here on the ship and catch up with you and see what you all have to say about what life's been like." So they came over and it was a fun time. Of course they were most impressed that she knew who I was, that I would take time to have dinner or lunch with them.

EM: The women in your command, when they were outside of U.S. cantonments, did they have to wear a hijab? Or what did they do? Were they in uniform?

SA: In Saudi Arabia they had to have their hair covered.

EM: If they are in uniform, just a cover, a hat, would be good enough?

SA: A cap would be good enough.

EM: Yeah.

SA: But then if they were inside they still had to have a cover, so that was a sticking point with them. After the war, when I got back up to Japan—of course now we are well into women at sea and stuff, so each time I went to sea, even though we didn't have women, women were not assigned to *Blue Ridge* yet; women were not assigned to the staff yet—I took females from Yokosuka who were in surface warfare trying to get their SWO [surface warfare officer] pins. I would allow the flagship to bring them to sea with us and to continue to get their qualifications so that they could get their SWO pins.

EM: So they were on the ship, but they weren't part of the ship's company?

SA: No, they weren't part of the ship's company, and that worked out pretty well. I probably dodged the big bullet in that we crossed the line, and so I said okay, we are near—

EM: The equator?

SA: Yeah, the equator, and so you know how those ceremonies go. So I said, well, I think I'm going to have to interject myself in here before something goes wrong. I got the senior female, the lady who was with them, and I said, "Okay, we are going to have this crossing the line ceremony. I want you all to be a part of it because this is a big thing in the Navy. But we are going to have to be very careful that we don't do something dumb." And I said, "One of the things that we do—and you have to tell me if this is going to cause any problems—a lot of the times what we do, we have what we call special cases.

King Neptune decides that he's got a couple of special cases that have to go through a more rigorous type indoctrination. One of the things we do, especially for senior guys, like I did for my XO when I had the *San Jose,* I made him get dressed up as a female. He had high heels and a skirt and lipstick and all this stuff." I said, "Is that going to cause any problems?"

She said, "No, we will just dress up in suits and ties." I said okay. That sounds fair to me. So we did it. Nobody made a complaint.

EM: Everybody had a good time?

SA: Everybody had a good time, and we all carried out our business. Of course, later on we had all sort of different incidents happening with the crossing the line ceremonies.

EM: During your term?

SA: After I was Vice Chief, yeah. But it worked out good.

EM: All right. What was the status of the Tailhook investigation when you took over as VCNO?

SA: The investigation had hit a point where the JAG officer on the Secretary's staff and the IG [inspector general], one of the guys on the IG staff, had been sort of drawn up in a scuffle with Barbara [S.] Pope, who was ASN M&RA [Assistant Secretary of the Navy for Manpower and Reserve Affairs], I think at the time was her job.

EM: On the SecDef staff?

SA: On the SecDef staff. I guess the JAG officer had made some comments in her presence that she had taken umbrage to as being not quite what you would expect to hear from somebody that had done an impartial investigation. So the noise had been made that the investigation was tainted, and the investigation was then turned over to [Derek J.]

Vander Schaaf at the OSD IG staff. So that's where I came in. The two guys that sort of had been canned were on their way out.

EM: Those were the—[John E.] Gordon was the JAG?

SA: Yes.

EM: [George W.] Davis was the—

SA: Who was it? It was—

EM: [Duvall] Williams [Jr.] was NIS [Naval Investigative Service].

SA: Williams, yeah.

EM: So they were canned?

SA: Yeah.

EM: Because she reported that they had—

SA: Yeah. She was part of that, and then it got caught up in the, I think in the DoD [Department of Defense] that she had made a plea to the DoD.

EM: Did you ever talk to those three folks that she ran afoul with, Williams and Gordon?

SA: I had talked to Gordon some, I know. I knew Williams from before. I'm not sure that we had any dialogue after I got back to be Vice Chief. But basically, I was now dealing with the DoD IG guys.

EM: It was kicked upstairs to DoD?

SA: Yeah. So my first visit, one of the first visits I had with them, I was trying to get my handle on all of this stuff. They came in, and the first thing they said to me was, "Your flag officers are lying to us." Lying to us.

I said, "How do you know this?"

"Well, we just know it."

I said, "Can you give me a definitive instance where you have black and white that says this happened, and you got this black and white that says no it didn't happen?"

"Well, no, we are not ready to share that yet. But we want you to tell them not to lie to us."

And I said, "Look, every flag officer who knows me knows that they are not to lie. I don't have to tell them that. They know they are not going to lie. And until proven otherwise, that's the way it's going to be."

I was not going to because Dan Howard had pulled everybody in before I got back and had a big, big séance on Tailhook and women in the Navy, and there were a lot of ruffled feathers that he had sort of overstepped his line, and he was part of the reason that Garrett had got the heave-ho so fast. So I said, "You can trust that I will, if you bring me evidence that one of my flag officers is lying, I will certainly take care of that. But no, I'm not going to go and issue a message that says, hey guys, don't lie. I'm not going to do that."

And they said, "Well, okay. We will see how it goes."

I said, "And tell me—you can come back to me anytime, and if you think somebody is not cooperating with you I will make sure they cooperate with you." So we left it at that.

Then they said, "Well now, the next problem is that we don't think Frank Kelso's being truthful with us."

And I said, "Whoa. Why do you say that?"

"He was out there with John Lehman, and he had to have seen all of this stuff."

I said, "Now, wait a minute. Do you really know Frank Kelso? Do you really know this guy?"

"Well yeah, we know him. We know who Frank Kelso is."

I said, "But do know him? Do you know what kind of guy he is?"

"I don't know, but we are concerned."

I said, "Frank Kelso's one of those guys that believes everybody is good. He's a kind-hearted guy, and he's a very smart guy. He's not a guy that I've ever known—and I don't know him that well, because I worked with him just a short time when I was OP-04 and now here I am as his Vice Chief—but I know his character. I've seen him enough to know to tell you that Frank may have been places where stuff was going on, and Frank would not tune into the stuff that was going on underneath the table. It's just not in his nature."

I said, "Let me give you an example. I've been going through all this trying to get up to speed on what you think you know and what you don't know, and what's available for people to know about what's been going on out there. I looked at some of this, and I ran across this thing that was going on with the Marine photo squadron that had the Rhino suite. There was this episode in there where the guys had a free leg shave for the ladies if they would come into the Rhino suite. And it just sort of caught my fancy, and so I said—Frank and I are talking about things and I said, 'Frank, you won't believe this. These guys were offering free leg shaves to anyone that would come into the Rhino suite.' And he looks at me and he says, 'Well, why would any lady want to get her legs shaved?' [Laughter] And I said, 'Well, Frank, it was just sort of funny.'"

The guy was naïve about these kind of things. I mean, he had not been in the culture that had seen this kind of stuff happen before. And I said, "Again, if you've got something definitive, I will help you work. I will do anything I can to work your way through it, but right now you guys are—you know, you've got to do what you've got to do, but I'm telling you that some of this is—"

EM: Did the fact that he was a submariner rather than an aviator, was that pushing him more toward the naïve side; that sort of thing just didn't happen in the submarine community?

SA: Yeah. I was never more convinced in my appraisal of Frank than when he passed away. Jennie Lou and I drove to Tennessee for his service there in his old hometown.

Well, I wish that Barbara Boxer and that clan of ladies over on the Hill would go over to Fayetteville, Tennessee, someday and ask about Frank Kelso. I mean, he was active in the church. He was active with the kids. He was Frank. He was not Admiral Kelso. He was old Frank. And the people just loved him and the stuff that he was doing there. And then try to tell me that Frank was a bad guy that was doing bad things against women's rights. I said it made me sick at the time, and I knew that he didn't know exactly what he had seen out at Tailhook, and I was never more convinced of that than when I went to his funeral and met the people of that town, and the people who came up and talked to me said, gee, you had the opportunity to work with this guy, and I said, yeah.

EM: What was Frank's reaction when he was asked to retire? He retired early, right?

SA: He retired early.

EM: Who asked him, by the way?

SA: I'm sure the Secretary did. I think Frank contemplated serving out his term until the Secretary probably told him that he wasn't going to support him for any more terms. I think he was on his second term then anyhow. I don't remember now. But Frank realized that he had now become part of the lightning rod. That things weren't going to get any better as long as people could point and say look, he's still CNO.

EM: Yep.

SA: So he said, it's time. Great loss. He was a good CNO. He was a good guy to work for. He was very good with his people. Of course, he had grown up in the building. He had all of the key jobs in the building, so he really knew how to get things done.

EM: Other than Tailhook, your relationship with Admiral Kelso—as a submariner, was he sympathetic to your support for naval aviation and your involvement in the aviation world?

SA: Yes. Yeah. Frank had had some good aviators working for him over the years. I don't know all of them, but he was very knowledgeable on aviation issues. He was very supportive of me relative to aviation issues. He knew that I knew how to present naval aviation, and he also knew that I was not bashful about—if there was something wrong that I wouldn't try to bury it. That I would try to fix it rather than push it off.

EM: Was there a time when he did something that favored the submarine community and you thought, boy, we got short-shrifted here?

SA: Well, there was always that dynamic with the submariners. The submariners are very, very good, and here's what they did that I tried to get the aviation community to do. I'm still upset about it over the years, but what the submariners did, and it goes all the way back to—I can't talk about John [M.] Richardson right now, the new CNO. But his predecessor—now, when I was moved from OP-50 to be OP-90, the programmer, my chief, my civilian deputy was Irv Blickstein, the SES. Don Pilling ran the shop, the bullpen as we called it, where all of the community guys were in there, and they did all of the POM scrubs and everything. The last CNO—

EM: Greenert?

SA: Greenert, Jon Greenert, the last CNO. He was a young commander in that shop under Don Pilling. Jon was to the go-to guy to do all of the computer manipulations of grinding this thing out, and what were the alternatives—if you added one destroyer what else did you have to take out, and if you added one sub what did you have to give up. So Jon was trained in that arena, and Jon was just an absolute dynamite guy. The submariners had programmed him for that job a long time before, and they had one of those guys working at every place all along.

EM: All prepped?

SA: All prepped, ready to go. Bill Owens, another very bright guy. Bill had the right credentials to do these things. And Bill Smith, who was OP-08 I guess before Dave Jeremiah. Bill Smith was the programmer, the big programmer when I was OP-04, another submariner. So the submariners always had a team of guys trained in how to put together a program. And they knew where all the dead horses were buried. They knew where all of the weaknesses were in other people's programs, and they were always able to protect the submarine community, as far as possible, from raids that the rest of the people sort of had to go through.

Frank Kelso had the same bent. I mean, there were times where he would make a decision on a submarine issue so that they didn't take their fair share of a cut. That's okay. But it was well thought out. It wasn't a natural biased decision. It was a decision built upon a lot of experience that says we don't need to take this cut. It can be taken someplace else. Now, when I tried to preach that to the aviation community, oh no, we've got to be out there flying airplanes. We've got to be out there doing this. We've got to be out there doing that.

One year in one POM, the aviation guys screwed up their POM, and they had given away about $2 billion that they didn't know they had lost. One of the guys over in the bullpen came to me one day and said, "Look, these guys down there in the OP-05 area just mishandled this completely. They don't even know what they've done."

So Don Pilling, my surface warfare guy who the aviation community hated for the most part—I said, "Don, one of your guys, the aviation guy over there just whispered in my ear."

He said, "Let me take a look." He came back and said, "He's right. What do you want me to do about it?"

I said, "First of all I want you to go challenge them and say, 'Hey, take another look at this thing.'"

And he did, and he said, "No, they think I'm trying to steal their money."

"Don, put it back in. Put the money back. Don't even talk to them anymore. Go put the money back in wherever you best see fit where they've still got the money."

EM: Now, this was when you were VCNO or before that?

SA: This was when I was OP-90. So I had him do that. Everything gets done. The POM gets submitted, and then all of a sudden this same aviator, worker bee over there in the bullpen, comes to me one day and he says, "The OP-05 guys are still trashing me about stealing their money."

And I said, "You've got to be kidding me. Can't they see it? Can't they find it?"

He said, "I don't know. I'm through talking to them. But you know, it's sort of disheartening that we worked up here to fix it for them and they don't even know it."

So I finally called the deputy for the 05 at that time, the 05B. I think it may have still been Bud Edney, I'm not sure. No, it wasn't Bud. It was the guy right after him. Anyhow, I called him up and I said, "Okay, this is going to end this dialogue, but you guys screwed up when you submitted your POM. You dropped about $2 billion on the table, and we found it and we put it back, and you don't have anyone down there smart enough to know that it's back in there. So you better get something going down there, and you better send another guy up here instead of beating on the guy that's up here now. You'd better send another one up here to get trained."

EM: Did they take that action?

SA: No. And I tried several different times to get the aviation community to understand how the submariners did their work.

EM: Does it go to the culture? I mean, aviation you have to be spontaneous, flexible, seat of your pants kind of decision making, rather than methodical?

SA: Yeah. And it's a funny thing, because part of the aviation culture is just as you describe it. The other part of the aviation culture is very precise. You've got to do X, Y, and Z before you can do W over here. Check-off lists, and the maintenance and those types of things. We are much like submariners to a certain degree, but then all of a sudden we shift gears and we start paying more attention and give more credit to the spontaneous-type stuff instead of the rigorous-type stuff. Submariners are rigorous all of the time. There is no spontaneous.

EM: So the Rickover training?

SA: The Rickover training.

EM: There's a couple more things about Admiral Kelso. Apparently he overhauled the aviation OCS [Officer Candidate School] at Pensacola; do you recall that issue?

SA: Yes. I do.

EM: Why was that?

SA: That's one where we had a little problem.

EM: All right. I thought you might. [Laughter]

SA: And this will take you back to the same thing that I was talking about with his personality. He said, "You know, the most useless thing that I encountered when I was a midshipman was this drill instructor type mentality. Yelling and screaming at us to do this and do that, and everything had to be done just exactly the way, or somebody was going to rip your hat off and throw it down and stomp on it. We don't need that crap. That's just childish stuff that has no place in our business. You know, there are still vestiges of that with the Marines down there at Pensacola, with the aviation cadet or the AOCS [Aviation Officer Candidate School] program? I don't support it. I don't think it needs to be there. I'm going to kill it."

I said, "No, no, no, no. There is a place for it. We've got to put people under stress to see what's going to happen to them."

He said, "I don't believe that. I think it's useless. I just do not believe that you need to stress people like that, and I don't want to be a part of it." And we argued about that several different times and he said no.

EM: That's it?

SA: "I've got a chance to make a change and I'm going to do it." So that was one of the issues.

EM: What was the upshot? Did you think it made the training less effective?

SA: Well, yeah, I do. I do. Of course in today's world, I mean, everybody's going that route—you know, everybody, nobody can fail, everybody's got to be given a pat on the butt. But it's just like when we had the 40th anniversary of our '72 cruise with my old squadron, and the guys are telling sea stories about when they first came to Pensacola and getting into the AOCS program, and the Marines shouting at them and hollering at them, and it's a fond memory to them. It was something to laugh about and talk about.

EM: I recall the same thing.

SA: And these guys that I'm talking about basically came right out of flight training and came right to the RAG and went to war, and flew in some of the hardest times in the war that we had, and they were good. I think that the fact that they could take all of that stress so early in their careers had to come back to being part of that training.

EM: Another thing with Admiral Kelso. During his time, I'm sure because of everything involved with Tailhook and the role of women in the services, he instituted getting women into combat aircraft. Is that accurate?

SA: Yes.

EM: Did he push that, or did he sort of go along with it?

SA: I think he went along with it. I don't think he was a big pusher of it, although he certainly didn't push back at it, no.

EM: Who would have recommended it? Did it come from you?

SA: No. Basically, in '93 the Congress had enacted females could go into combat.

EM: So it came from way above?

SA: Yeah. So now we were faced with how we were going to do this, and where were we going to do it.

EM: Yeah.

SA: Of course aviation was an easy choice. We had lots of discussions about it. I would always push back and say, "Well, if it's good for aviation, it ought to be good for submarines as well."

"That's a whole different issue, you know; we can't do it."

I said, "Oh, yeah, remember the old movie the pink submarine, that had the pink submarine, with Tony Curtis? You've got these boomers. You've got blue and gold crews. So why couldn't you have a blue and pink crew? I mean, you know."

"Oh no, because they've got to be trained on the attack subs before."

EM: You'd have an entire crew would be female?

SA: Yeah.

EM: Why not integration?

SA: Well, at that time people couldn't—submariners would say you can't, there's not enough space. There's not enough space on the submarines. We sleep on torpedoes, and all of this other stuff. But I was always on the side of, "Look, when we brought women into aviation, we brought women into aviation because we were looking forward to the demographics of the future, which said that we will never have—there is a point in time where we may not have enough qualified male candidates to fly all of our airplanes. There is no reason that we can't mirror what was tried in World War II with the WASPs [Women Airforce Service Pilots], where we had female pilots to fly all of the non-combat missions that we have, since in those days, no one ever thought of sending a woman into combat."

EM: It was just beyond the pale?

SA: In those days there was a rule that you [women] couldn't fly, so there wasn't a thought that, "Well, why don't we teach them to fly combat airplanes?" It was, "You can't do that because they can't fly in combat." But we realized that we needed more qualified people to fly all of the airplanes in the missions that we did have. So that's sort of how aviation walked its way into having female aviators to begin with. So where now we've got this big push to start integrating women into combat, then it was, gee, we can't really do that.

EM: Why not?

SA: That was sort of where I was. I said, "It's certainly a lot more comfortable for the families to know that their guys are at sea with just the guys and no girls. But that's probably not going to be in the future, so let's figure this out." In my talks with the people as we were trying to do this, I said, "I have no problem with females flying combat aircraft. Flying an airplane is a monkey skill. Some people have it, some people don't. You can't predict it—I mean some of it's predictable, but because she's female doesn't mean she can't fly a combat airplane. You know, can the body take all of the Gs and stuff? We've got to find out. But the fact is that women can fly airplanes just as well as a guy can, any day, any time."

EM: So you had no problem with that push by Congress?

SA: I had the problem of how are we going to integrate it, you know, give us time to do it right. Don't force us to do something dumb. We'd already started to see that every time we tried to reach a little bit further that we would do, something dumb like the first women that went to sea, you know. We didn't have enough female leadership above them, role models to follow. So I said, make sure that we know how to figure this out, how to do the heads [restrooms] and everything else so that we don't end up with bad things happening.

EM: Yeah.

SA: But I said the concept is fine with me. We can do it. They can do it. We just don't know how far it can go, so don't rush us into doing something dumb, but, yes, this can happen.

EM: Yep.

SA: So that's sort of where I was, and Frank was on the same boat, except he couldn't see them in submarines, yet.

EM: Only recently?

SA: Only recently.

EM: That one that came from SecDef.

SA: Yeah.

EM: All right. Anything else you would like to say about Admiral Kelso, your relationship with him or any issues you might have had?

SA: No. I had no idea—because of the short time I had worked with him when I was OP-04 before I went out to Seventh Fleet, I really hadn't got that close to him. We still didn't know each other that well. I think I probably knew him more—well, I don't know if I knew more about him than he knew about me. I don't know. But I found him a lot easier to work with than I'd thought. I wasn't sure our personalities—that we would find a relationship that would be easy for us, but it turned out to be very easy.

EM: Did you have occasions to have social engagement with him?

SA: Yeah. Oh yeah. We had lots. One of the things he always did was, when they had foreign visitors and stuff like that, he hosted many a dinner over at his place. And we were always invited to those, and we always had a good time.

EM: Did he have a sense of humor?

SA: Oh yeah. He had a great sense of humor. My wife and Snuffy Smith, I don't know if you've been around Snuffy, he tells a joke. Well, almost every time we went to one of these dinners, they split the husband and wife teams up, you know. So invariably Snuffy and Jennie Lou were together at some table across the room, and it wouldn't be 10 seconds into the meal before there's rip-roaring laughter going on over there.

EM: From that side.

SA: Frank would look at me and say, what's going on over there? I'd say, you've got to learn to keep those two apart. [Laughter]

EM: Oh, where did you live, by the way, when you were VCNO?

SA: I lived at the old Naval Observatory, right across from the State Department.

EM: Oh, that's right, sure. Until the VCNO got bumped out of there.

SA: Oh no, over at BuMed [Bureau of Medicine and Surgery]. Yeah, at the old BuMed. When I was OP-04, I lived up at the Naval Observatory across from the Vice President's house.

EM: Sure, right. [Laughter]

SA: I had some fun there.

EM: All right, Let's see here. In the first investigation that was done by the Navy in regard to Tailhook, some 50 naval aviators were found fault with, some of whom I guess retired and others kept on, but they had a letter of reprimand in there. Do you think that punishment was appropriate or not?

SA: Well, we worked hard with that list and we ended up with Paul Reason, Admiral Reason down at LantFlt being the—what did we call him? Something authority. Anyhow, he heard the hard cases. We nailed it down to just a few hard cases. We took the rest of them and, like I mentioned back when I was talking about counseling one of the flag officers, I took most of those back into my world and dealt with those. Yes, I think for the most part that they were valid. The problem was—and you know we will probably talk about Bob [Robert E.] Stumpf and others later—

EM: He's the next on the list here, yeah.

SA: When I was a junior officer in the squadrons, and even when I was CO of the squadron and we all did dumb things, but one of the things, we did always sort of designate one guy to be the junior officer control officer. And the job was, when you see things getting really out of hand, step in, grab the guy, and take him back to the ship. And they weren't doing that out there.

EM: Yeah.

SA: They just weren't doing it out there. It was let the boys be boys. Let them do their thing and then, you know, Monday morning when we are all back in the squadron, everybody would be all burnt out and ready to settle down and behave. So as I counseled the guys that I had to counsel and talked to Bob Stumpf later on, when they were trying to get him back on the promotion board, I said, "Look, here's the thing, you knew—in one case you left a party because you knew what was going to happen next. You knew that once you walked out the door, things were probably going to get wilder than they already were. What I'm expecting is that you would have taken that moment to turn around and say, 'Joe, I want to make sure that nothing bad happens here. You're in charge.'"

EM: Yeah.

SA: I said, "That's the connection all of you have got to make from here on out. You just can't say, well, I wasn't there, if you saw something that was headed in the wrong direction."

EM: And that was the case, there were too many who just saw what was happening and knew where it was going and didn't take any action?

SA: Yeah. It got away from them. So from that standpoint, I think they were valid. Maybe a little too harsh in some cases.

EM: Admiral Dunleavy was DCNO Air at the time, and he was reduced to rear admiral. Was it just because he was DCNO Air, or had he been at Tailhook?

SA: He was there. Basically by fiat, he is the designated head air guy. In other words, it happens under his watch.

EM: The fact that he was there, but also because he was the captain of the ship.

SA: The captain of the ship, yep.

EM: Do you think that Williams—Admiral Williams was head of NIS at the time—when his report came out and Pope, Assistant Secretary of the Navy Barbara Pope, said it was a whitewash, was that an accurate characterization or not?

SA: I wouldn't say it was a whitewash, but it certainly didn't measure the real amount of mischief that had gone on there. I think it downplayed it to the point of—well, it's just as the saying goes, boys being boys. It didn't really hit the real issue, and I think he was not alone. In other words, I think his report more or less reflected what most people thought at the time. And so Barbara was hard over on women's rights and stuff. I mean, she was an activist in her way. She was a very nice lady, and I think the fact is—what bothered me about the whole thing was that a lot of people knew things were not going right. There had been enough warning signs in years ahead, and before that that there were things that looked all right to the people involved, but for an outsider to see them, they didn't look all right.

EM: Yeah.

SA: And people were slow to grasp that we were rapidly moving into a new era where what people saw outside was as important as what the people saw inside. And so the thing that got everybody sort of across the breakers in all of this was that there were females who willingly participated who all of a sudden said—

EM: Enough?

SA: Enough—which, you know, should have been listened to. And there were enough guys who saw things and knew that it was going too far, who should have said enough and didn't. So both sides took some blame, but the most blame goes to the guys who had been in that culture.

EM: Well, at one point DoD took over the investigation after the Navy did, and you were told hands off; let them do their thing and don't try to influence it. Is that accurate?

SA: That was pretty much it. They were in charge, and of course then I had—as I spoke earlier about them coming up to me and asking me to do certain things for them, which I thought was sort of inappropriate at the time as far as, you know, I was there to facilitate. I was not there to issue their edicts. If they wanted, if they could bring me information that said somebody had done something bad, then I wanted to have it and deal with it.

EM: Dan Howard was kept on as Under Secretary by Secretary O'Keefe. When the IG said they thought he didn't press the investigation enough—the Navy investigation, I guess—do you think that was accurate? I mean, did Dan Howard not do enough? What was his role in this?

SA: By the time I got there he had already done his thing, and then after that it was in the DoD's chain. Now, Dan had previous DoD experience. He had been on the DoD staff before he had been Under Secretary of the Navy. Dan had taken a very proactive approach, and it happened before I got back there. He'd called all of the flag officers together, I guess in the building, and did a real rip-roaring reaming session on them up there in the theater, which had not gone down well. Dan had a very prickly personality. I mean, he was a tough, tough talking, tough acting sort of guy. We got along well. He had very positive positions on everything, and I think he really—a lot of people were very upset with him. A lot of people thought he had gone overboard, and I had no way of really knowing because I wasn't there.

EM: Yes.

SA: And somehow or another, I always had the impression that he and Barbara ended up at odds with each other, which was sort of uncomfortable because she would be touting one position and he would be touting another. But on the other hand, Dan always had several female assistants close to him that he trusted implicitly. I had worked with him on some very sensitive issues back in OP-04 days, and he had a couple of female assistants that were really dynamite people. Really good, sharp, and he trusted them to carry his word and his thoughts wherever they went. So, you know, some people sort of got to the point where they thought Dan was anti-female. Not at all that I could ever see. He picked and chose along the way, but I know that I worked with one for years and she was just dynamite to work with. As far as Dan was concerned, if she spoke, she was speaking for him.

EM: Yeah.

SA: So I don't know. I think some of that is probably brought about by the things that Williams and the JAG got involved with that perhaps Barbara felt excluded from. And I think the DoD IG probably thought that Dan seemed to be more sympathetic to what the investigators had done than what Barbara was saying.

EM: Okay. We probably should have dealt with this earlier, but Secretary Garrett, now he resigned before his time was up. Was that more related to Tailhook, or did it relate to the A-12 issue? What was the cause of his early retirement?

SA: I think it was more Tailhook, part of the Tailhook.

EM: Right.

SA: Yeah. That was again something that he wasn't going to be able to walk away from.

EM: What actions did you take to deal with sexual harassment in the Navy?

SA: We had lots of meetings. We instituted lots of training. Gosh, we had all sorts of training episodes that we had to do.

EM: I recall. [Laughter]

SA: Some of them a little too much. It seemed like we couldn't move fast enough. We just couldn't outpace the demands. I remember—God, I forget what kind of meeting this was, but it was one of our standard meetings that we had, and one of our female flag officers was in the meeting—

EM: Bobbie [Roberta L.] Hazard?

SA: It wasn't Bobbie, no. I think it was Marsha [J.] Evans. I think it may have been one of the meetings; I had a lot of meetings on medical issues. We had all sorts of medical problems over those days, and during the time I was in OP-04, Bud Edney was the Vice Chief and I was his main guy for the Blue Ribbon Panel on Navy Medicine, to the point that they dragged me into the hospital and I had to watch open heart surgery and all of that other stuff. So Navy medicine was a big thing that we kept dealing with. We kept having problems, so I think it was one of those meetings. And in the middle of the meeting, Marsha and I got into an argument about something. She started arguing about, you know, "if I go to the hospital, I've got to be seen for my annual physical, I've got to be seen by a male doctor, and I think that's just wrong."

I said, "But Marsha, we've got Navy flight surgeons that are female, and if I have to have a flight physical and that's who the flight surgeon is for the day—"

"Well, it's not the same."

"Well," I said, "I think it is the same."

"You just don't understand." I said okay.

EM: Agree to disagree?

SA: Yes. We will agree to disagree on that, but it was things like that that would just pop out of the middle of nowhere, and all of a sudden it was a cause célèbre. I can be more philosophical about it now, I guess.

EM: Before we move on, do you have anything more you would like to say about the whole Tailhook episode?

SA: No. It was a sad thing to happen. It tore away a lot of good feelings about the Gulf War from the aviator standpoint. We lost some good aviators that had done some great things, and so it tore away at that. The thing I feel bad about is that somehow we didn't get out, and we didn't get in front of it soon enough. There are two things—my reflection was, when I got back there, that one, the OpNav staff was slow to understand what had happened and so had sort of shuffled it off. And then of course we had a bungled investigation. But hindsight was that there had been plenty of signals back to my OP-50 days, when Chuck McGrail and I were back there, and Ed Martin had come back and said, boy, I've got to do something. Somehow or other that didn't translate into some changes being made earlier.

EM: Do you think the Navy changed as a result of Tailhook?

SA: Oh yeah.

EM: In what way? Positively or negatively?

SA: No, I think it changed positively. I mean, there are a lot of negative things that had happened within the process, but I think when it was all said and done, the Navy became a much more aware organization that realized things were not as simple as having something down in black and white and expecting it to be okay. That some of these things had to be worked. You couldn't assume that just because you said we are going to integrate the ship with females—you had to follow through, and you had to have somebody monitoring some of the fallout. The one negative that I think came from

Tailhook is the fact that people still had a great fear, now, of getting involved, that it might not be safe to get too close to anything new and different. That something was going to happen by fiat. Just because you were there, you were going to get hung for it. [Break]

EM: I wanted to just get your impression of some of the folks that you worked for at one level or another. Sean O'Keefe, of course, Secretary of the Navy from '92 through '93. What interaction did you have with him?

SA: We had good interaction. He was a budget guy. Sean's big strength was the budget.

EM: He came from the budget community?

SA: Yeah. We were sort of apprehensive at first that it would all be a budget-driven agenda, but Sean had a very outgoing personality. Very engaged with the process. He drew the best out of people. If he had a problem, he called people in and talked to them. Really enjoyed the job, and you could tell he was enjoying the job or else he was very good at faking it, but he seemed to enjoy the job and I thought was very effective. We had him for a short time.

EM: Did you have any personal contact with him on certain issues?

SA: We did. We talked on different issues. I can't recall all of them, but he would sit down with you and talk over issues. Let's see if the timing there was right. Of course we had a lot of discussions over the Tailhook issue and all. I was a little apprehensive at first. I wasn't sure how the interface would work, but as it turned out it worked very well. I had more interface with him after I left the Navy, and he was teaching up at Syracuse and had that course up there, a government course.

EM: How did you interact with him after he retired?

SA: Well, he had me come up there a couple of times to be one of their guest speakers and presenters, and so I got a better appreciation for him, and he was a family man. I enjoyed the association with him, and then afterward I've enjoyed the interfaces we've had. I talked to him a little bit after his episode up in Alaska, when he was in the plane crash up there with the senator from Alaska.

EM: Who was killed?

SA: Sean and his son were both on that airplane.

EM: The son survived too?

SA: Both of them survived.

EM: Wow. Stevens, Senator [Theodore F.] Stevens [Sr.].

SA: Yeah, Stevens.

EM: Was Tailhook the big issue when he was SecNav, or were there others?

SA: Yeah, well, that was a great part of the thing. Plus he was there for that POM build that year, and then he was gone, so he only put together the one POM. But he was very engaged and had a clear, pretty much a clear picture of where he wanted to take the Navy.

EM: Where did he want to take the Navy?

SA: He was moving forward, and he wanted to tighten up our budget process, of course, being his background. But the thing was that he made you a part of the process. He wasn't a guy that squirrelled himself away and then all of a sudden came out with these pronouncements that caught you by surprise.

EM: He wasn't a sphinx?

SA: No. He was very engaged, and he had a great personality.

EM: Which is kind of interesting, what I've heard this morning about you involved in the budgetary process. You were a flier—I mean had you had any experience with POMs before OP-04? Is this something that came naturally to you?

SA: No. When I went to ICAF, McNamara, the POM was all new business, all new-wave stuff and everybody tried to figure it out. I couldn't figure out why we had all these lectures at ICAF about building a POM, and I said this stuff is not of any interest to me. Why do I need to know about this? So then when I came back to be OP-50, first thing that happened, I'm building a POM. Well, this is the first POM I've ever built, you know, and I'm thinking, why didn't I pay a little bit more attention to this process when I was in the ICAF? And then of course I went from OP-50 right straight to OP-90, and OP-90's job is to put the POM together to present on up to the staff and to the Secretariat. So thank goodness I had people like Don Pilling and Irv Blickstein and all the boys in the bullpen to get me smart on what a POM really was.

EM: Did you do any midnight reading or anything else?

SA: Oh, yeah. [Laughter] I had to teach a course in the building about the POM. So for all of the new guys who came to the building, one of the things that OP-90 had to do was go down and tell them how to build the POM. So I had to get spun up pretty quick on that.

EM: On the job training.

SA: Well, that's the beautiful thing about the Navy. You get to do these things that you never thought you would ever have to do, and you learn some good stuff.

EM: That's always puzzled me how the Navy would put somebody in a position for two years where they are just getting into the job, learning the job when off they go to something else.

SA: I know.

EM: Clearly the intent there was to give them broad background in all aspects of the Navy. But doesn't that interfere with efficiency?

SA: It does. It does. There's no doubt about it.

EM: Wouldn't it be better to have, like, four years for a tour?

SA: It would be. [Laughter] It was the same thing with your squadron tour. A year as exec and a year as CO, you know. To make real change, to do it in a year is a hard job.

EM: How did the Navy come to that procedure? And why do they stick with it?

SA: It is the grooming of the force. In other words, you've only got so much time that you can stay at one place, or you are not ready to go to the next place. You have to move people so that there is an opening for the younger people to come through. Now what we could do, and probably what we should do in today's environment, is let the force age a little bit more and not want to move everybody every two years. But we haven't got there yet.

EM: Anything else you would like to say about Sean O'Keefe?

SA: No. I thought he was a good Secretary of the Navy. He and Frank Kelso got along very well. Sometimes you would have to weave a fine path between the CNO and the Secretariat, and with Sean and Frank there it was very easy.

EM: And he dealt with Tailhook satisfactorily?

SA: Yes.

EM: How about John Dalton, his successor?

SA: John Dalton, of course, was completely outside the structure. Came out of Texas, I guess, and was unknown to us and pretty much from our standpoint, the initial blush was, you know, this is a truly political appointee. I mean, somebody that hadn't been on the Hill. Somebody that hadn't been in the building.

EM: This was the Bush Administration? Bush senior?

SA: No, this was Clinton. So he was a really an unknown quantity, and we were very suspect, I think, at the early stages because he didn't really know the building and really didn't seem to know much about what we were doing. But over time we had a lot of interface with him because again, Tailhook, we had to sort of clear everything. The final clearance of all of that was under his watch, when we finally got out from underneath the umbrella. So, worked very closely with him on that. Worked very closely with him on the Rebecca Hansen episode. His bent was political. He loved being Secretary of the Navy. Believed in the mission. Wasn't as comfortable with how the building worked and some of the hard things that had to be done, but as far as showing the flag and being supportive of the Navy and the Marine Corps, John certainly was.

EM: Did both he and O'Keefe get out to the fleet periodically?

SA: John did. I can't remember for that short period with Sean. He did get out some, but not a whole lot that I recall.

EM: Sean was a Republican appointee?

SA: Yes.

EM: Okay, so change in administrations. Anything stand out about Dalton? Any programs that he is tied to? What contribution did he make?

SA: John was really into the people programs. We were trying to get off of dead center on the women-in-the-Navy issues, so he was very engaged with those. Had very little tolerance for things going wrong in that area. He was going to make sure that things like Tailhook were not going to happen. When it came time for handing out the punishments and the disposition of the Tailhook things, he was very much on the side of, let's get it done. Get it behind us and let's get moving in the future. I'm trying to think what big program might have—well, I guess we had the E/F stuff was going on during those days.

EM: What's E/F?

SA: The F/A-18E/F program was still coming of age during that time.

EM: Oh, okay.

SA: I think Mike [Jeremy M.] Boorda was pushing, and LCS got started under his watch.

EM: Littoral combat ship?

SA: Yes. The littoral combat ship. Mike Boorda and Dalton got along pretty well. They got along well. We didn't have to be too careful weaving between them.

EM: During your entire time in Washington, do you recall where SecNav and the CNO did not get along?

SA: There were issues, in every instance I think we had some issues.

EM: How about with Secretary Lehman?

SA: In Lehman's time, I mean there were lots of issues then, because he was so energetic and so forward-thinking. You are going to do the oral history with Joe Prueher. Joe can tell you a lot about those days. He was a very much a Lehman protégé. John Lehman thought highly of him, and of course we had Admiral Watkins as the CNO, so there were lots of friction points every now and then between Watkins and Lehman. Totally different personalities and very strong personalities. Watkins was a very strong personality, and John Lehman was a very strong personality.

EM: But during your time, the SecNav and CNO seemed to get along?

SA: During the time I was the Vice Chief, they normally got along pretty well. There were some friction points every now and then. Both CNOs that I worked for, if there were some sticky points that they thought I could do better than—could calm the waters better than they could, they'd let me play in it a little bit. So I had two bosses that both basically let the Under and the Vice Chief—if things were going to come to a head, they sort of said, "Vice Chief, you go talk to the Under and let's see if we can get this thing so I don't have to go next door and throw a fit." But every now and then you'd get a call, because they had connecting doors, the CNO and the SecNav, in those days.

EM: Did they use those doors?

SA: Every now and then they would get used, and then you'd get this call saying, "You're going to have to backfill on this thing. I just had a visitor." Or, "I just made a visit and here's the issue." But they were nothing really big or I would have remembered them. I hoped I would remember them.

EM: All right. How about some of the SecDefs you served under too. Dick Cheney, of course.

SA: Oh, SecDef, yeah, Dick Cheney. Of course I had dealt with him some during Desert Shield/Desert Storm, because he always came out with Colin Powell. And every now and then I would have him in my helo and take him from one ship to another.

EM: He was always coming out with recommendations for various things, even though he's not in the operational chain of command. Anything that sticks in your mind from the Gulf War that you thought was maybe it is not the best thing for us to do?

SA: Well, the one that I was most upset about was trying to re-wicker Navy doctrine on, you know, cohabitation on ships with the husband wife doctor team on the hospital ship. But other than that, he and Powell were very close. They always came on scene together. So they had their script pretty much prepared between the two of those; so you weren't going to drive a wedge between them, so Cheney was a good backup to Powell. Powell put his emphasis on something, then Cheney would sort of take an opportunity to back him—yes, that's exactly what we are thinking about, and so you knew that they were aligned.

EM: How would you characterize Cheney's personality as SecDef? Was he a forward, an aggressive-type SecDef, or—definite points of view?

SA: No. As a SecDef, he had definite points of view, and he wasn't bashful about expressing them. And sometimes, you know, the fact that he and Colin Powell both came out in the theater together all of the time, it was pretty evident that the two of them had made the decision on where we were going to go, and they were there to make sure everybody knew that not only did the Chairman—this is his view, but this is the SecDef's view. And if that's the SecDef's view, then that's the president's view, so that's where we are going to go.

I remember before we started combat operations, we had a final gathering in Riyadh. And prior to that session, Schwarzkopf called me and said, "Look, you are going to have to make a presentation, and this is what the Secretary and the Chairman want to hear. You are going to be able to sink any Iraqi ships that enter the theater at any time."

And I said, "Whoa, whoa," you know, "wait a minute. Law of the Sea and all this. We've got to be very careful about this stuff."

He said, "There's going to be Iraqi-flagged ships coming down through the Red Sea and out into the Indian Ocean, and you're going to sink them."

I said, "Well—" [Laughter]

"If that's what's in the rules of engagement, that's exactly what I want you to tell them."

EM: Schwarzkopf said this?

SA: Yeah. Apparently this is what Powell and Cheney wanted to hear, because when I got up to give the brief, then they both chimed in and said yes. Of course that didn't happen, because the first time I sank a ship, everybody was upset that I had sunk the ship. [Laughter]

EM: Which one was that?

SA: Oh, that was a guy that was anchored up there by Bubiyan Island that was transmitting the radar—

EM: Oil spilled from his tanks?

SA: Yeah, the oil spilled from his tanks, and all that. Be careful what you ask the guy to do.

EM: So he was the fellow who canceled the A-12—Cheney, from my understanding. Is that accurate or not?

SA: Cheney, he started, but it finished up with—yeah, that's right, and I think the final disposition and all happened under Perry. As I recall. [Les] Aspin was there for a short time.

EM: I think you talked about the A-12 and why it wasn't a viable program before.

SA: In my mind the A-12 was a good program. It ran into those cost overruns, but where they were trying to go with the A-12 was the right thing.

EM: It was going to be a replacement for what?

SA: The A-6.

EM: Okay.

SA: And it was going to be a survivable version of the A-6, and it was going to keep us in the long-range strike business. So when they got rid of the A-12, then we were left with no long-range strike capability.

EM: That must have made the Air Force very happy?

SA: Sure.

EM: Because they like that.

SA: Yeah. They like that a lot.

EM: So did he side with the Air Force, or were there other reasons?

SA: You know, I don't know how much the Air force influenced that program. I think the thing fell on hard times between the contractor and the systems command. I thought it was a salvageable program. I thought it preempted a viable program. To me it would be interesting if somebody would go back and reconstruct that, and tell me whether that was really true or not because we, in my mind—and it had been for a long time—if I couldn't reach out 1,000 miles from the carrier, then there was something wrong. Reaching out 1,000 miles from the carrier with the F/A-18 is not an easy task.

EM: Even the E/F?

SA: Even the E/F. And even with the F-35. We don't have the range that we need to have. We just have to have that capability.

EM: Which means you have got to get closer into the danger.

SA: You have got to closer in, or you've got to have a string of tankers, and so all of those things are giveaways. All of those things start tipping your hand and you can't mask your tactics, because either you are getting in too close or all of a sudden they are seeing tankers.

EM: Lining up?

SA: Lining up, and the Navy has done it to themselves—when we bought into the concept that we would no longer have a robust onboard tanking capability.

EM: When did the Navy buy into that?

SA: That was under the rubric of jointness, when Watkins was CNO and the Chairman was the submariner.

EM: Submariner?

SA: Yeah, that came out of the Mid-East Force.

EM: Oh, Crowe?

SA: When Crowe was Chairman, Watkins was CNO, and the Air Force Chief of Staff brought to the tank the concept of the air component commander for the 150th time since World War II. They decided that under that rubric of the air component commander the Air Force would be in charge of tanking, and if we needed long-range tanking capabilities, the Air Force would provide it for us. We no longer had to buy something like the A-3. We could have our short- range tanking around the ship—you know, recovery tankers and stuff like that, and some escort for a little bit of distance—but for really long hauls, we were to be relying on the Air Force for that tanking.

EM: And that caused some problems during the Gulf War, right?

SA: Yeah.

EM: Tanking?

SA: Yeah, the tanking was an issue. It turned out to be more of an issue because of the fuel in those days, when we had to have JP5 and they had to have JP4. If you had a mixed load of fuel on your airplane, if you tanked from an Air Force tanker and took JP4 on, then when you recovered on the ship, we had to dump all of that fuel over the side because it was not compatible with shipboard operations.*

EM: It's flammable, right? More flammable?

* While the Navy has been especially environmentally conscious for many years and does not routinely put fuel or other foreign matter into the sea, this was a wartime exigency. The JP4 used by the Air Force was considered more volatile than JP5 and therefore a danger on board a carrier, where fire is always a major concern.

SA: More flammable. And it would light off on a steam catapult—JP4 would light off just like that if you spilled any on the steam catapult.

EM: All right. How about Les Aspin, SecNav from '93 to '95?

SA: SecDef, yeah.

EM: SecDef.

SA: Yeah. He didn't have much impact.

EM: He was a political appointee as well, right?

SA: Yes.

EM: Former congressman from Wisconsin.

SA: Wisconsin, I think, yeah. Pretty benign.

EM: In contrast to Cheney, I guess?

SA: Yes. Particularly in contrast to Cheney, yes. Then Perry. Perry had an extensive background in think tank stuff around town, and jobs that he had had in the building before. So we went from a very brash, direct guy in Cheney to Aspin, very quiet, reticent sort of personality. More interested in the political ramifications. And then you get Perry who had a pretty extensive background in thinking of the future, and what the future was going to bring. So it was sort of a quiescent period.

EM: You can't remember any programs that he really pushed through?

SA: I don't remember anything that he really made a big difference on.

EM: Did you have any personal contact with him?

SA: I had some; very little. I sat in again with the JCS meetings, and we would sit in on different meetings with him. In my mind there was nothing that—at least that I can remember now, that sort of got me up on the step and excited about.

EM: He left when he got ill; didn't he have a heart attack?

SA: There was a health issue I think, yeah.

EM: All right. Okay. Now, having gone through those you worked for in one capacity or another—you brought this up last time—the transition from Bush 41 to the Clinton Administration. What impact did that have on what you did?

SA: We had this dead period where we didn't have the Under, and the political appointees that came in with the Clinton group were not of the same ilk as what we had dealt with in the Bush years. So it was really a transition more from a business oriented atmosphere to a more political atmosphere, and a way of doing business that was not as precise as what we had been used to. Where we had meetings and decisions were made and, *boom,* you are off and running.

EM: Right.

SA: These decisions kept getting revisited. "Well, we want to talk a little bit more about it," and so there was a little bit more; it was a different pace, and it was hard to adapt to. There was within the new administration—this is just my reflection—I always felt that there was an awful lot of competition within the appointees themselves; some were not where they wanted to be. They wanted to be in some other job. So it seemed to me like things weren't hooked together quite right. That we were always sort of struggling with who was going to be in charge. Who was going to do this? Who was going to do that? And trying to match personalities with different people.

EM: Do you recall anyone who came in with the Clinton team that stood out, that really wanted to make an impact?

SA: Well, let's see. You know, John Dalton I think aspired to bigger things along the way.

EM: Yeah.

SA: Nora Slatkin. Nora came in, and she had been a Hill person. So she knew the business of the building because she had been on the committees, working for committees over there that did a lot at the time with the building. She took over the job from Jerry [Gerald A.] Cann, who we'd worked with on the Goldwater-Nichols legislation, the old Mel Paisley job and the technology. He had the acquisition side, and another Under or Deputy SecNav had the technology side.

Well, under Goldwater-Nichols it all came under one guy, RDA, Research, Development and Acquisition. Jerry Cann was the guy that put that together. Jerry and I had long discussions about that job when I was OP-04, because when that transition took place, part of my job at that time was to be one of the key players in making Goldwater-Nichols work within the Secretariat and the Navy staff. That was sort of the bailiwick that I was given. So Jerry and I had lots of discussions at the time that one—that job was almost too big for any one person; it was a very big job. But he was convinced that with the Goldwater-Nichols, it was the right thing to do, and the fact was very clear that the OpNav staff was no longer in the acquisition business. We had no voice. Our job was only to write the requirement. And once we wrote the requirement, we were to take our hands off of it and never look at it again. It was his job to make it happen.

EM: Yeah.

SA: We argued about that long and hard that I didn't think that was right. I thought that we had to build an interface between the OpNav staff and the acquisition guys that still let some give-and-take take place. All of the other services maintained that.

EM: Yeah.

SA: We didn't. We made an absolute split. When I came back to be Vice Chief, just at the final days, a few months were left of Jerry Cann's tenure as RDA, we had this discussion. He came and he talked to me and said, "I think in the end that you were probably right, that we've gone too far. That we've pushed the OpNav staff so far away from us in the acquisition side of the house that I don't think it's as healthy as I had hoped it would be."

So when Nora took over, she was very gracious. She came down to see me and wanted to talk about what our interface should be and how I wanted to do business, and tell me how she wanted to do business. I told her at the time that I had talked to Jerry some between the start of RDA and now the ending of his tour, and I said, "He has backed down a little bit to my way of thinking at the time that there should be more interface between us and you guys."

She said, "How can I best do that?"

I said, "It would be good from my standpoint if you had a military deputy that could sort of be the interface that would be the kind of guy that could, although he's attached to you, he would still have good contacts with the OpNav staff. And when he could see that the OpNav staff and the acquisition side of the house were not happy with where they were on a particular program, he could probably help you weave your way through the process without disturbing what everyone says is Goldwater-Nichols, and this is how Goldwater-Nichols is."

I pointed out to her that all of the other services did that. They all had that interface. She took that under advisement and eventually did do that. John Lockhart went over to be her military deputy, and that was a step in the right direction. You'd have to talk to John about whether he liked doing that or not.

EM: Do you think it improved the connection?

SA: It did improve the connection, yeah. It showed us where some of the real problems were and helped us game our way into the future. But Nora was a very interesting, hardworking young lady. Very strong-minded and had the complete confidence of John Deutch, who was DepSecDef [Deputy Secretary of Defense]. So she was backed up by John Deutch.

EM: Did she continue in government service, or did she get out?

SA: She was still there when I left, and John Deutch had already left. So then John Deutch ended up going to CIA and he pulled Nora.

EM: Yeah, right.

SA: He pulled Nora over to CIA with him after, and then they both got in trouble. [Laughter] With the same thing that we've got with [David H.] Petraeus, and all of these days with he was taking super, super secret stuff home. Nora was walking around the building without her pass over there and giving people grief when they would stop her and say, where's your pass? She would fire back at them, "Don't you know who I am? I don't need a pass." That kind of stuff. Then she left and went into industry from there.

EM: Okay. How about your dealings with Capitol Hill? You would have to go over there and testify periodically, right?

SA: Yeah, yeah.

EM: Which committees? Armed Service Committees?

SA: Armed Services Committee mostly.

EM: Appropriations Committee?

SA: Appropriations sometimes.

EM: Yeah. Who were your friends and who were your enemies?

SA: I had good friends with Inouye and Stevens. Both of them were very gracious to me anytime I went over and had any dealing with them.

EM: Were they supportive of the Navy's programs?

SA: Yes. They were, both of them. So I did some work with them. I think I told the story when—I'm thinking I was still in OP-04, but it was really when I was Vice Chief that I had to go over and talk about the atomic clock to our good friend from West Virginia.

EM: Senator Byrd?

SA: Senator Byrd, and tell him that no, we weren't going to build a third atomic clock in West Virginia. [Laughter]

EM: No need for a third clock?

SA: There was no need for a third clock. I went over to testify, or I went over to meet with [Senator] John [F.] Kerry on shipyards when we were doing the BRAC stuff and the shipyards up in the Northeast, and we were trying to close one of the shipyards up there.

EM: Portsmouth, right?

SA: Yeah, Portsmouth. So he wanted me to come over and talk a little bit more about what we were doing there.

EM: What kind of reception did he give you?

SA: It was a for-show effort. It was something that he had to do. He called me over and thumped on the table a little bit to say, "I'm in support of Portsmouth and hope you guys can see your way clear to do that." It was one of the more standard of the things that you went over to do.

EM: Well, Portsmouth survived, so it must have had some impact?

SA: Yeah. It survived because we still hadn't figured our way completely out of how we were going to do what we had to do with our ships. We still needed that capability. I wish I would have kept a log, but you were always getting called to go over on the Hill, and that was one of the things that was good. It was much better for the Vice Chief to go over on a lot of those things than it was for the CNO to go. So I could go over.

EM: Were you called over there in regard to Tailhook?

SA: You know, I can't remember ever going over to the Hill on Tailhook specifically, no. I went over there, of course, with the Rebecca Hansen deal. But no, I don't think I ever got called over there on Tailhook.

EM: Were there congressmen or senators or representatives who had the Navy in their crosshairs, that you recall?

SA: There were some folks that weren't actually too friendly to the Navy over there. Of course, John [W.] Warner was our big friend. I used to see him quite a bit. He was always very supportive of the Navy, of course.

EM: Was [John S.] McCain [III] a senator yet?

SA: McCain was a senator. Our relationship with McCain was off and on. There were some programs that he was in favor of that we weren't, and there were programs that we were in favor of that he didn't want to deal with.

EM: Can you name a few? What was he against? Which programs?

SA: Well, let me see, what was the big one that we had? One of the aircraft programs he was against for a while; it may have been the F/A-18, I'm not sure now.

EM: What did he fly in Vietnam?

SA: He flew A-4s. John was a little hard to deal with. He would get himself entrenched in a program and make up his mind, and then he just wouldn't listen to anything else. Just no way to reach him. We always tried to coopt as many people as we could on the staff, so we would always have somebody on the staff that you could talk to so that you could sort out and weave through things.

EM: Right.

SA: John [H.] Glenn [Jr.] was never really a friend.

EM: Really?

SA: Friend to the Marine Corps. Not so much a friend to the Navy. So he was sometimes cantankerous.

EM: What about Senator [Edward M. "Ted"] Kennedy?

SA: Oh, well, he was very seldom a friend.

EM: Very seldom a friend?

SA: Yeah. I got into a big argument with him on one of the hearings on sealift. The union guys had gotten to Kennedy on sealift and had given him some bad numbers. Now, I'm not sure whether I did this in OP-04 or as Vice Chief. I can't remember the timing, but it was testimony on spares. Kennedy and Glenn both came down hard on us on spares. Then Kennedy came down hard on me and called me over for a hearing on sealift, and then he gave his little speech up there about how poorly we had supported sealift, et cetera, et cetera, et cetera, and when was the Navy ever going to do its job, and do what it's supposed to do about making sure we had a robust maritime force behind the uniform side.

EM: What was the union's big push? What did they want to have happen?

SA: They wanted more ships being built in the U.S., and parts of more crews for the ships. So when we were building the preposition force and stuff like this, they wanted us to do more of that and use American-flagged ships and not foreign-flagged ships in anything that we did. So we got into this numbers thing, and finally I challenged him on the number that he had presented and said, "Senator, that's not right. You know, in the last five years we put $2.5 billion into sealift, and you are giving us no credit for it."

He just came unglued. Somebody had taken a picture from behind me looking at him, and he was up there—you know Ted Kennedy had a little trouble with the bottle, so I think he had been imbibing a little bit already that day, and his face got so red. I mean, it was just like a red torch up there, and he was pointing at me. Some OLA guys had taken it and put a big caption on it.

EM: It would be great to find that picture.

SA: Yeah. I would love to find it again, yeah. I'm sure it's someplace. I will have to call Mike [Michael L.] Bowman. Mike's probably still got a copy of it.

EM: Did he give the Navy credit for sealift's contribution during the Gulf War?

SA: Yes, yeah.

EM: All right.

SA: But he wanted more.

EM: He wanted more. All right. Any other issue that got his goat? How about the personnel side?

SA: He was in the personnel side a little bit, but I don't think he and I ever had any dialogue in the personnel side.

EM: Do you recall any individuals who worked on staffs over at Capitol Hill who were very useful for what you had to do?

SA: They were there, and I'm trying. Larry, let's see. He was one of the staff guys over there that was always helpful to us. Came back into the building, came back like so many staffers come back into the building. Gosh darn, he lived in Annapolis. Caviotta. Larry [Lawrence] Caviotta. Great guy, and we ended up at Lockheed Martin together later on.

EM: All right. This was the period of the post–Cold War peace dividend.

SA: Yes.

EM: How did that affect the Navy? Did you guys get a dividend?

SA: We didn't get a dividend. We got a humongous cut. I don't remember the numbers now, but we took a hit, really big hit. Then we could see another one. We thought we saw another one coming that was really going to be catastrophic. Back in my OP-90 days, Don Pilling and I and Irv Blickstein, we all sat down one day or night after work, and we had put together the POM, and we were talking about what did we see coming down the track. And we were convinced that we were going to see some really monumental cuts coming. So I had asked Don to form a real close-knit group over there that could keep quiet with his gang and build me a fake POM that would reflect what would be the worst number that we could think of. For us to just look at and see what had to fall out so that we would know, if this thing ever started to really take place, where we should go first instead of just continuing to do the cut by 1,000 knives, you know. Where to finally—

EM: Cut a major program instead of a bunch of little programs?

SA: Take a major program to give us time to buy into where we were eventually going to floor out. It was a very good exercise, because when I came back as Vice Chief and some of the stuff was still going on, and we could see things going up to Capitol Hill, I at least had a framework in my mind of things that were very important for me to be seen as supportive of and things that I should not.

EM: Well, based on your worst-case scenario, what were the major areas that were going to get the axe if it came to pass?

SA: We were going to take huge personnel cuts and the 600-ship Navy that basically, when I was OP-90, I had given the POM to John Lehman that told him which ship was going to be number 600. It was going to go down to sort of where we are now.

EM: 285.

SA: Yeah. Which was just unheard of in those days, I mean, here we were, we worked all of these years for 600, and now we were going down. When we had John Lehman there—and the team that John Lehman had were really sharp business guys—they were teaching the Navy staff how shipbuilding should be done. In other words, you know, figure out a way to get the Hill to let you buy more lead time and not have to pay up front the full cost of the ship. It takes you X number of years to build the ship. Why should you have to POM the whole number in one year, and this type of thing.

So they had started making great progress in that line, and that team went away, and then the smart people who knew how to do that were gone, and so we were starting to get boxed in, or we could see the *Nimitz* class coming along—how long could that class go before we had to go to the next class of ships? When could we do it, and, oh, you've got the boomers, the *Ohio* class. And we are right at that intersection now, today.

We are down a couple of carriers, but we need to do the *Ohio* replacement. How can you correct that minus number on the carriers and build the *Ohio* class at the same time? We had seen that coming, and we were trying to sort our way out to how to get the Congress to—and you can see that the CNO has already tried to persuade Congress to consider a new way of funding the *Ohio* replacement.

EM: Which didn't go down well with the current Secretary of Defense?

SA: No, it did not. So from the standpoint of when people say, "You guys never see over the horizon—"

EM: You saw that coming?

SA: We saw that coming 20 years before it happened. But the result has not changed.

EM: During the time when you were in, were there big cuts in particular weapons systems?

SA: Of course the A-12 was a big setback. Then you had no replacement for the F-14. Just tie into the F/A-18F instead. We started to see the end of the DDGs [guided missile destroyers] up at Bath Iron Works [shipbuilding company], and how to—what were we going to do now? That's when we started moving forward with the *Zumwalt,* starting to talk about the follow-on, which became the *Zumwalt* class. Here we are today, still having to resolve. We got maybe three *Zumwalt*s authorized, and the first one has just sailed and is still not turned over to the Navy yet.

EM: Yeah.

SA: We've got a carrier that's built, or close to being finished, and we are going to play Ring around the Rosie with it for three years getting it tested before we will let it go to sea.

EM: Why is that? They are just dragging it out for the cost factor?

SA: Dragging it out just because of political incompetence as far as I'm concerned. And Navy incompetence.

EM: Which ship is that?

SA: *Ford.*

EM: The *Ford,* yeah. So it's pretty well constructed, but it needs some testing.

SA: It's going to need testing, and here's the thing that happens. You get shifts in administrations, and you get people. So you had [Donald H.] Rumsfeld come in with transformation. Everything is going to be transformation. I'm out of the Navy by this time, but I'm called back to be part of the industry team that Rumsfeld wants to have present their view of transformation. So I'm a part of the industry team. It's industry and SecDef people working this thing for Rumsfeld on what should be transformation. Well, one of the key things that I took into it, since I was the Navy-experienced guy, was I didn't want the carrier to be considered—that it had to do something different than what it was just to be transformational. The carrier was still a platform that could be evolutionary. You walked your way through the technologies so that you were actually transformational from the day you started, because as new technologies developed, each time you came in for a major overhaul period, you could bring those technologies aboard.

EM: Sure. The basic structure was the same, but it could be adapted.

SA: So I didn't want to get into the game of trying to do away with the carrier program that we had, and say okay, we are going to build this transformational carrier. I was pretty successful. The carrier hardly got mentioned. Of course some people in the Navy said, why didn't you get the carrier mentioned? I said I didn't want the carrier mentioned.

EM: Don't draw attention.

SA: You know, I want it to be assumed as a part of the future without making something different.

EM: But that's not the case now, because there are those saying that you don't need carriers.

SA: Yeah, yeah. And we've gone through that argument every 20 years. So when *Ford*—and I was also on a panel with Don Pilling. He was still—no, Don was out then too I guess—anyhow, we ended up with CNA on a study on the carrier of the future. Where we were going, and this was *Ford* and basically we said that there were technologies that needed to come along with the carrier.

And so the *Ford* class was first designed to bring on the first technology, electromagnetic catapults. Then part of that was a big difference in the electrical system on the ship. The nuclear interface and the electrical system was completely different than what we had on the other nuclear carriers. Then the second technology we wanted to bring on was the electromechanical or—

EM: Magnetic?

SA: Electromagnetic arresting gear. Same thing from the catapult side of the house. Now bring that technology back to the arresting gear. The third technology was the new radar, whatever the new radar was, but the radar after Aegis. So instead of the carriers having a destroyer-type radar on them, we were going to go back to where we were with *Enterprise,* when she first came out with a phased-array radar; that was one of the unique things about *Enterprise.* So the thing we wanted for the carrier was the follow-on to Aegis. Where were we going to be with radars for this next 50 years? Well, when the Rumsfeld team got ahold of all of that, they dictated to the Navy that the *Ford* class—no, the Ford class was not going to do incremental integration of new technologies. They were going to do it all on the first of the class. So we were stuck now with *Ford* the first of the class. We've got problems with the catapults. We've got problems, bigger problems with the arresting gear, and we've got major problems with the radar that it hasn't been developed.

EM: Everything is on hold until they get each of those taken care of?

SA: So now were trying to work our way through, and of course as soon as you slow down a program, the dollars just go—you slow like this and the dollars go like that.

EM: Yeah.

SA: So it's a terrible situation, and basically where we are going to end up is, now the ones that have always been against the big deck carrier are going to be very successful in arguing that the price is going out of sight on the very first of this new class, so you can't afford the rest of the class, so why are you doing it?

EM: All right. Any other issues or programs that you recall that you had to go up to the Hill to talk about more than once?

SA: No, I think we are good.

EM: Let's talk a little bit about Kara [S.] Hultgreen.* One of the first women; a carrier-based fighter pilot. She was in EA-6 Prowlers first and then F-14s. She failed her first carrier qualification. Was that typical of folks in the aviation training business?

SA: Not unusual.

EM: Not unusual?

SA: Not unusual.

EM: Okay.

SA: In fact, I looked very closely after it became a cause célèbre that she had flunked her first thing. I went back and looked at that class that she was in. There were guys behind her that had flunked twice. I think even one flunked three times before he succeeded, so it's not unusual. It's an issue that you need to look at and make sure that whatever caused the first failure, there is a path forward that says it's not a fatal flaw.

EM: Yeah.

SA: I personally took a look at the records of those in the class and hers, and what I had seen in other RAG classes when I was a RAG instructor—basically it was, yeah, it's one of those that she did.

EM: I'm sorry, one of those what?

---

* Lieutenant Kara S. Hultgreen was the first woman to qualify in the Navy's top fighter, the F-14 Tomcat. Like many of her male counterparts over the years, in 1994 Hultgreen was killed while recovering onboard an aircraft carrier, an inherently dangerous maneuver. Some critics charged that to satisfy feminists the Navy put an unqualified woman in the pilot's seat. Others said that a woman should not have been exposed to such risk in the first place. In the end, the Navy determined that it mattered not in the least whether the pilot of that F-14 was male or female.

SA: One of those episodes where, yeah, somebody's not going to make it through their first carrier qual in the new airplane.

EM: But then she did subsequently?

SA: Subsequently she did just fine. So, you know, a lot of angst over that. There were people that were first and foremost against women being in the fighter community. There was no doubt that that existed.

EM: Aviators or people on the outside?

SA: Aviators.

EM: Yeah.

SA: Aviators. It was pretty well ingrained in the culture there at Miramar. Tom Cruise and the great movie, you know, sort of cemented their view of the world. This is who we are, and this is how we behave, and so they were a little chagrined that they were the first ones to have to give up a seat to a female.

EM: Did they give her a tough time?

SA: I've heard different things along the way. I talked to her mother at great length during the workup to her funeral and some events after that relative—Elaine Donnelly and all. Kara was pretty tough skinned; as far as I could tell from the reflections that I got, she took it all in stride. Knew it was going to be there, and knew she was going to have to handle it. She did. She didn't let it—she did it much differently than, say, Rebecca Hansen did relative to her circumstances.

EM: You attended Kara Hultgreen's funeral?

SA: I did. Got a lot of criticism for that.

EM: By whom?

SA: From some of the aviation community, particularly from out west. I thought it was sort of petty. I mean, the Vice Chief attended a lot of funerals. I attended more than a lot of guys, probably because when I was OP-04 I had a window right behind my desk. The OP-04 office was on one of the corners of the Pentagon, so the window right after the corner sat right behind my desk in OP-04. So if I stood up from my desk and turned around and looked out the window, I was looking straight at the cemetery. I can't tell you the number of times I'm standing behind my desk working my way through a problem and I'm looking out. I'm looking over at Arlington, and there is a bugler, the color guard, the preacher, and maybe one of the ladies from the Arlington Auxiliary attending a funeral of one of our service members. NavPers always had designated somebody to be at every funeral so there were people there. But I thought if there's somebody you know, or somebody that there's an attachment to, there ought to be more people. There ought to be more people to see this thing.

So I remember we had a seaman, female seaman down in Norfolk who was murdered, and she was going to be buried in Arlington. I checked to see if anyone was going to be there for the service, because her dad had been a chief petty officer. What I got back was, "There will be the normal rep there." This is when I was OP-04, not Vice Chief per se. I decided, well, you know, I've got time. I can go and tell the chief that we understand.

EM: Were there others there? Was it a small group?

SA: No, it was a very small group. It was just the chief and his wife and the ceremonial people. So when Kara Hultgreen came along, I said, here's our first female aviator that we put into a tactical airplane. She's the first of a kind. Why wouldn't senior aviation leadership show up at her funeral?

EM: Right.

SA: I had been there for the guys killed in space, when we lost all of those guys in the space capsule at one time. I went to that funeral. I said, you know, this is something that I ought to do. I am the number one aviator, so to speak, and she was number one of something that we were trying to do.

EM: Yes.

SA: I should be there. I know her mother appreciated it. We had a very good—

EM: Were you asked to speak?

SA: No, I didn't speak. Well yeah, I did speak, I take that back. But I know her mother appreciated it, and I thought it was the right thing to do. I was surprised that some people said, that's just playing into the hands of the politically correct. It wasn't politically correct; it was just the right thing to do.

EM: You thought she was qualified to fly F-14s?

SA: I think she was qualified. I don't think that she was going to be the best F-14 pilot. I am absolutely 155 percent sure that she wasn't going to be the worst.

EM: You've seen some of the worst?

SA: Yeah. I had seen some, and I had seen records of some. I knew that she wasn't going to be the worst. I didn't think she was going to be the best, but I thought she would be in the middle of the pack. That's more than you can hope for.

EM: And you explained about her accident. You don't need to go into that, but that wasn't an unusual occurrence, that type of accident?

SA: No. That type of thing could happen. It was a characteristic of the airplane that if you got the needle out of balance and got a sideslip going, you could stall the air to the compressor, and you were going to have a compressor stall.

EM: Yeah.

SA: Theoretically—and they practiced recoveries from it. Basically she didn't do everything she should have done, which, in those circumstances, you never, nobody—I mean not everybody gets it right, and she didn't.

EM: So you got pressure from women's rights groups, right, over that issue, and Elaine Donnelly?

SA: Elaine and I had lots of discussions over time. We finally didn't end up where we should have, I don't think. She was a big supporter of me during the Hansen episode. So she thought I should be a big supporter of hers, with her trying to show that we had cut corners on the training for Hultgreen. And I kept saying I don't think it's the same, you know. Elaine was pretty much on the side that women shouldn't be in combat.

And I understand that, and quite honestly, you know, you take a look around, and it's like the Disney Marathon that was run I think the weekend before you came down here. They ran the Disney Marathon. The guy that won the marathon this time around was three times in a row. He's a Brazilian who lives here in the vicinity. He came in under two hours on the Disney Marathon. The first woman who has run the race for the women's side of the house I think two or three times herself, came in just under four hours.

EM: A lot of time to make up?

SA: A lot of time to make up. And you've got people like the present Secretary of the Navy saying there are just absolutely no differences. Well, we all know there are differences. There are differences for good reasons. Now, do those differences make a difference in some things and not in others? Like I mentioned before, I'm convinced there is no difference in how well a female can fly an airplane and a male can fly an airplane. It's an inherent monkey skill that people have or they don't have.

EM: A monkey skill? You've used that term.

SA: Yeah, that's not a good term to use anymore, but it's one that I've always associated with, you know, who makes a good pilot? You've got all of these things, the standards that you've got to go through to get there, but once you get there, there are people who just absolutely—they were made to fly an airplane. And there are other people who may have been the best athlete in the whole universe that get there that can't get the airplane in the air. Something doesn't connect.

EM: Well, let me ask you this. What do you think of the current—the Secretary of Defense has mandated that women will be in every operational specialty in the services, including SEALs and other Special Forces and submarines. Should every billet be open to women?

SA: No. I don't think so. I don't think in general that women all want to be in every billet. There are onesies and twosies. But onesies and twosies doesn't work if you've got to fill out a whole SEAL platoon or something like this. Are there jobs within the SEAL community that the females can do? Certainly. Are there jobs in submarines? I've always thought submarines are a good fit for women, as long as you can figure out how to make the accommodations so that you don't have male and female both afraid to go to the bathroom, for heaven's sake, or take a shower.

But to do certain things it takes a different set of muscles, and you have to figure out what those certain things are. Like I say, if we were back in the days of the Indians, where you always had a scout who ran back to the tribe to say okay, over the next mountain is a bunch of cavalry, you want the guy that can run the mile, or 26 miles in two hours, and not the one that can run it in four hours. So that would be a discriminating thing.

EM: So it's capability that's key.

SA: It's capability, yeah. And somehow or other you get people like the Secretary of Defense and Secretary of the Navy, they don't talk in those terms. They talk in, "everybody has to have the chance to do it." Well, why do you want to give somebody that has a chance to fail? Why don't you want to give somebody a chance to succeed? So that's the argument. There is a place for women anywhere in any of the services, but there are certain things that a male can do better that they should be—that's where the last dollar should be to pay that guy to do that job, and the last dollar to do that job should go to the female who can do that job better. But they want to get everybody at the same—it's the same thing that we are seeing throughout society. Nobody can be told no. Everybody's got to be told yes. It's a waste, a waste of resources, human and monetary resources.

EM: Can we talk about Rebecca Hansen and that episode?*

SA: This is a classic example of the first thing that I learned in the Navy, and I think that anybody who joins the military does too, don't volunteer for anything. Wait to be assigned. [Laughter] This was a classic violation of that basic principle. You could see this one coming down the road. She had failed. The training command had failed her out, and then she went to Congress, and Congress was now interested. Why did she fail? What are we going to do, and she should be back and flying? Why aren't you putting her back in a flying billet?

---

* In 1994, to spare the Navy further unwanted attention over gender issues, Arthur withdrew his nomination to be Commander in Chief, Pacific. Arthur had endorsed a subordinate command's ruling that Hansen was unqualified to fly naval aircraft. She argued that sexual harassment had been the cause of her failure. Congressman David Durenberger of Minnesota promised to put a hold on legislative proceedings unless hearings were convened on the issue.

And it was starting to bounce back and forth, and I could see it coming, and I talked to the CNO. I said, "The Secretary is going to get it, and the Secretary, Dalton, is going to wrap themselves around this program, and we going to get issued a directive: do this." I said, "and if we get a directive that says do this and it's the wrong do this, then you and I are going to be embarrassed when the next shoe falls and things don't turn out right." You know, again, I'm the senior aviator. I've been an instructor pilot. I've got 30-plus years of watching people fly airplanes. I've seen a lot of accidents. I know what can happen. I said, "Let me go to the Secretary and make a deal, and say let me make the decision on this case or provide you the decision."

EM: So it was normally handled at a lower level, whether she should fly or not?

SA: Yeah, and it had been reviewed two or three different times, finally up to the two-star level in the training command that no, she's not going to go back to flying.

EM: At every level they said she should not be flying?

SA: She should not be flying. So then it came to the Hill, and then the Hill came to the Secretary. So I tried to intercept this and say okay, look. I went down to the Secretary and I pled my case. I said, "Look, I'm the senior aviator. I've been an instructor pilot, and here's what I'm willing to do. Rather than you getting in the middle of it, let me come back to you and tell you what I think. And here's what I will do. I will go back, and I will go through every one of her flights. You know, each flight is written up, everyone's graded per episode. I will go through all of her flight records." Although the reason she's not flying is the fact that she claims that she was sexually harassed and she's not happy with how that finally ended, even though the guy got bounced out of the Navy and everything, she still thought that she, that she should—

EM: That he did her in in his evaluation?

SA: That he did her in, and that he had lobbied other peopled to make sure that she didn't get through flight training. So I said, "I'm not going to decide whether—I'm not going to go back and review that case per se. We already know that yes, he sexually harassed her. He was punished for it, and now we are going to go from that point on to see, one, if the results of that punishment fed back into the system anywhere. Whether there is significant evidence within her flight grades that shows that she can't be a pilot. And when I complete the thing, the last thing I will do before I make a decision, I will interview her." I said, "The reason that I want to interview her is that it's amazing what you can see when you sit down one on one with somebody and how they respond to certain things. Whether that reinforces what I've already seen in paper, or not. If I think that she should go back to flying, that's what the recommendation will be. And if I don't think she should go back to flying, I'll tell you why."

Good. He was happy to get rid of that problem. So sure enough, I called for all of her records. I brought them in. Lots of interesting things. Some that I didn't pick up at the time, but I found out later. One of the things that was never really mentioned in all of this was the final, final, final, final check ride that she had that was going to be—this is the last opportunity ever to get a chance to go fly an airplane. Her instructor for that flight was the ops officer of the squadron who, never declared by anybody, was a female.

EM: Right.

SA: You know, never had that come up.

EM: She made the final recommendation that she should not fly?

SA: She was the one that said absolutely this girl should not fly.

EM: Wow. That's not well known.

SA: No, it isn't well known. So I went through every one of her flights, and when you are an instructor, the thing that you look for is—almost everybody gets a down somewhere along the way. Very few people ever get through the flight training program without a down. Some do, but not many. But even with marginal flights, what you are always looking for is, what was the problem on the previous flight that got low marks? Did that same problem show up on the next flight or the flight after that? Did—all of a sudden, did the trend back down again to the same thing? Well, her record was rife with those types of things. She would go out and fly a flight, and she would get airspeed and altitude control things below average. She would get a warm-up flight to go back and review the techniques and all that, and on the warm-up flight she would be back up into the green area, so to speak.

So then she progresses to the next flight, which now is emphasizing other techniques, other procedures, and lo and behold, on that flight she would be marginally doing her thing on the new objectives, but all of a sudden back here in these old objectives, she was getting—the problem resurfaced. And it was consistent throughout her training program that she always retrograded back at some point fairly near to where it had been discovered. Which tells you, at least in my mind, it tells you that given a set of circumstances where she's uncomfortable, where she is in distress, she is going to revert back to the lowest common denominator, which is going to be fatal.

Seen it. I mean, I don't know how many times I've seen it. I've seen people that I knew real close that we always knew—sister squadron, fighter squadron. One of their lieutenant commanders, great guy, great, absolutely everybody loved him. Everybody loved him in the air wing. He was such a gregarious type of guy, but he had this propensity to drop his nose in close. He was flying Crusaders, which, in Crusaders if you drop your nose in close, it's not a good thing. And as long as everything was going right, he would pretty well control it, but if you get a big black night with a little moving deck and stuff, and sure enough it would be there, that same tendency was there, and they would have to call him: don't drop the nose. Don't drop the nose. So I'm the recovery tanker, so I'm the last guy to come aboard this night, and he's the last F-8 to land. I'm behind him. So his lights are up in front of me, between me and the carrier and we are on the same frequency so I'm hearing the LSOs talking to him. I can see he's getting in close, and all of a sudden I hear the LSO saying, "Power, power, power!" [i.e., add power]—and then, *boom.* He hit, you know.

EM: Ramp strike?

SA: Ramp strike and didn't survive. I've seen several of those episodes. For some reason I'd always end up behind somebody that was doing that. So I knew what happens when people sort of tend to—because of their other factors, they know that something is not right. They know it needs to be fixed, and they keep hoping that it's going to get better. And it doesn't get better.

EM: And you saw that in her record?

SA: And I saw that in her record. So I said, you know, we don't need another accident. Of course she's going to be flying helos, so she's not going to be by herself. There's going to be other people on the airplane with her. So I looked at that, and I said, "From the record, she doesn't know. They all made the right call."

EM: Is that when you called her in for your person-to-person meeting?

SA: Yes. So then I called her in for a person-to-person meeting. I'm not sure what I was going to find. She brought a lieutenant along with her who was her lawyer, and we sat down. I had a sofa and a couple of chairs there in my office so I didn't sit behind the desk. I came over and sat in the chair, and she sat down on the sofa with her lieutenant. We talked a little bit, and I sort of got a feel for who she was and what her personality was like. Quite honestly, my impression with talking to her just reinforced what I had seen on paper.

EM: How so?

SA: There was this—and I understand, you know, young JG and a four-star admiral, but there was this hesitancy, there was a little bit of hesitancy about answering certain things the way I would have liked to have heard them. And on the other hand, there was a little overconfidence in some things that I would have said, that overconfidence is a little out of place for that particular answer. So my feeling after I had that interview with her was that yeah, what I saw on paper matched with what I felt when I had a chance to talk to her. I didn't come away feeling that there was some disconnect between performance and what she appeared to be.

EM: Did you make your decision then, or just wait until after the meeting?

SA: I think—I don't know, I don't remember now whether—

EM: You didn't tell her right out that you've failed the program and I'm going to stick with that?

SA: I'm trying to think if I did then, or whether I did the written letter first. Now I can't recall. I sort of think that I told her at the end of the meeting, because I know she said later, "I think he had his mind made up before he even talked to me." It wasn't made up. I knew what I saw on paper, but I didn't know what I would see in person. But in the end the person matched with what I saw on paper. Later I was to find out that if I would have pulled the records a little bit further back—I did not and I should have at the time, but I did not go back to what her aviation scores when she was accessed into the aviation program.

EM: What did those records show?

SA: Those records showed that she should have never been allowed to enter the program. Her scores were below what we were accepting at that time. That there was an exception made for her. Now, she was an attractive young lady, I mean, she was poised. She was trim. She had all of the attributes you would say of a young person, would she look good in a uniform with a set of gold wings on? Yes, she would have projected the right image that you would have wanted to see. She presented herself well other than the little ticks that I saw, with a little bit of this and a little bit of that, but overall if you met her, you would say this is a well-spoken young person and seems to have a good grasp of the world. So she had all of those positive attributes.

Her college records showed that she had been a fairly athletic person in college, so she had some of the extra things you look for in aviation. Did they play varsity sports? Did they do these types of things? So she sort of checked off those boxes, but in the ASTB [Aviation Selection Test Battery]–I think it was called the ASTB in those days—the score was below what we were accepting at that time, for anybody. And so the question, when I found that out later—of course everything had been done—was, who in the devil made the decision to not adhere to the standards that we had put? During those days, we had more talent than we could accept. So when that happens, what you do is you drive up the scores so that you are only taking the best of the best.

EM: Who made that decision?

SA: Somebody in the recruiting command.

EM: Did you have any contact with Congressman [David F.] Durenberger from Minnesota over the issue?

SA: Oh, I sure did. [Laughter] Once I talked to the Secretary of the Navy, then the first thing I did was go over to see Senator Durenberger. I told him who I was. Told him my background, you know, and why I was doing it, because I thought if we wanted to settle the issue about whether she could fly an airplane or not, I felt that I was as capable as anybody he was going to find to make that decision. I told him, "What I volunteer to do is, I have not gone back to see if the sexual harassment stuff was handled satisfactorily or not. I know it was handled. I know that there have been several judgments made that it was handled properly, and that the individual that sexually harassed her was dealt with. I know she still has some reservations about whether it was enough or not, but the issue on the table here, and the issue that I want to resolve, is whether she should continue in flight training or not. The other issue has been—"

He said, "That's fine, yeah. That's what we want to know is can she go back to flight training."

So I said, "Well, that's what I will do, and when I'm finished, then I will tell you one way or the other."

EM: So you hadn't made a decision yet?

SA: No, I hadn't even started the process. I said, "This is what I'm going to do, and this is how I'm going to do it. Then we will tell you the results." So that was fine. Then we went through this thing and we made the decision. I sent him the letter for the Secretary of the Navy, and then the dialogue started again.

"You need to answer this question."

"I've answered this question. What question is it that you want me to answer?" There was somebody on his staff that was pulling the strings on this. I'm not sure, and it wasn't of interest, really, to me who it was, but there was somebody obviously pulling the strings on this thing, and he was sort of the dummy in the middle.

EM: He didn't seem especially antagonistic?

SA: He was trying to be forceful relative to the issue. "I want you to do," and this type of thing. But I told him what I was going to do, and that I was pretty sure that I was capable of doing it. The thing kept getting wrapped around about—he kept saying, "I've asked questions that haven't been answered."

"Well, every question you asked me was answered."

I even had everybody come back in that was around the thing, and the lawyers and all, and say, "Okay, in all of our correspondence with Durenberger, and my meetings with him, is there any question that is left unanswered?" No. We don't know what he's talking about.

EM: Did the two of you meet, just the two of you, or did you have others present?

SA: He had his staff present.

EM: You were by yourself?

SA: I usually took—well, I took one of the OLA guys with me every time I went, so I always had one of the OLA guys with me. I kept getting confused about this fact that, you know, he called the Secretary of the Navy and said I hadn't answered his questions. The Secretary would call me. I'd say, "Mr. Secretary, I don't know what questions he's asking that I haven't answered. You see all of the paperwork. We've had the lawyers review all of the paperwork. Everything has been answered. I'm still believing that he's thinking there are some issues relative to this sexual harassment. That wasn't what this was all about. That has been closed and done with and has nothing to do with this." I said, "I can tell you right away that the sexual harassment has nothing to do with what she's done in her flight training. Now, if she's afraid there's going to be another instructor like the one she had way back when, that's another issue, but that hasn't happened."

EM: Was there any evidence that the instructor, who eventually was kicked out of the Navy—did he influence others to give her poor evaluations?

SA: I couldn't find it, because with the evaluations, she doesn't fly with the same person on every flight. They rotate around, so there are certain areas of flight where you have the same instructor for a period of time, then you shift to a different instructor. There was a consistency in her grades that could not have been—especially when you take a look at the last person or the last check ride, she validated everything that had gone on before. So I mean you would have had to have a cabal of X number of people that had all got together and said okay, these are the grades we are going to give. Even that would be hard to do because it's a long grading sheet. I mean it's not just three or four blocks. There are 20 or 30 grades on every sheet. So I couldn't find any evidence that anybody there—

EM: It was Admiral Boorda at this time, is that correct?

SA: Yeah.

EM: What was he saying while this process was ongoing?

SA: He was happy it was out of his hands, and then when all of this started again, then he decided—Mike had this belief that there was never a personnel issue that he could not solve. That he was the best people person on earth and that, you know, if there were any people-like oriented issues, if he could get his hands on it he could fix it. And he was pretty much right. I mean, Mike was a master.

EM: So, he thought that you should be able to fix this?

SA: As far as I know he was very content with that. Mike and I were very close. We had been close for several years in the building when he was Chief NavPers and I was OP-04. So we had been around each other for a long time and had a great respect for each other, and he knew I was doing the right thing for this thing. But then he got caught up in this—that he could charm her to quell the debate. He was going to offer her another job and see if that would be enough.

EM: He had a meeting with her and offered that?

SA: By this time, now we had sent her out to Great Lakes, to a job out at Great Lakes. On the way, when he was on the way out there, all of a sudden the EA—I think by this time it was Jim [James] Metzger who had relieved my previous EA Jim Hinkle—he got the word from somebody on Mike's traveling team that he had agreed to meet with her out at Great Lakes and offer her another job. I'm saying, what in the world is this all about?

EM: And Admiral Boorda recommended that?

SA: Admiral Boorda was the one that was going to meet with her.

EM: Oh, okay.

SA: Unbeknownst to me. So, that was my first clue that Mike was now dabbling, going to solve another people problem. And all I know from that meeting is what I read later on from her accounts of, you know, "He offered me this job and that job, and I didn't want to do that." So it all got tangled up in that. Then, by this time I'm nominated to go out to CinCPac.

EM: By Boorda?

SA: Well, Boorda had to sign off on it. So it had already gone up through the chain over on the Hill.

EM: Who was your champion for taking that job?

SA: I don't know who proposed me for it, but I assume—well, Mike had to sign off on it, and Chairman of the Joint Staff had to sign off on it. Secretary of Defense had to sign off on it. So it was now over on the Hill.

EM: Had you expressed a desire to take the CinCPac job?

SA: The first I heard of it was, they told me I was nominated to go to CinCPac.

EM: Right.

SA: Of course I wanted the job, being a Pacific sailor all of my life. That was the top job out in the Pacific. As far as I was concerned it was probably the best job in the Navy, since the Navy had held on to it for that long. So I was thrilled to be going back. My daughter lived there. We had two grandsons out there, and we had had a previous tour in

Hawaii. So I was a happy camper. But then Durenberger put that hold on there, and that's when things started to go.

Admiral [Charles R.] Larson, who I would have relieved, had been picked by Boorda to come back to the Naval Academy for the second time. He had been there as a two-star and come back as a four-star to solve some of the problems we were having over at the Naval Academy relative to women issues and professor issues relative to—some of the professors were a little off the range on a couple of issues. So Larson wanted to come back and save the Naval Academy. He wasn't willing to stay and wait at CinCPac. He wanted out of there soonest, so he could get back and get ready for the new year at the Naval Academy.

EM: So there was some pressure coming from that end.

SA: There was pressure from his end; he said, "I want a relief out here. I don't want to wait. I want to come back to the Naval Academy and get started." So then the timing got such that Mike and I—he'd been down to talk to the Chairman, I guess it was, and came back.

EM: Who was who at that point?

SA: Let's see. The Chairman at that time must have been Shalikashvili.

EM: Okay. [Break]

SA: And, based on the version that I got from Mike, I said, "If the structure is not willing to hang with me on this thing, and you don't think that it can be resolved in time to meet the demands of making sure a Navy guy relieves Chuck Larson, then you need to tell me, because I'm not going to stay here and demand to wait for people to decide whether I'm cleared to go or not go, and then us lose that job out there." That's a very key job.

EM: That a Navy guy wouldn't get it, you mean?

SA: Yeah. We'd had it for so long. It's a true maritime CinC and so I said, "You just need to let me know what the deal is."

So he'd had a meeting with Shalikashvili and I think SecDef also, I'm not sure, and came back up and said, "We don't think we can wait."

Then I said, "Well, I will retire."

"I don't want you to go right away. I want you to stay as Vice Chief."

I said, "Well, yeah, I don't have to go right away. When's the best time to retire? But by announcing my retirement, then that takes my nomination off the table and you can put somebody else in as CinCPac." So that's what we did. Then he—you know, he hung on to me for, I was there for almost a year after that.

EM: So Durenberger didn't press it after you withdrew your nomination.

SA: Yeah, after I withdrew my nomination.

EM: You say you were there a year after this decision was made?

SA: Yeah.

EM: How did that affect you doing the job?

SA: I just did the job. [Laughter] It was a little awkward. Jennie Lou was really upset, and rightfully so. She said, "Why should I participate in all of this stuff to show the flag if that's the way they're going to treat you?"

EM: Did she think the Navy had somehow done you dirty?

SA: Yeah. She was not happy. She was not happy with the Navy on that issue. She was very unhappy with Mike. She and Mike had been sort of buddies. They got along real well.

EM: Yeah.

SA: But she thought he ought to have been more supportive.

EM: What did you think?

SA: Yeah, I thought they should have, you know, fought a little bit harder. But I don't know what the conversations were like down with the SecDef and Shalikashvili. I don't know if there were other issues.

EM: Yeah.

SA: There had been, you know; I'd had a couple of little warnings along the way—hey, you're getting too heavy, and stuff like that over the years.

EM: I was going to ask you—it seems like some people brought things out that I almost laughed about when I heard them. I mean, you were criticized for failure to censure officers over the '83 *Ranger* fire? That came out during this?

SA: Yeah.

EM: Was that a valid thing? Or something that they just dragged up?

SA: Well, somebody dragged it up, and Admiral Watkins had trouble with that. When my JAG report came back to the building, I understood that Admiral Watkins was upset that here was a report that blamed two low-level guys down in the fire room for the fire.

EM: Yeah.

SA: Then, nobody—you know, the captain of the ship wasn't held responsible, or anybody else. In part of that investigation, what I found—again, that I couldn't make part of the investigation and didn't make part of the investigation—was that one of the two guys down there who was involved in that, the perpetrator who really left the cap off of the sounding tube that the oil came out of and then was ignited, had started a fire in a theater at Great Lakes when he was going through training at Great Lakes.

It was a classic example of one guy that's supposed to be a two-man team down there, one guy letting the other guy go off on his own, unwatched. So that guy leaves the cap off of the sounding tube, and the other guy doesn't double-check it. They all knew the procedures. Their training records were up to date. Everything was up to date. The only way it would have been prevented is if there had been a third guy there that caught it. And so it was not one of those things where these two guys were down there unsupervised and didn't know what they were doing. It was two guys who basically, one didn't do his job of following up on the other guy, and the other guy deliberately, as far as I'm concerned—even if it wasn't deliberate, it was still a violation of protocol. He left the sounding tube open.

EM: Yeah.

SA: So as much you would like to say, okay, the engineering officer was responsible and the chief and everybody else, hey, there was no reason for any other person to be there but those two. Both were qualified. Both knew their jobs. Both of them made a mistake.

EM: But people were thinking about bringing that issue up again.

SA: Yeah, they wanted to bring that up again.

EM: The other issue that caught my eye is that you recommended that Lieutenant Paula Coughlin, who was part of the Tailhook thing, repay an $18,000 bonus after she left the Navy. Does that ring a bell? She left the Navy in a huff, and she had an $18,000 bonus. So what was it; you recommended that she pay it back to the Navy?

SA: Yeah, yeah, bonus. I think I probably signed off on a document that said that under the contract she owes the Navy $18,000, and I probably did sign off on that. Standard, you know, a standard thing.

EM: Anything else you want to add about the Rebecca Hansen issue?

SA: Well, later on with her hearings and stuff that she wanted to do, we just did an awful lot of dialogue on both Hultgreen and Hansen. Hansen, you know, came before Hultgreen, and so we wanted to go back and revisit that. But I was very comfortable. I was happy, when I got ahold of her records and could go through them in detail, that I found it easy to make the decision. I mean it wasn't even a close call. I didn't have to search my soul to say, well, are you leaning this way or are you leaning that way. It was so clear cut that if you knew what you were looking at, it was just as clear as could be. So in my mind, I would say, "Rebecca, you ought to thank me because you are still alive today." I know that one would never like to hear that.

EM: She never really accepted your decision?

SA: No. She just thought that this guy just took the easy way out. But I put a lot of time in on it. But I was happy that when I had a chance to go through it, it wasn't a close call. It wasn't one that I had to agonize over. Do I lean this way or do I lean that way? It was easy. Then later on, after I got out of the Navy and found out what her aptitude scores were for aviation, it all made sense. That if, in fact, those aptitude scores were what we thought they were, they were enough to say she shouldn't have been there in the first place.

EM: How did you get those scores?

SA: They were given to me by—remember the guy on PBS [Public Broadcasting System] that did an interview with me?

EM: Oh, Ted [Edward James Martin] Koppel?

SA: No, it wasn't. It was a PBS guy [Peter J. Boyer].

EM: Actually, I listened to that.

SA: After this thing was all over, he said, "How much do you really know about her?"

I said, "I didn't try to find out much about her. I didn't want to know much about her. I wanted the record to talk to me, and I wanted her to talk to me. I didn't want other people to tell me what their impressions were." So, I said, "I've steered completely away from going back and trying to find people that knew her here or there or anything else. I thought that was not what I should be doing. That I needed to look at the official records," and so, I said, "I know very little about her."

He said, "Well sometime you ought to go look at some of the stuff about her. She's been involved in these types of things several times before. This claiming of discrimination and claiming people didn't treat her right."

EM: But since you were retired, you couldn't go back to her personnel record, right?

SA: No.

EM: How did he get that information? He talked to people?

SA: Yeah. He said she and her mom had been into this thing before in other episodes. Then he said, "Oh, by the way, you should know what her—"

I said, "Well, I should have looked at what her score was, but I didn't."

He said her score was below what was being accepted in the Navy at that time. And sure enough, it was.

EM: Well, it's history. [Laughter]

SA: That's right. [Laughter] That's right.

EM: I don't have any more questions with regard to Rebecca Hansen.

[End of 19 January 2016 interview]

Interview Number 8 with Adm. Stanley R. Arthur, USN (Ret.)

Date: 20 January 2016

EM: This is the 20th of January, 2016, the interview with Admiral Stanley R. Arthur and Edward Marolda from the Naval Institute doing the interviewing. We had left off with a decision not to go for the nomination for the CinCPac job.

SA: Yes.

EM: And the last CNO for whom you worked was Admiral Mike Boorda.

SA: Right.

EM: And how would you describe your relationship with Admiral Boorda?

SA: Mike and I got along very well. We had been around town together for some time. Back in my OP-04 days, Mike was Chief NavPers, so we had interfaced in the building during that period of time, and then when he became CNO we stayed pretty close. Mike would come down to the office. I had a basketball net and a Nerf ball down there, and we would shoot baskets while we tried to work our way on whatever the problem of the day was. So we got along quite well.

EM: Was your relationship different than that with Admiral Kelso?

SA: It was a more casual interface with Mike than it was with Frank. Mike would have these light moments where he just wanted to goof off and laugh and scratch. Frank was a fairly serious guy, as we talked about—a really charming guy, but when he's in business, he's focused. Mike had these periods of time where he was almost like a kid laughing.

EM: Get away from the pressures?

SA: Get away from the pressures, and I always had something going on. I had some electric cars down there in my office that I used to race up and down the E ring. [Laughter]

EM: Is that right? [Laughter] Down the main corridor?

SA: Down the main corridor. [Laughter] Just to relieve tension every now and then. So we got along quite well. People were his big thing. Mike was very, very attuned to people and spent a lot of time thinking about what we could do for the sailor. He was very, very engaged in those topics.

EM: Because he had come up from the enlisted ranks?

SA: From the enlisted ranks, and then as time over at Chief NavPers. So he was really tuned into a lot of the people side of the house, promotion boards and stuff like that.

EM: Were you involved in his Seaman to Admiral Program?

SA: Just from the standpoint that that was his design and we were to execute it.

EM: So you didn't have any hand in the development of it?

SA: I didn't have any real hand in the development of it, no. That was his brainchild.

EM: Who did he work with on that? BuPers?

SA: BuPers. I think Zap—Ron Zlatoper—was over there at that time. I'm pretty sure he was there at the time we implemented that.

EM: What was the intent of that program?

SA: Well, it was sort of to mirror Mike's career, where he came in as an enlisted sailor and he caught somebody's attention as he was coming up, and they mentored him to a point where he switched over and got his commission. He thought that there should be a formal program to identify young sailors who showed the potential to go much further than they were.

So I think really what he wanted was a sort of flag to say, hey, I should not be the only one that has done this; there is a path for others. And so this program put the signboard up that said you can shoot for the moon. You're not tied to—if you enlist in the Navy there is a way to get a commission, and if you get a commission, you can continue on. You don't have to come out of the Naval Academy to be the CNO.

EM: Right. Did that program continue after his passing, in some form?

SA: I assume it's still out there in some form. Of course, I left before it was really under way and we had some time to see what the results were going to be, because it takes a long time. My only worry was false expectations. In other words, we needed to be careful not to say that everybody has this opportunity, and all of a sudden guys come in and a year or two later they don't think they are moving fast enough and are disenchanted and say it wasn't what it was supposed to be. But that sort of showed his interest in giving people an opportunity to take a look forward to what could be.

EM: Right. I wonder how many availed themselves of that and actually got through?

SA: I don't know. You ask a good question. I just wonder if it's still alive and well today. I haven't followed it.

EM: Probably in some form.

SA: Probably, but I don't hear it mentioned anymore.

EM: Maybe not that title.

SA: I have not seen it.

EM: Another program was the C4I [command, control, communications, computers, and intelligence systems] enhancement program—command, communications, what, intelligence?

SA: Intelligence, yeah.

EM: Do you recall that, beefing up that sector?

SA: Yeah. Art [Arthur K.] Cebrowski was sort of the architect of where we were going in that arena. He made a pretty impassioned plea that we really needed to get started in that, and then we needed to probably think about more of a career development path for the folks that really could understand where we were going with computers.

EM: Sort of a techie billet?

SA: Techie, and Art was a very much a techie, and he followed—oh gosh, the names are—they race across my head and then they disappear. Aviator, good friend—we were at PG School together and then served a lot of different places on the way. It will come to me here in a minute. Oh, Jerry Tuttle. Jerry Tuttle was pushing that way very hard during the last part of his career, and he's continued it out in the civilian world. And Art

Cebrowski then came along, and he was a very cerebral guy and sort of pushed it over the top. He did good work for us. The Air Force was way ahead of us.

EM: So this is creating an operational specialty in IT and that sort of thing—communications?

SA: Yeah.

EM: Well, we had communications technicians before.

SA: We had the communications side of the house and that's sort of where everybody came up, and then the intel guys took it a little bit further, and this was an attempt to bring it all together, because they were going on different paths. If you wanted to talk real secure communications, you were probably talking to the intel guys. If you wanted to talk broad-based communications, you were talking to the N6 communicator guys. They weren't all tied together as neatly as they could have been.

EM: Did that work? I mean, has that eventuated?

SA: I think so. I think that that process has worked. They've taken it another step forward. They now made that sort of like a fleet job for the top person in that position.

EM: Right. Jerry Tuttle, didn't he run afoul of Tailhook or for inappropriate jokes or something of that sort?

SA: No, I don't think so. Not Tailhook, and I don't think he got caught up in jokes. The Vice Chief had lots of fun with missteps at all levels. The one that was toward the end of my time there was Barney [Robert J.] Kelly out at PacFlt told a frog joke at the—

EM: A frog joke?

SA: Yeah.

EM: About French people?

SA: No, it was about a talking frog, but it had some off-color connotations, and he told the joke at his morning lineup meeting out there at PacFlt, and someone put him on report for inappropriate language in front of a mixed audience, and so I had to call Barney up and tell him no more frog jokes.

EM: That was the extent of it?

SA: Yeah. It faded away. But it caused quite a stir.

EM: Was there a letter of reprimand, or just a verbal admonishment?

SA: No. It was a verbal. [Laughter] Hey, Barney, no more.

EM: Come-to-the-shed meeting?

SA: No more jokes.

EM: Do you think the culture has changed in that way, that folks are less willing to come out with off-color jokes?

SA: Oh, yeah. Folks are scared to death to do almost anything.

EM: In the current Navy?

SA: Yeah.

EM: Yeah. To the point where it's changed the—

SA: It's changed the culture. You don't see them socializing as much with each other, because they are afraid. You're completely away from the Navy business, and you are hunting or fishing or something like this, and somebody hears something and the next thing you know, you're on report. So basically no one goes to the clubs.

EM: Can you talk about a related aspect—alcohol? Alcohol was a big part of not only the Navy's culture, but naval aviation in particular. Has that changed since your time?

SA: Oh, yeah.

EM: Well, had it changed during your time?

SA: It had to change during my time, and it has gone drastically further since I've left the Navy. I mean, to the point where if a guy goes out and has a couple of beers and is stopped on the way home, not for a DUI [driving under the influence] but for a stoplight or something, and there's any indication that they ask him if he had a drink or something, "Yes, I had a beer."

"Alcohol level is fine; it's not a problem. Fix your taillight." If that individual doesn't report that, he's probably going to get canned.

EM: Wow. That is tough. That is the current approach?

SA: That's the current approach.

EM: Wow.

SA: So if you self-report an incident that ends up not having any real—I mean, he wasn't drunk, didn't cause an accident or anything, but not self-reporting it will get you canned.

EM: I remember at one point the Navy closed down clubs. In fact, during the day at the Washington Navy Yard, we would have lunches and people could go to the bar and get whatever they wanted. At some point that ended, and people were told you are not supposed to go to lunch and drink.

SA: Yes.

EM: Was that during your time?

SA: Started during my time. Yeah.

EM: Did you impose some of those restrictions?

SA: They were sort of imposed upon us from the standpoint of you've got to de-emphasize the use of alcohol. It became such an issue that, basically, through the club network, it became, hey, you have got to be careful. Then the folks in charge of clubs recognized the fact that they were going to get shut down if there was an incident, so they clamped down. It was the de-emphasis of any alcohol use. We've gone too far, I think. My first squadron—well, it was actually my second squadron. We had a CO that demanded that everybody attend Friday happy hour.

EM: Happy hour, yeah.

SA: But you could buy your way out of it. So if you knew you were going to take the family out for dinner that night or you had visitors or you just didn't want to go, you put 25 cents or 50 cents, or whatever the fine was, in the mess kitty, and that was your ticket not to attend happy hour. The happy hour turned into people who wanted to drink and people who didn't want to drink. You had Coke and stuff, and we had plenty on each side.

EM: So the intent was just to get people there interacting.

SA: To get people there and interacting in a more casual thing, and I have to tell you I learned a lot in those sessions. But there were some tragic outcomes from some of those sessions, too. Not particularly in the ones I attended, but ones that I learned about later—guys taking off and driving through fences. In my younger days, we had some alcoholics among us that were pretty bad. They were true alcoholics.

EM: They were aviators?

SA: Aviators, yeah. Luckily I don't ever recall anybody crashing an airplane because they were drunk.

EM: They couldn't have been at their best effectiveness.

SA: They couldn't be at their best, no. I know of guys who completed the flight and climbed out of the cockpit and fell down, so—

EM: Oh, jeez.

SA: But yeah, we had to back away from some of the glamorization of the alcohol. There's no doubt about it. My take is now that we've come down so hard, if somebody has an alcohol problem, you're not going to find out about it until something dire has happened.

EM: People try to keep it hidden.

SA: They keep it completely hidden, and that's why there's not a whole lot of association amongst themselves, because there are some that maybe want a few more drinks than others, and they don't want anyone to be in a position to tell on them. I don't know where the happy medium is, but I think we've gotten to the point where we've sent people—

EM: There aren't many happy hours anymore, are there?

SA: No.

EM: Has that affected the clubs? I mean, there aren't that many clubs.

SA: Oh, yeah. The clubs are closing left and right, and most of the clubs now are combined clubs, and that has its own self-fulfilling prophecy.

EM: You mean joint service clubs?

SA: No, officer and enlisted.

EM: Oh, right.

SA: So if the officers want to have a big gathering, then they take over and that becomes the center of activity at the club, and the enlisted folks sort of feel uncomfortable. And the reverse—if you've got a big ship gathering with the enlisted having a birthday party or a promotion party or something like that, then the officers tend to stay away unless they are physically, individually invited. So there's not this freedom to pick up the family, and let's go to the Sunday brunch at the club. It just doesn't have the same feel that it used to have.

EM: All right. Getting back to Admiral Boorda. What do you think were his major accomplishments?

SA: Well, I think Mike did a lot in the way of communicating with the sailor. I think he showed a side of himself to the sailors in the fleet that—here was one of them who had made good and was doing well. He was a very energetic guy. He was approachable, and he was sort of the champion of the sailors. He was really well thought of, I think, among the sailors. For those of us who'd worked with him in Washington, he was an easy guy to approach and talk to. He had pretty strong ideas about things. Once he was latched on to something, it was sometimes pretty hard to back him off of it.

EM: Like what?

SA: Well, the LCS program, the littoral combat ship, started under his leadership, and he was convinced that this was a ship that we needed to have, and he put some pretty tough markers on the table for the acquisition community as far as costs and things. Some were sort of unrealistic, pretty unrealistic, but he wouldn't let the program move without tying it to those numbers.

The numbers were not reachable, and people kept trying to tell him they couldn't. "Well, how do we get to that?" Well, we get to that by making the ship differently than we've ever made a ship. In other words, we won't follow the NavSea specs for ships. We will follow the commercial, so we can get a commercially designed and built ship for use by the Navy, and we will make some accommodations to some safety items, and all that we've always had in our ships that some of the commercial ships don't have. So that was the only way that we were going to get to the price marker.

EM: Was that a valid approach?

SA: It's a valid approach depending on how you intend to use the ships. The big argument that you see in the papers now—and I'm so far away from it I don't know how much of this is real or imagined or whatever—this ship is not a combat ship. You can't take it into combat or you are going to lose it.

EM: The LCS?

SA: Yes, the LCS. So you've got this argument now going on that this ship needs to have more robust—

EM: Antiair?

SA: It needs more antiair, anti-surface, and it needs some more robustness about the infrastructure inside the ship. The first thing that happened when they first went to the drawing board on this thing was—the ladders between decks on Navy ships need to be at a certain pitch, and that pitch was designed by NavSea for the fire and rescue guys to haul air-breathing equipment and all that, and fire extinguishers up and down ladders. The commercial ladders were much steeper. So the initial design was steeper ladders. Then somebody said no, that's not going to work for us. Okay, then change the pitch of the ladder. Now you don't have enough room at the bottom from the compartment. You have to move the bulkhead from the compartment farther away, and that just dominoes through the ship.

And the same problem happened to us in the aviation business when we tried to go what we call green hulls. If you try to take a commercial design and adapt it to the military standards, you find all sorts of things that won't connect. So that was the start of the LCS, and that's why the price of the LCS started to creep up, and they got to the point now where people say why are we building something that we are not sure that we can really take into a hard combat area?

EM: What was Boorda's motivation for the LCS? He obviously was pushing it.

SA: There were a couple of things. One was the ship count. Make sure we had enough ships. Two was command opportunity for a more junior officer—sort of back to the experiment we had where we had the lieutenant commanders as squadron COs and lieutenant commanders as destroyer skippers instead of commanders. It was to provide more command opportunity earlier in a career, and also to have a more robust force relative to some of the missions that the Navy did close into the littoral areas.

EM: I know you supported his program. You worked for him. But were you a little bit anxious about that approach?

SA: Oh, I was. I sat in several meetings with him and we discussed some of these issues. But it was his program, and as long as he was there, it was going to go the way he wanted. Now, I don't know how much changed in the program after Mike departed. I followed it a little bit from my Lockheed Martin days, but that was not my business area, so I didn't follow it as closely as I would have.

EM: Did Admiral Boorda have impact on some of the major aircraft acquisitions?

SA: No. I think from the aviation acquisition side of the house, I don't remember him fussing much about that.

EM: Was there an obvious focus on the surface community because he came from that community?

SA: Well, it was an obvious focus there because we were right in the middle of—we had to do something to the FFG-7s, the frigates were coming to the end of their life, and what were we going to replace them with? That's where Mike sort of came in with this idea of the littoral combat ship. I think it was just that juxtaposition of the timing of we've got to look at the next small combatant. What's it going to be? Mike loved ships and he loved ship handling, and we would talk about it a lot. That resonated with him, and he could see his opportunity to put his mark, the Boorda ship, on the fleet.

EM: Were there things about naval aviation that he didn't understand and you had to bring him up to speed?

SA: No. Mike was pretty well-rounded. He understood the aviation community pretty well. It certainly wasn't his primary interest. He was always going to be a surface sailor, and that was it. But he wasn't super-critical about the aviation community. He was still concerned about the recovery from Tailhook and where we were going in the leadership and that type of thing.

EM: Did he defer to your judgment about naval aviation matters?

SA: He did, for the most part, yeah. If he had a concern or issue, he always bounced it off of me.

EM: Did you feel he had confidence in you in that regard?

SA: Yes.

EM: What were some of his frustrations?

SA: Mike, being a people person, his frustrations were some of the dumb things that people get involved with and some of the dumb things that leadership would do that would make the problem worse. Mike was a big believer in a one-to-one presence. He was convinced he could convince anybody of anything where he wanted to go. In other words, there wasn't a people issue that Mike couldn't solve himself. So he sometimes got out in front of the rest of the team when we would have a personnel issue come up. Maybe I will think of a real example. Say somebody, a young sailor out in the fleet, all of a sudden found himself—he'd gone on liberty, got a call when he was on liberty from the hometown. His folks or his mother or father were ill and he just took off and went home to see them. Then he shows up a week later and they would take him up to Captain's Mast, and then somebody would write the congressman and say that this kid just went home to check in on Mom and Dad. Why is he in the brig?

EM: He would get personally involved?

SA: Oh yeah. Something like that was something that he could sink his teeth into in a minute. He would sort of go around the system and say, "Now we need to fix this," and off he would go.

EM: I would imagine he was looked at favorably by many sailors, enlisted sailors? How did the officer corps treat Mike Boorda?

SA: I thought pretty well. I guess there was some residual question of how did he get there. But most of us who were in the flag ranks at that time had interface with Mike somewhere in the building. He had been around Washington enough that everybody knew him. He had had the right jobs out in the fleet, and he was such an easy guy to meet and talk with. I didn't see many people trying to ignore him.

EM: There wasn't resentment with the fact that he wasn't a Naval Academy graduate? That he had come up the enlisted way?

SA: I didn't see it. No.

EM: What about the wives? Maybe that's an old wives' tale, but I've heard that some of the wives didn't accord him the respect that might have been due him.

SA: I don't know of an incident of that. The only thing along that line that I would think of was that some thought that he was sometimes a little too casual. That he should have been more formal, I guess. That was Mike. He was one of those energizer bunnies. He was always at full tilt. He was always going and he was always—usually had a smile on his face, but he could bite really quick if something that was dear to him didn't go the way he wanted. He would bite back. But socially he and his wife entertained some. She was very uncomfortable with that. So I would guess you may have heard where some of the wives, who had been close in a social circuit with the CNOs' wives, would feel that it wasn't the same, because she was not the same. She was pretty reserved, did not like to be out in front, and he didn't try to push her out in front.

EM: All right. Well, you know all about the medals issue. Can you go into that? Your feeling about—was that contributory to his suicide, or was there more to it than that?*

SA: Here's my feeling on the thing. That medal, the Navy Achievement Medal, I think was the one that was of concern. The criteria for a Combat V on that ribbon bounced back and forth two or three times during the Vietnam War. The fact that Mike had a Navy Achievement Medal with a Combat V on it that he didn't deserve—you know, wasn't awarded with the Combat V, but was just awarded as a straight Navy Achievement Medal in a combat theater—that would have been an honest mistake. And it shouldn't have caused any concern one way or the other whether it had a Combat V on it or not. It was a very low-level award. And like I say, during the Vietnam War, the criteria for whether you could award it with a Combat V or not was changed.

EM: And he was offshore?

SA: He was offshore. He was in the combat zone on a ship, so it was an easy mistake to make. Mike would have known that if challenged. I mean, he would have understood the history.

EM: Well, didn't he remove the V at one point?

SA: I think he did. That all occurred after I retired. I have never believed that that had any contribution whatsoever to what was going on with him. Now, what was going on, I have no idea.

EM: Did he seem like he was under undue stress toward the end of your time?

---

* Admiral Boorda, the Navy's first enlisted sailor to rise through the ranks and serve as Chief of Naval Operations, committed suicide at his residence in the Washington Navy Yard in May 1996.

SA: You know, Mike called me up a couple of times on aviation issues. There were a couple of accidents that happened. I think one of them was an F-14 that crashed into some civilian houses, if I remember. He called me up a couple of times on issues like that in the aviation world and said, "Hey, I'm being told this. Would you agree?" I talked him through what I thought about the subject. From that I understood that he was feeling some pressures, especially from the aviation community, the old retired aviation community. The old retired aviation community went after Mike after my retirement.

EM: Over what issues?

SA: Over my issue. They didn't think that Mike supported me as much as he should have. And he was trying to weave his way through some issues there to make sure that he wasn't contributing to more angst amongst the aviation community. So that's why he would call. There was something else going on. I was with Loral at the time, and I was on the corporate airplane with Frank [C.] Lanza, the president of Loral, and a couple of our guys that I worked with, headed out to the West Coast. I got a call patched through to the airplane, and it was the Navy staff calling to let me know that Mike had died. The phone on the airplane wasn't that good, and they said that he got shot at the Navy Yard. We had lots of problems right before I retired with people coming over the wall at the Navy Yard and trying to break into the houses and stuff like this. So my first inkling, I said, "Oh, somebody got into the Yard?" They said no, no, no, no, it was Mike that did it. I said, "Oops." Last person in the world that I would have guessed, because of this propensity that he could fix anything. There's nothing that's relative to—

EM: So maybe he found something or some things that he could not fix?

SA: Yeah. Something.

EM: But it's speculation, really?

SA: Yeah. It's speculation. I have been told that there are a couple of people who really know what the story is.

EM: What is the story?

SA: And they won't tell.

EM: They won't tell?

SA: So I don't know. I just don't know. But in my mind, the medal had nothing to do with it. Mike knew enough to understand that it wasn't like somebody had a Silver Star they shouldn't have had, or a Navy Cross. It was a Navy Achievement Medal, and over the years it did or did not have Combat Vs authorized.

EM: Yeah.

SA: So, it was that easy.

EM: Anything more you would like to say about Mike Boorda?

SA: No. He was a fun guy to work with. I enjoyed working with him. He had his things that he was very protective of that he wanted to do. Most of the things where he would sort of jump off the range and we couldn't catch up with him were—he would get a bit between his teeth about some personnel issue and he knew he could go fix that. And away he would go. You just waited to pick up the pieces if there were any left behind.

EM: Okay. I'm going to switch now to the gender issue. We've talked about that in many ways, but, specifically, there was a heightened interest in female development in the Navy during your time. What actions did you take to further that advancement of women in the Navy?

SA: We were always looking for opportunities where we could put women where they had not been before. The thing that we tried, the message we were trying to portray out to the fleet and to the folks that had to implement, was, one, this was not going to be a situation where we took people and put them in jobs that they weren't qualified for. That that was going to be a prescription for disaster; they had to compete for the job. They had to have the record that supported them going to the job.

In the past, a lot of people were not concerned whether there was a female in that job or not. It just was a thought process that males had always been there, that females had other opportunities. This is reserved for a male. We had to break that paradigm. We had to change that. So we spent a lot of time—we had training courses and all of this stuff to make sure that people understood that it was a new world and that women were going to be a part, a bigger part, of the operational Navy. They weren't just going to be in supporting roles—a pretty hard sell to a lot of folks. They just could not grasp the idea that there would eventually be a female captain of a ship and a female squadron CO.

EM: Where did most of the opposition come from, the chiefs?

SA: The chiefs had a lot of opposition to it. The retired community had lots of problems with it. It was just a mindset that people hadn't been used to thinking in that direction.

EM: During your time as VCNO, what billets were women in? Were they commanding ships at that point?

SA: I don't think we had our first—maybe we had, I think we did have our first female CO of a ship during that time. Some of the issues we had that stick in my mind were a couple of incidents on submarine tenders—females in jobs and people were upset about it. It was just that culture of not wanting to accept it. But I think we did have our first female ship CO during that period of time. I think it was out on the West Coast.

EM: What about squadron commanders?

SA: We had female COs of non-carrier-based squadrons. The things got out of sync. We still had the combat restriction, and yet we had this new pressure that said you've got to bring females into the operational and they've got to be able to command carrier squadrons and ships and all of this other stuff, and we are saying that, one, you've still got the combat restriction, and so why are we spending money on somebody that we can't send to combat? We've got plenty of jobs that—you know, they aren't the career-enhancing guy that's got a squadron, or female that's got a squadron on a carrier, but we've got career paths for them and we can expand those. But first of all, if you really want them to be a part of the front line, you've got to figure a way to work on this combat exclusion, because as long as we've got the combat exclusion, we're headed down the path of all of a sudden getting to a point where we've got a squadron CO that we can't send to combat. So fix that—and oh, by the way, why don't you fix the draft legislation so that an 18-year-old male who has to register for the draft and has to give up a spot that he particularly wants for a female who doesn't have to register for the draft, and basically she can't go to combat and he can—so let's make all of this, if we want equality, let's start with the basics. Let's make equality right from the start. Females, 18-year-old females, sign up for the draft just like an 18-year-old male signs up for the draft.

EM: That still hasn't come to pass.

SA: It still hasn't come to pass. The talk this year is the first time I've heard it start circling up again after all of these years, and you've got to address this combat issue, because you keep drawing us down and you want us to give females more opportunity, which we understand. But if you are telling us to put them in the direct line to go into combat and they can't go, then what are we going to do? We can't all of a sudden create out of thin air a brand new CO to take the squadron into—

EM: Where were you getting this pressure, from Congress?

SA: Congress. The Hill was just all over this. But they wouldn't do their part. They liked to stand on the Capitol steps and throw swords at Frank Kelso, but they wouldn't go back in and vote.

EM: Did you take the step anyway of preparing women for these combat roles?

SA: Yes.

EM: Even though you had the combat—if the balloon went up, you would have to pull those people out of those billets and put in green people?

SA: Yep.

EM: That's not a—

SA: That's where the antagonism was.

EM: The combat exclusion ended when, in the mid-'90s, or just at the end of your tour? I mean, you had women aviators flying in Desert Fox dropping bombs. So that's what, '96, '97? [1998]

SA: Yeah, probably.

EM: That's my recollection. That was the first time they were actually in combat.

SA: That was the first time, and that was an exception for the aviation side of the house, that the aviators could fly in combat, but for the ground folks there were still combat exclusions, and we are still fighting through some of those.

EM: Right.

SA: It ate up a tremendous amount of energy. From the female side, you could understand why their angst was the fact that "Hey, I joined up to make a career in this outfit, and now all of a sudden I find out that I don't have the same opportunities to make flag rank or captain as my male counterpart because of these restrictions. So I'm not getting the opportunity to achieve the career that I should be able to have based on my talent." And that was a very valid claim. But we had to get all of these other things out of the way to make it a clear, clean path.

EM: Right. How did you deal with the pregnancy issue? Would a woman who got pregnant before deployment not go on deployment?

SA: They didn't go on deployment. What probably caused more consternation amongst the fleet than anything was the fact that if a female got pregnant and couldn't go on deployment—one of the reasons why it was so long for them to get into the submarines was this pregnancy issue. What are you going to do with a pregnant submariner out there, all of a sudden have a troubled pregnancy and can't get her off the sub? It was a very real problem. We never handled it exactly right because, again, people didn't want to hear about it. The Hill didn't want to hear about it. "You are just finding another reason to slow this process down. We don't want to hear about it." So the fact that we had major problems with pregnancies and having to fill in behind them never really surfaced.

EM: What about women who became pregnant while they were on deployment? Were they airlifted home?

SA: Yeah, they had to come home.

EM: Is that right? They couldn't stay with the ship?

SA: Depending on what the schedule of the ship was. If the pregnancy occurred right at the end of the cruise, she'd come back with the ship. Those were the kind of questions that were, you know—when did it occur? When was it known? How much time do you have before you have to react to it, and then what do you do for a backfill if you've lost a very key person? Where do they come from?

EM: Was there concern from any circles about fraternization with women out in the fleet?

SA: Oh yes. Lots of concerns, because you are dealing for the most part—in the fleet, 80 percent of your folks are between the ages of 18 and 21. [Laughter] And, you know what 18-to-21-year-old young people like to do. [Laughter] So it's going to be an issue on how do you deal with it. It was sort of the same thing that we had to do during the drug war days, when we were really trying to clamp down on drugs on ships and stuff. You had rules about when spaces could be locked. Could you be in a space with the door locked or not locked? Could you be in a fan room if you weren't from the R [Repair] Division? Why were you in the fan room? We would set up a whole set of rules so that somebody who shouldn't be there couldn't be there—or if they were, they were going to be hauled in for Captain's Mast. We had to go back and revisit those. Do you get two 18-year-olds, male and female, in a fan room and they happen to be yeomen in R Division space—what are they doing in there? Whether that was allowed or not allowed.

EM: Do you think the Navy has adapted with women in the service?

SA: It seems to me like they've done pretty well. I know there are still some issues, and again it's, what are the qualifications? Does everybody meet the same qualifications? The younger people today don't have the same view of the world as the group of folks we were dealing with at that time. They are much more comfortable with the women in any role, but there's still this undercurrent—and you see it now as the new focus is to get them into these hard-combat roles. Are they as qualified as their male counterparts, and should they be? Until we resolve that issue, there is always going to be a little bit of rumbling. But on the main, I think things are pretty good in that regard. Pregnancy is still an issue.

EM: You had brought up this very specific case here, but now that the Secretary of Defense has mandated women will be in all billets across the services, what do you do with a pregnant woman on a submarine? A boomer is down on the bottom of the ocean for months at a time. How are they going to handle that?

SA: I don't know, because—I think the term is etopic pregnancy?

EM: Ectopic, I think.

SA: Ectopic.

EM: Yeah. I mean, they would probably come up and get the person off?

SA: Yeah, but it's a high risk and it's also a break in the security of that mission for a boomer.

EM: Oh yeah.

SA: You know, for a boomer. But I recall—I think it was when I had the carrier group on *Ranger*—we were someplace and there was a female midshipman on another ship, a surface ship. Not part of our group, by the way, but I think it may have been when we were steaming down to Central America or someplace. In any case, there was a frantic call to get this young lady off of the ship to a hospital because it was a life-threatening episode. We participated; I think we got a helo and COD to get her to where she needed to go. But it was one of those touch and go things where she could have lost her life, and people weren't prepared to deal with it at that time. I assume over the years people have gone through all of these what-if opportunities that occur and have backup plans on how to deal with them. There are real issues—if you are going to really protect your people, you've got to make sure you've thought through all the way.

EM: Can you name a few female naval officers that you came across in your career, when you were VCNO, who really stood out as making a contribution?

SA: Well, Marsha Evans was one. Marsha went on, after her Navy career, I think to Red Cross. I think she was at Red Cross for quite a while. There was a female captain who was a civil engineer and worked for the Secretariat at one point in time. Ended up being CO out at Great Lakes. Really very talented—very, very talented.

EM: Well, Roberta Hazard was CO of Great Lakes at one point.

SA: Yeah, Roberta was out there at Great Lakes at one time, too.

EM: Do you recall Rosemary [B.] Mariner?

SA: I've known Rosemary for a long time. I was in the bureau when we put Rosemary and her first class into flight training.

EM: How well did they do?

SA: They did well. Of course, they were ahead of their time and all of the opportunities that the females have today weren't open to them. But they set the stage for the next, for the follow-on bunch. One was killed in training, I think, in training command. But Rosemary did well. She had a squadron out at Point Mugu somewhere along the way and then retired. There were several. There were a lot of good female officers on the OpNav staff. We had a ton of really good, solid talent.

EM: All right. I wanted to talk about racial issues. African Americans and Hispanics and Pacific Islanders in the service. It was a major issue back in Admiral Zumwalt's time, and then Admiral Holloway. What was the status of racial issues during your time?

SA: The things all came to a real head back in the days when I had command of a squadron. We were at war, and some of those issues didn't surface quite as big then, but right after the Vietnam War they became classic issues. So by the time I had command of both *San Jose* and *Coral Sea,* we were heavily involved in the racial awareness programs. And that had really sort of started while I was still back in Washington with BuPers. The Zumwalt years. I remember one of the things that we had to do when we were in BuPers—they had noontime showing of movies that were going to go to the fleet. And you had to have officers from different branches of BuPers to go down for a noontime showing, so you would go down with your sandwich and sit and watch this movie, and then you would grade this movie, if it was satisfactory to send to the fleet. If there were any issues in there that would be—

EM: Were these instructional movies or Hollywood movies?

SA: No, these were Hollywood movies. [Laughter]

EM: Oh, all right.

SA: So you had the movie watch. You would go down, and I forget which movie it was, but it was about the old South and slaves, and it was brutal as could be. I mean, it was bloody and gory, and every bad thing that you ever heard about the treatment of slaves was in this movie. I happened to be there for that one. Those of us in the viewing room looked at each other and said, "Who in the world would have even thought of sending this movie out into the fleet?" We all gave it a big fat no. There were the really tough issues. It was really a culture shift, and some people made it worse than it really was, but on the other hand, there were a lot of subtleties that those of us who thought we didn't have any problems with it had sort of not caught up on.

EM: You weren't totally attuned to what might have been grievances?

SA: Yeah. There's a lot of things that were grievances from one side of the aisle that from our side, from my background, I looked at it and said that that's not a good grievance. I got into a big argument with the HR guy that was giving us our—we all had to attend this sensitivity class, and I'm CO of *San Jose,* and we are going through this exercise where they show an episode and then they say, what went wrong here, you know, what things?

Part of the episode was that a white guy walked into a space, and the greeting wasn't what the instructors thought—that this was all wrong. This individual walked into this place and didn't introduce himself, and he should have introduced himself. So that was a big faux pas that led to all of this other stuff happening. I said, "Wait a minute. Why is this a racial issue? I don't see it as a racial issue. I see it as a cultural issue. I moved around a lot as a child, so I moved from big cities to farms and back and forth. The place that my folks were raised, and where I spent the last few years of school, was a very rural, small town community. The fact was that you didn't go into any place and broadcast who you were. You waited to be invited. You didn't walk in and say hey, I'm Stan Arthur."

EM: Right.

SA: No. You walked in and kept your mouth shut, and somebody would say welcome or can we help you or something, and then you could—but if you walked in and preempted—

EM: Considered too forward?

SA: Much too forward. So, I said, "Why is that a racial issue?" We couldn't come to an agreement on that. He said no, there's no place like that. I said, "You and I are on different wavelengths, and we are not going to get together. But now I understand what you are saying. There's two sides to that story, so everything that is perceived to be a slight is not necessarily racial, but certainly is cultural. We've got to understand that we've got all sorts of cultures in the Navy, so some poor kid from some farm in

Tennessee or southern Ohio—don't expect him to understand why he has to walk into some place and say here I am. He's been taught to walk in and wait to be recognized. Not to press it." So we had lots of problems.

EM: In 1991, the Navy Affirmative Action Plan, the NAP, had been around for a while. There was a complete revision, from what I've read, and it called for 6 percent black population in the Navy and 3 percent Hispanic, which wasn't the same as the proportion of blacks in the overall society, which is like 13 percent. Apparently Secretary Dalton came out with a 12/12/5 plan.

SA: Yes.

EM: In other words, blacks 12 percent, Hispanics 12 percent, Pacific Islanders 5 percent, which according to historian Bob [Robert J.] Schneller [Jr.], my good friend, was a bombshell. Do you recall when Dalton said we need to go to this 12/12/5?

SA: I do remember that, yeah. It's déjà vu all over again with the current Secretary and things that are going on today. Those edicts were always—and we see them all the time, to kick-start a program. The problem with it is how do you get there from here. The goal—

EM: That was a goal?

SA: That was a goal. And yet they want to start measuring the day after the goal is established. That's always been the dilemma, and if you mentioned that you've got to give us time, then it was, "Okay, you are trying not to do it. You don't want to do it, so you're just buying time. We are not going to give you any time. It's got to be done now." It's that forcing function that creates a lot of animosity. The goal was a good goal. How do you get there was another issue. Sometimes those goals take extra resources, and there doesn't happen to be anyplace to get those extra resources. That's just one of the facts of life in this business—that people give you unrealistic goals along the way, and you've got to figure out how to do it.

EM: Do you recall when Secretary Dalton came out with that plan? Did you work directly with the recruit command to change things?

SA: That's where it had to start, and that was in Mike Boorda's alley, and Zap was still over at BuPers. How to get the recruiting command to shift into that mode. What kind of resources did they need? For the officer accession program, we had to go out into the traditionally black colleges to establish NROTC units. Could they get more candidates from the historically black community colleges? All of those things had to be orchestrated, and boundaries had to shift to reflect where we could go to find the people to get that kind of a quota.

EM: Did you register any kind of success during your time there?

SA: I don't know what the numbers were. We were making progress. All I remember is that we were making progress. Not as fast as anybody wanted, but we were making progress.

EM: You weren't at that 12/12/5 level?

SA: I don't think so, no. I think we are well past that now. I'm not sure.

EM: Do you remember any specific actions you took to improve the racial climate?

SA: Oh, over the years, we started recognizing things like Black History Month and—

EM: That began when you were VCNO?

SA: I think it was already started by the time I was Vice Chief. We then had, in the human resources area, all sorts of new occasions where people gathered and had celebrations or had decorations in the offices for Black History Month and women issues and Asian Americans.

EM: Was that appropriate?

SA: Some of it was superficial, but it still served a purpose. It reminded people that you've got to be thinking about this. This shouldn't be a one-day event; this needs to be something that you understand and help us make progress in this area. After I got out of the Navy, at Lockheed Martin we were doing this stuff in spades. We had Diversity Week. We had car shows, with diversity as a background, that the more diverse you are, the more talent you can bring to the job—these types of things. It was déjà vu all over again. The military was so far ahead of industry, and I got across the wires several times in my industry days after I retired, by saying that you guys are so far behind where we came from that, you know, I'm back revisiting things that I was trying to do 20 years ago. They would look at me and say, "Oh, you're full of you-know-what." I said no.

EM: Well, you look back at the history of integration, 1948, Truman said it will happen. It started there. It didn't go well for quite a while, but that was the start.

SA: Yeah.

EM: What black leaders do you recall, guys that served with you?

SA: Well, we had our first four-star, Paul Reason, and Paul was a surface nuke guy and he did quite well and was a delight to be around. He was a very interesting guy. Very smart guy. When I was in BuPers there was a lieutenant commander who worked for me, and later on, when he got out of the Navy—he was sort of behind the—he just couldn't get pushed up far enough before he got out. But he went over on the Hill and was the staffer for several years who did a lot of good stuff with feeding back into the Navy from the Hill, and also took it in the other direction to help soothe some false expectations over there. An aviator that took command out on the West Coast, a two-star.

EM: Admiral [Kendell] Pease?

SA: Gosh, I can see him as clear as day. Walt [Walter J.] Davis was a fighter pilot, and he became CO of *Ranger* and then made flag.

EM: What about some chief petty officers who, throughout your career—do you recall any African Americans who stood out?

SA: From my earliest days in the Navy, there was always a black petty officer or a black chief petty officer. Not that many black officers in my early days, but there were always, in the enlisted side of the house—we always had a very talented group of people. Always. There were guys I can remember—I knew his name at the time. We weren't in the same squadron, but we went to survival school together. He was an enlisted aircrewman, and we all had to go to survival school in those days. I'm leading this team on a forced march for two days or something, and we had to forage for our own food. We are up in the mountains behind San Diego where there are just tons of rattlesnakes, and everyone was saying, you know, you've really got to eat that rattlesnake. You find a rattlesnake and catch it and eat it.

EM: Eat it raw?

SA: Well, no. We had to build our own fires and stuff. What we were doing, one of those techniques they taught us was if you are on a march with a group—I think there were four or five of us—you always had to replace the leader, because each person had his own pace. If you let just one guy dictate the pace, he's okay but none of the rest are. So you have this rotation going on where you rotate from back to front. This kid, it was his time to be the lead, and we always gave the lead—we'd found a big branch with a fork on it, like Daniel Boone days, so we could trap a rattlesnake if we saw it. None of us were going to trap it, but it just so happened that when he got to the lead, a rattlesnake went racing across the trail. [Laughter]

EM: Chasing after him?

SA: No. It went, but none of us were going after that rattlesnake. It was headed one way, and we all headed in the other direction. We laughed about that. We all said we were never going to be hungry enough to go chase that rattlesnake. But he happened to be the lead.

Years later, it was water survival school when I had the CarGru command, and all the people that would fly with me on the helos to go visit ships and stuff—I had a team with a radioman, a yeoman—and they had to go to water survival along with me. We were at the edge of the pool. The first thing you had to do to start the process was, you had to step off the edge of the pool, go down in the water, and surface yourself and start to swim. You couldn't dive off and swim. You had to step off and then get yourself back to the surface and then start swimming. It was the first thing that you had to pass. Well, it was time for him to step off the edge of the pool. He stepped off the edge of the pool and went right to the bottom. And he's down there standing at the bottom of the pool.

EM: He was black?

SA: Yeah. And no movement. He's just standing down there. So they always had a rescue guy in the pool. He goes down and pulls him up and gets him out of the water. "Everything okay?" Yeah, yeah. "You know, you are not supposed to stand down there at the bottom of the pool. You push yourself back up and then you have to swim."

So he gets up, steps off again, and [*whoosh*], down to the bottom of the pool he goes, and he's standing down there again. He was really a lean guy, so there was no floatation to him at all. [Laughter] And they drag him back out of the pool, and by this time I'm over there saying, "What's wrong?"

He says, "Admiral, I don't know how to swim."

I say, "And you stepped off the edge of that pool and went into the water and you didn't know how to swim?"

He says, "Yes, because that's what we are supposed to do, right?" [Laughter]

I say, "No, you are supposed to tell us that you don't know how to swim first."

EM: Isn't swimming required now in their boot training?

SA: Yeah. It was then too.

EM: How did he get through it?

SA: I don't know. I don't know.

EM: Wow.

SA: But it just showed you that here was a kid who knew what we expected out of him, and he was going to do it whether he knew how to do it or not. I said, you could have put a gun in my back and I wouldn't have stepped off of that thing if I didn't know how to swim.

EM: Well, that's dedication I guess.

SA: So over the years, I mean, it was just a constant reminder that, jeez, there's just talent anywhere you want to look, no matter what the color. And sex too, it's the same thing, no matter male or female, black, yellow or white, good talent.

EM: All right. Okay. Onto base realignment and closure, BRAC.

SA: BRAC, yes.

EM: There were two major ones while you were in, '93 and '95. You were involved with both, I would assume? What do you recall about that evolution?

SA: Well, I think it was the '93 BRAC; it may have been in '95. The big thing then was the recruiting depots. We were going to close down—we had three recruiting depots, here in Orlando, we had Great Lakes, and we had San Diego. Marines had San Diego and Parris Island. We were all told that we had to get down to one or two. One, I think it was. So from the Navy's standpoint, we wanted to get out of Great Lakes, at least those of us that were—

EM: Why?

SA: Because it was so expensive to maintain. I mean, all of the cold weather up there, the infrastructure. My dad went through Great Lakes as a sailor back in the '20s, and most of the infrastructure was still there from the '20s. We had this brand new recruit depot down here, or recruit training center down here in Orlando that was fairly new. It had all the infrastructure with it. It had its own property and was a great place for people to come see their kids graduate from recruit training and have a little vacation in the sun at the same time. And of course gorgeous San Diego, which was 20 structures and stuff there at the recruit depot in the training center there in San Diego. So we thought we would probably end up with—what we would have liked to have done, at least what I thought we were going to try to do, was we were going to try to hold on to San Diego and Orlando and dump Great Lakes. The final analysis was we were going to dump the two new ones, or the two nice weather ones, and stay in the old one at Great Lakes.

EM: Who made that decision?

SA: You know, it was one of those things that went all the way down. I'm still convinced that Charlie Nemfakos made the final decision on this one, but I'm not sure. But that's the way it shook out.

EM: What was his billet?

SA: He was in the comptroller side of the house in those days.

EM: But he had a lot clout?

SA: Oh yeah, yeah. He had a lot of clout. The most important part to me, though, was the aviation repair facilities. There had been a movement in the previous BRAC when I was N4 to consolidate with the Air Force all of our aviation depots. The OSD staff was convinced that we ought to combine them, although they didn't have a complete appreciation for how each depot was organized and what they had to do. But they thought one aviation depot is just like another depot. Not anywhere close. So we were trying to protect our ability to depot our own airplanes and engines. The Air Force, of course, saw this as a grand opportunity to grab onto, you know, get more control of air, whether it be the airspace or airplanes or maintaining airplanes.

EM: But you would have what—rotating Navy and Air Force commanders?

SA: That would probably be the case in the long run. Maybe you would do that, but maybe not. So we worked very hard to try to figure out, we needed to get rid of some of our excess capability. There was no issue with that at all. We had too much. Then it was a decision on where do we take it away from, and so all of these start to intertwine—if you take away the aviation, the NARF [Naval Aircraft Repair Facility] at San Diego and the recruit training command in San Diego, and the Marines also take out the recruit depot in San Diego, that's too much out of one place at one time.

EM: It becomes a political issue.

SA: It becomes a huge political issue and also was—you know, close Miramar. Well, what we did was, we closed Miramar for the Navy and let the Marines come down from El Toro to Miramar, so it became Marine Corps Air Station Miramar for the second time in its career. The Navy went north to Lemoore. So all of the BRAC process is really a painful process and a very political process.

EM: The end result is you give a list to Congress and they yea or nay that list?

SA: They end up with a yea or nay.

EM: In the '93 BRAC, which bases did you close? Do you recall?

SA: I'm pretty sure that's the one—again I'm not sure if it's the '93 or '95 that we closed the recruit depot here and the recruit depot in San Diego and put everything at Great Lakes.

EM: Was there political pressure there? I mean on its face it seems as if the—San Diego and Orlando, the more modern ones—

SA: Yeah. For the Navy, the decision on the two was because the infrastructure in Great Lakes needed huge, huge input, an infusion of stuff, so we didn't make out as well there as we should have. I was surprised when I came down here to Orlando, when Lockheed Martin sent me down here, because I'd never been here. The recruit training is gone. We had some nuclear power training here at the time also. We had a brand new hospital. I get down here and somebody says, "You know we've got a Navy Exchange here?" I said why do we have a Navy Exchange here? Oh, it was an SOP after the BRAC, that the Navy would leave a small Exchange and Commissary here for the retired community. [Laughter] So there's a Navy Exchange and small Commissary over on the east side of the airfield, over there at Orlando International which was a BRAC closure years ago when the Air Force lost their SAC [Strategic Air Command] base there and handed it over. So I think we closed NARF Alameda on that BRAC. We kept North Island. North Island was still there. We did a couple of things around Oceana. There was a big movement to close Oceana then, but we didn't do it. Oh, there was a long list, but those were the big items that I recall.

EM: Did you periodically meet with your counterparts from the other services to decide?

SA: Yes. That's where the Vice Squad—we spent a lot of time together dealing coins back and forth on what can we do. Of course it was really important for the Marines, because Cherry Point was on the thing. We toured almost—as the Vice Squad, we went around to almost all of the air rework facilities and showed each other what was going on there. Went out to the big Air Force one out in—

EM: Edwards [Air Force Base]?

SA: No. We did go to Edwards too, because we were talking about anechoic chambers. There was a big anechoic chamber in Edwards. There was an anechoic chamber in Pax River.

EM: What is an anechoic chamber?

SA: That is where you can put an aircraft in and start electronic tests on all the antennas and stuff. It's a quiet room for electronics. They are all padded cells so you can test things like radar signatures.

EM: It's not a wind tunnel?

SA: It's not a wind tunnel. It's a foam-filled building, spikes of foam sticking all out of everything in there, and then you can direct electronic stuff, the antennas, and measure reflections, and so you can test your stealth capability and stuff.

EM: The list that you finally sent forward, did it include bases where Air Force and Navy were combined? The depots and—

SA: Yes. It ended up we combined some, like out in Hawaii, where we had Hickam and Pearl Harbor right next to each other. Now they are Joint Base Pearl Harbor Hickam. We had Joint Base Dallas Fort Worth, where we had Air Force reserves in one place and Navy reserves in another airfield, and we combined them too.

EM: So that occurred during these two BRAC periods?

SA: Yes. During those two BRAC periods.

EM: Okay, Joint Base Andrews—did that go back that far?

SA: I think Joint Base Andrews was in that same sweep-up. I think it happened at that time. The problem of trying to remember which BRAC—because each BRAC, the same things would come up, and then some would be discarded. But they were never dead, because the next BRAC they would come back again and you'd deal with them again. But yeah, the Vice Squad—we spent a lot of time together trying to convince each other why they should fold and we should keep what we had. I know we came down here to Eglin, and there was a big move to combine Pensacola and Eglin.

EM: Let me ask you this: did congressmen try to get to your level to try to influence the deliberations?

SA: [Laughter] Steny [H.] Hoyer and I spent an awful lot of time together. [Laughter]

EM: From Maryland.

SA: Yes.

EM: Over Andrews?

SA: Over Pax River.

EM: Pax River. So there was similar interest expressed?

SA: Oh yes. We were—yes.

EM: And what about local political leaders, mayors and—

SA: Yeah.

EM: Okay.

SA: And, you know, you sometimes used them—you'd approach them and say, okay, you need to do something here to help us stay. It ended up, after I got down here to Orlando with Lockheed Martin, one of the BRACs came around again, and it was Governor [Ellis "Jeb"] Bush [Jr.] who wanted to get our base up in Jacksonville opened up again—the big air station that we used to have in Jax [Jacksonville, Florida]; not NAS Jax, but the one west of there, NAS Cecil Field. Which had been a big attack squadron base during the Vietnam era and afterward. So Oceana had had the big crash of the F/A-18 up there, into the housing area, and they were looking for another field for bounce, carrier bounce, because you couldn't use Oceana at certain periods of time. Too much noise and stuff, so there was a big movement to close Oceana again, and Governor Bush wanted the Navy to come back to Cecil Field.

EM: Yeah.

SA: So I joined his team. Bob [Robert J.] Natter and I—Admiral Natter—Bob and I were part of his team, and we tried to argue with the Navy to open up Cecil again. Because what had happened was, in the time that the Navy had walked away from Cecil, industry had taken over the airfield and brought industry in, and part of the industry was servicing Navy airplanes again. So there was a facility run by Northrop Grumman up there—I believe it was Northrop Grumman—and the Navy was flying airplanes in to get reworked by Northrop Grumman. So I went up to visit the facilities.

Well, the last time I had been to Cecil Field in the Navy was when I was an instructor at VA-122 out in Lemoore, and I flew an A-7 across country to Cecil to see my old buddy Jack Calhoun, who was my counterpart in the squadron there at Cecil. And the airplane that I flew was going into a rework at the field there at Cecil; we had a rework facility for A-7s there. Jack had taken me on a tour around Cecil, because I had never been there before. So here these many years later, I go up and take a look at Cecil, and holy mackerel, it's a whole different place. It had new housing. It had new hangars. It had gorgeous runways out there. I'm thinking, they really have done a good job, and basically, in my mind, it was twice the complex that you could ever develop out of Oceana. So we tried to help Governor Bush get that pushed through the Hill, but it didn't work.

EM: Where was the opposition? From Senator Warner?

SA: Well, the Virginia delegation was very strong, and the Navy was strong. The Navy wanted to stay at Oceana because of the closeness there to Norfolk, even though they knew they had some problems, and I'm not sure how they are going to survive with the—

EM: Those planes still come over every five minutes, it seems like.

SA: Well, wait until the F-35 gets there.

EM: Is it louder?

SA: Oh, they make the E/F, you know—the E/F was a major jump in decibel level between the A/B/Cs. The F-35 is a whole new adventure.

EM: They are going to be based at Oceana?

SA: Yes.

EM: Oh. [Laughter] You'll have some local opposition there I think.

SA: You bet.

EM: Anything more you want to say about BRAC—the process?

SA: Well, it's a very political process. I mean, it's an agonizing process. It's a necessary process because we still have too much infrastructure, and you can't get rid of the infrastructure politically without that yes/no vote from the Congress. So it's a painful exercise, but it's a necessary exercise because we still own too much turf that we don't use.

EM: Are we in need of a new BRAC?

SA: Yes. We will have to have them. It's just like shooting yourself in the head with a gun, but you've got to have it. [Laughter]

EM: All right. You had on your list the last time I was here, "uniforms." Were you involved in the change of uniforms?

SA: Well, the one uniform item that came out after my joint experience with the RDJTF and the Gulf War was a jacket. When I joined the Navy, we had this old thin khaki jacket for when we were wearing work khakis. And that went away; then we never really had a common jacket. People would have windbreaker jackets and stuff like this, and of course the aviators had their leather jackets and we had the foul weather gear jackets. But none of it was something that you would want to wear with a blue uniform, or even a white uniform or a khaki uniform.

Well, as I spent more time with the joint guys, especially the JROC trips and the Vice Squad—and so I'm back after all of this time that I'd been with the joint world, where everyone had a nice black jacket. The Air Force had blue, but the Army had black. And of course we'd had our old pea coats that were black—navy blue, they called them, but they were black. So I'm saying, gee, these jackets—why can't we have this jacket? The Army's done all of the testing of it. They've had it for years, it works great, it's a nice-looking jacket. It always stays looking nice. It will fit our blues. It will fit our whites, and it will fit our khakis. All we have to do is tie into their contract and we've got a nice, beautiful jacket to wear.

EM: What color would it have been?

SA: It would have been black. And it was just a little bomber jacket, and by this time the Army had developed a version that had a zip-in liner, so for colder weather you could zip in a liner, and we could get rid of our old black raincoat that we used to wear with the liner in it. So I thought this was an easy thing. And everybody in the Navy knew we really needed a new jacket, but they hadn't thought of this.

So I come back and say, here's what we are going to do. We are going to take this jacket, and we are going to make it our own. I checked with Frank Kelso. He said yeah, that works good. So I had a trial one that I wore around the building, and everybody liked it. Tried to get the Supply Corps to get this jacket, just tie into the Army guy. "Oh, no. We've got to go to the Uniform Board."

Jeez. Oh, God almighty. So the Uniform Board wants to get in the middle of all this, and I'm saying no, no, no, no, no. So finally I just sort of overrode the system and said this is going to be the jacket.

EM: Yeah.

SA: Everybody is going to be happy. Well, everybody was happy except the people that had to get it, because they didn't want to tie onto the Army contract. Eventually the Supply Corps got their way. They did their own contract and did their own spec for the jacket, and so rather than us being able to just go down to any Army Exchange and buy a black jacket, now we had our own jacket. Looked identical. Probably made by the very same people, but it had a Navy tag in the middle of it.

EM: Probably cost three times as much.

SA: Probably does. So at least I was successful in getting us a decent jacket to wear.

EM: Is it still in use?

SA: Yes.

EM: Oh. Well, that's a success story. [Laughter]

SA: My only success with uniform things.

EM: You also mentioned you had some issues with TransCom.

SA: Oh yes.

EM: Transportation Command.

SA: Well, with Goldwater-Nichols, there were lots of accommodations made between the services of who is going to do what with this and that. One of the things that kept bubbling up when I was N4 was that TransCom, when they stood up, the next thing that they wanted to do was, they were not content to just handle the air side of the house. They wanted, especially as we started working Desert Shield—"Why are we out here flying airplanes to, say, Charleston, and then from Charleston we have to train stuff to Jacksonville, or fly it into Jacksonville? We don't know the schedule of the ships."

"We've got MSC. Just tell us what you want and we will be there."

"No, no. We think we ought to schedule those ships."

"No, you know MSC belongs to the Navy. We are Sealift Command and you are the Airlift Command."

EM: Who was commander at TransCom during this time?

SA: The first guy was—not [General Alfred G.] Hansen—during the episode that I'm leading up to was Ron [Ronald R.] Fogleman.

EM: Air Force.

SA: Air Force was TransCom at the time. And Ron and I had worked together when I was OP-50; he was sort of my counterpart over on the air staff. So Ron and I had worked, and when I was OP-50 and 90 I was president of the POAC, the Pentagon Officers Athletic Club. They always had a president from one of the services down there, although it was an Army run facility. So when I left I was looking for the next guy to take over the job, and I sold it to Ron Fogleman. Shortly after he took over, the Army instigated an investigation on their guy who was the keeper of the flame, the manager, and he had been doing some illegal contracting. The first thing that Ron got faced with was POAC in trouble with illegal contracts, so he thanked me for giving him that job. We knew each other well, and so now Ron's out at TransCom. I knew where this was going.

MCS would end up reporting into TransCom as part of their network. But there were certain things that I wanted to make sure we had, and of course we had a three-star over at MSC, and we had a list of things. I knew where it would finally end up, but I was going to fight it tooth and nail to the end, and I told Ron, and he understood that. I said, "Look, we both know where this is probably going to end up. What I don't want to have happen is for the final decision to be announced without us having been told ahead of time. So I'm asking you—"

EM: So you thought it was valid that TransCom would direct when ships should go and what time?

SA: Well, I knew that the accommodation—that there had to be a bridge made between TransCom and MSC. I knew that was—

EM: There was no bridge prior to that time?

SA: No, no. Well there was an informal, but not a formal type. So I knew in my mind where this would eventually end up. And he did too. So I said, "Look, the one thing I want is, I don't want this thing to be announced without me having a chance to talk to my guy at MSC."

"Okay. You've got my word." Well, I don't know how much longer it was after we had this discussion, all of a sudden here's the announcement.

So I call Ron Fogleman, and I say, "Ron, you know, we had this discussion and you promised me."

"Yeah, I know."

I was really hot. I started swearing at him, and I said, "That's the last time we are ever going to do any business with you," and slammed the phone down.

A couple days later, out in the office I hear Ron Fogleman's voice and they open the door. Here he comes in, and—I had been out in St. Louis for a Navy League function sometime earlier during the middle of all of this stuff, and they had taken me to breakfast at a hotel in St. Louis that had the best sticky buns I'd ever had in my life. I had mentioned this to him. So here he comes trudging into my office with this box of hot sticky buns. [Laughter] I said, "Okay, okay, we will speak again." [Laughter]

EM: This occurred after the Gulf War, right?

SA: Yes.

EM: Okay. My impression was that MSC worked fairly well during the Gulf War.

SA: Oh, MSC worked wonderful. Yeah, they did.

EM: But TransCom wasn't happy with the interface?

SA: Well, no. They thought that, hey, it's a transportation network, and they are the godfather of transportation for the military now. They ought to have tighter control.

EM: Oh, okay.

SA: Now I think we've had an admiral out at TransCom. I know we've always had a deputy out there. I don't know if we have the top job now or not. But since that time we've had a Navy deputy.

The other thing with Ron Fogleman was, then he became Chief of Staff of the Air Force when I was still there. I don't know if I told you this story about the Christmas party that—their first female Secretary of the Air Force, Sheila, [Dr.] Sheila [E. Widnall]—anyhow, as all of the secretariats do at Christmastime, they have a Christmas reception in the Secretary's office, and people from the staff go by and the Secretary talks to them, and they have a little something when they go by.

So over the years I had—from my aide in the N4 days, who one Christmas bought this Christmas tree suit for himself, then gave it to me to use for the rest of the day. He'd gone down and surprised Frank Kelso in the office by standing in the corner dressed as this Christmas tree. So then I ended up taking over the thing, and every year I would get out the Christmas tree suit. When I came back as Vice Chief, I would get this suit and wander—we always had a day before Christmas where we sort of had to relax, office parties and stuff, and I wandered the halls in this Christmas tree suit.

And so I had this invitation to go down to the Air Force reception. I said, gee, I will just go down there in this suit. Of course the Air Force, very formal about how they do things, you know. They had everybody lined up out in the E ring, and they had airmen standing there checking people off of the list. So I had to weave—I had to dig into my thing and show them—they didn't know whether to let me in or not, but I convinced them. I'm walking down the line, and of course Ron is before you get to the Secretary, and he introduces people that he knows to the Secretary. He looks at me and has no idea who it is.

EM: Was your face covered too?

SA: Face was covered, yeah. I mean, I looked exactly like a tree. There was none of me that you could see. So I said, "Hey Ron, you know who I am."

He said, "Oh my God." [Laughter] He turned around to the Secretary and said, "Madam Secretary, this is the Vice Chief of the Navy." [Laughter]

EM: What was her reaction?

SA: Oh, she laughed.

EM: I can just picture that. [Laughter] That's great. All right. How about the development of strategy during your time? Because there was *From the Sea* and then *Forward . . . From the Sea.* Were you involved at all in the development of these strategic approaches?

SA: Yes, in the final chop on the things, but those were done primarily in the N6 organization. Snuffy Smith with *Forward . . . From the Sea.* So the CNO and I, we were all in the loop, but we weren't in the working groups. They did some really nice work on those, I thought.

EM: How did those approaches differ from the Maritime Strategy?*

SA: They had taken advantage of some of the lessons learned from Desert Shield/Desert Storm, and so they now reflected more the regional types of stuff than we had from the Maritime Strategy, which was a Soviet Union–focused approach. You had the mention of the littorals and stuff like this, that started showing up much more in those two

---

* Admiral James Watkins, Chief of Naval Operations, and John Lehman, Secretary of the Navy, championed the Maritime Strategy as a means for maximizing American sea power in the Cold War confrontation with the Soviet Union during the 1980s.

documents. So it was really a refresh into the changing world situation from the Cold War into being able to respond to the regional conflicts and still have the ability to revert back to a major conflict, but not at the same level that we had with the Maritime Strategy.

So they were great documents to say that the Navy recognized that they were in a new environment, and they made that statement, that the Navy recognized that they were in a new environment, and they were going to have to change some of the ways that they did business. But they weren't going to forgo the traditional strengths of the Navy of presence and forward presence.

EM: So it was more of a regional focus, rather than just focusing on the Soviet Union?

SA: Soviets, yeah.

EM: What about the joint aspect?

SA: They recognized that we were part of a joint team, so we had those tickers in there so that nobody could say we didn't know how to play joint anymore. They were very timely and very necessary documents to help us with the political realities of defending our budgets over on the Hill.

EM: Right. Are those approaches still valid, or are we into a different era or different environment?

SA: The environment has changed. I notice the new CNO just dropped a new document on the street, and from all outtakes it's not at the same level as those strategic documents, but it's a nice look at how he sees the Navy and what the Navy internally has to do. I suspect there will be another, more strategic document coming out sometime during his tenure before too long.

EM: What about the adversary? Is there a different focus on potential adversaries?

SA: Of course China now is a great thing, to deal with its coming of age. It always seems like China lurks in the background, and then we let it go for a while. Then it comes back. It's interesting to see what they are doing now. I don't know why.

EM: What is your take on the anti-carrier missile? Is it all that it is cracked up to be?

SA: No, I don't think so. Again, I have no access now to the closely held programmatics and intelligence, but we've played around with this thing, I mean, ever since I was Vice Chief. We were worried about the carriers. We were worried about the carriers out in the Gulf, and some of the missiles that could be aligned against us there, which was a place where we couldn't hide. But in what I can read, and we've always done this. We always give the enemy a tall stature, part of the programming PR, so that we can find dollars to go do more research and development on stuff that we want to do.

EM: You mean, play up the threat to get more money?

SA: We play up the threat to get—you know, let's be honest about it. We've done that forever. It seems to me that we have given this thing capabilities that even in our mind we can't do, but we say this can. We've given our capabilities that can do certain things no credit at all. So the thing that worries me is, if we were playing this thing up for budgetary reasons because we want more things, what we are doing right now is sort of playing to the hand that says—I mean it's like cicadas, every 17 years or whatever, we have this anti–big deck carrier movement, and each time we knock it down, and, you know, I spent my life on the small decks, so I know what small decks can do, and I know what the difference is between a big deck and a small deck.

EM: Which is preferred?

SA: Big deck.

EM: Yeah.

SA: Big deck.

EM: Why?

SA: You just have so much more flexibility with what you want to do. The small decks are hampered by sea, although the old *Essex* class was a pretty rugged seagoing machine, but nothing like riding the *Nimitz* class. The big deck just gives you so much more flexibility, and of course the airplanes, you've got a lot more capability. Now, the one problem I've got with the big decks and the Navy strategy is that we have not populated that deck to the extent that we should.

EM: Too few aircraft?

SA: Too few aircraft, I think. But that's a budget problem. We've already killed some air wings. We had 80 or 100 airplanes on the small decks, and the big decks are 60, 70 airplanes, something like that. So we can do more with them. They just give you so much more flexibility, and they are so much harder of a target than a small deck will ever be.

EM: Harder of a target? Why is that?

SA: Well, they are harder to penetrate. They are harder, they can resist. They can absorb more—

EM: Damage?

SA: Punishment than the small deck.

EM: Okay.

SA: And keep on going. So, you know, you've got to always put that in the equation. Now, what I'm afraid we are going to see in the next, probably five years, we are going to see a big push by the Marines to use the big deck amphib as a carrier with the F-35Bs, and in certain roles that will be just fine. But if people try to equate that an F-35B big deck amphib is of the same capability and structure as a *Nimitz* or a, now the new *Ford*-class carrier, that's a big mistake, because somewhere along the way people will find that there is a limit to how far you can push the large-deck amphib into a heavily defended area. And it's an actual evolution for the Marines to do that. I'm not saying that they shouldn't do it. I'm just saying that people are going to take advantage of that for other purposes, and that's open even further—

EM: To say why do we need the larger carriers? We've got these big amphibs?

SA: Yeah. That F-35B looks just like an F-35C over here on the big deck. Why aren't they the same? We can build two of those instead of one of these. You get into that argument, but nobody wants to take it to the final conclusion. Who can take the first hit? The big deck can take a pretty substantial hit.

EM: Yeah. Okay. During your time as Vice Chief, Operation Sharp Guard took place in the Adriatic, right? '93 to '96, so it was toward the end of your term. Do you recall that, Adriatic operations again Serbia?

SA: I do, but I don't remember much about it now.

EM: You weren't supporting it?

SA: No.

EM: CinCUSNavEur was probably into that.

SA: I don't remember anything coming back our way.

EM: Wasn't there a shortage of P-3s or some PGMs?

SA: Oh, PGMs.

EM: Or Prowlers, I may be thinking of.

SA: Prowlers, yeah. The Prowlers. Yeah I think that may have been the first time that we had to base some—I'm not sure. Snuffy was out there at the time, I think. I don't remember me having to do anything. [Laughter]

EM: Okay, [Operation] Southern Watch. Not just Southern Watch, but the entire period after the Gulf War through 1995 at least, what was your view of Southern Watch and the embargo patrol? Was it effective against Saddam Hussein?

SA: Yeah, I believe so. I think it still kept the pressure on him. The things that developed right after we ended the war and cleared out of there, then we found out that we made a mistake relative to his use of helos.

EM: Right.

SA: So that was our first indication that things weren't going to go as well as everybody expected. The sanctions still were, I think necessary to continue to watch and keep him from really doing as much harm as he wanted to do, although we had given him a little too much leeway, I'm afraid. The presence out there and keeping that force there I think was very important, because we kept our touch into the area and into our coalition folks in the Gulf countries, and we ended up with different facilities for things that came on later. We had more access in the region than we'd had at the start of the Gulf War.

Usually what happens, you bail back out and you lost all of that touch and associations, and so I think that Southern Watch was good from the standpoint that not only did we keep the associations, but some of the countries were a little bit more forthcoming. In other words, a lot of our facilities that we used in Saudi Arabia during the Gulf War we gave up, and we were able to do some command and control in some of the other countries that were really more convenient for us to use. So I think all in all it was probably pretty good.

EM: The fact that the Navy had to maintain this patrol over many months, was that taxing on the support system getting enough pilots out there, and aircrews and other resources?

SA: It was—it was a resource-intensive effort, but it was sort of like we gave up presence other places; basically we were out of the Mediterranean, for all intents and purposes, for most of the time. So we had to give someplace to make that happen. It was very hard to maintain, especially when we started looking at the capabilities of ships, certain ships that hadn't been upgraded. We deemed that we didn't want to use those: "we want to use the ones that have got this upgrade." So it stressed the whole system to maintain that presence, but we were in pretty good shape to do it. We didn't gap much. We gave up some areas that we didn't think we needed to have as much presence as we'd had in the past.

EM: Were there any material shortages that come to mind? Aircraft or weapons systems?

SA: No. I don't recall us really having any real—there's something ticking in the back of my mind that we had a shortfall in something, and I'm trying to recall what it might have been. I remember fussing around about something that we needed more of out there, and I'm trying to think. But it wasn't a hard fix. It was just a momentary blip. Some of the stuff that was going on at that time was, we were still trying to reconstitute the Maritime Prepositioning stuff and upgrading some of that, so there was always that dynamic going on. How do we upgrade, and how much the Air Force—as I recall, the Air Force had laid a new requirement on us for some more capability that they didn't have during the war.

EM: More stores on ships? Prepositioned?

SA: I think it was more ammo prepositioned, and as I recall, they had some new requirements for some access there at Diego Garcia for new, I think, B-1s [bombers].

EM: Back in the States?

SA: Yeah. So some of that went on.

EM: This will be after your time, but was there a growing fatigue that this Southern Watch, or keeping watch over Saddam Hussein, was extending from '91 all the way through 2003? Were people getting tired of doing this, and maybe wanting some quicker action? What do you recall from that period?

SA: It had fallen into the routine category: this is where we've got to go. I mean, the Persian Gulf is not a fun place to drive around in the summertime.

EM: Right. I don't think we lost a Navy aircraft or a crew throughout that period, did we?

SA: During Southern Watch?

EM: Right. Or even the strikes into Baghdad. I know they were shot at.

SA: I think we had an operational loss during that period. Seems like we had an operational F-14 lost during that period for some. But no combat losses or anything like that.

EM: Okay. I'm going to your former EA, Jim Hinkle. He came up with this list and some of the things we've gone over, but some other things we missed—at least that I missed. He said you had no control over your schedule. [Laughter] Is that true?

SA: That's true. The JROC took priority over anything, and if the Secretariat wanted you for something, that took priority.

EM: So you had a schedule, but it kept getting bumped, right?

SA: Yeah. I basically was—it was out of my hands. One of the things I did do was, I refused to let the staff come in on weekends unless we had a real flap.

EM: Boy, that wasn't usually done in the Pentagon.

SA: No. But I made it a law.

EM: Well, that's good.

SA: You don't come in unless we have a real flap going on.

EM: What about during the workweek? The hours would be open-ended?

SA: Yeah, they were open-ended.

EM: So some guys would work until 10 or 11 at night?

SA: Yes. Some of them would. If we had to. But I tried to get everybody out early, and I tried to make sure they knew that if I was there, they didn't have to be there. That if they had certain things—but I almost always snuck in on Saturday morning early, and had three or four hours, or sometimes four or five hours that I would spend in there and clear the desk off. I could get a lot of work done, and then they would have all of the stuff that they had piled up. What happens in the Pentagon is, everyone wants to have their weekend clear. So starting at about 1100 on Friday morning, and running through about 1600 on Friday, everybody finally empties the things that have been on their desk since Monday, and it gets to the Vice Chief's office. [Laughter]

EM: So they want you to stay for the weekend?

SA: So at the end of Friday, a ton of stuff just happened to wander into the door. So rather than tie the office up, the guys would get the stuff that I should see as fast as I could. They would have that on my desk before they left on Friday. Then usually on Monday morning, they would find out that I had dumped all that back out their way.

EM: Jim Hinkle said that you had an unusual relationship with Under Secretary Danzig—several instances where you heatedly reminded him to do his job and stay out of yours. Can you elaborate on that?

SA: Well, I think we talked a little bit about that earlier, when we talked about the relationship with Danzig. We had an understanding that I had put a few markers on the table, when he was coming in, of things that I hoped to see from him, and that I was willing to accept things that he expected to see from me. Basically, that worked great. We were able to have discussions and, like I say, he always had the ACMC and myself up to his place, very frequently, at the end of the day. And we would go over things that were on the back burners or front burners, whatever. But every now and then—the lawyer in him was never going to go away—so there were lots of times that we'd think we had things settled, and then all of a sudden he would want to revisit them.

EM: Do you recall a particular instance?

SA: No. The thing about what goes on in there, in that office between us, there's just so much nitty-gritty, little stuff that you would think would just disappear, but they would pop up, and they are always associated with some—probably a bigger issue, but the things that go on at the time are probably pretty minuscule, but they get your temperature up. So he would want to come back and revisit ,something and I would say, "I thought we had settled this thing."

"Well, you know—"

EM: It's like a North Korean negotiation. [Laughter]

SA: And I would get frustrated, and luckily he didn't hold it against me when I would try to back him out.

EM: Jim Hinkle seems to think that Senator Byrd might have had something to do with derailing your nomination for CinCPac?

SA: Well, Senator Byrd, yeah. Jim has always felt that there was somebody other than Durenberger pulling the strings in the play. I knew there was somebody below Durenberger and his staff that was poking him. For what reason, I don't know. But anyhow, I knew that somebody was feeding him lots of information. Jim always felt that my episode with Byrd with the third atomic clock—probably there was still some residual bad feeling about me delivering that message. I don't think so, but you never know. And I really don't know why. Jim has had good contacts over on the Hill for a long time, so he may know something that I don't know. [Laughter]

EM: Okay. Who was Kernig? Or Koenig?

SA: [Harold M.] Koenig was the Surgeon General of the Navy. We had lots of issues with medical during my time. Different things that went on.

EM: The hospital ships, maybe?

SA: No. Well, we probably had some a little bit on the hospital ships, but not so much then as Vice Chief, because they were back, although we did send them off for some episodes. But the structure of Navy medicine—as I mentioned earlier, when I was in OP-04 I was part of the Blue Ribbon Panel on Navy Medicine that was headed up by Admiral Bud Edney, that was trying to look at the path forward for the hospitals. Again, part of that was the hospitals kept coming up in BRAC. The Navy had a different version of how hospitals were run compared to the Army at Walter Reed [Army Hospital]. There was always that pressure to combine Reed and Bethesda [Naval Medical Center].

EM: That's come to pass.

SA: And that's come to pass. The issue of a teaching hospital—the difference between a regular hospital and a teaching hospital, what were the advantages of keeping Bethesda as a teaching hospital, or were there advantages, or was it something that we could forgo? So there were all of those things that we sort of fiddled around with in that Blue Ribbon Panel. One of the things—and one of the reasons that as part of that episode, the Surgeon General at the time had me come over and witness an open-heart surgery at Bethesda for me to really understand, one, the teaching aspects of a teaching hospital being involved with an operation of that nature. And also what it really required to provide an operating room that could handle that kind of process.

EM: Were you persuaded?

SA: I was persuaded to the point that I believed in teaching hospitals; that the teaching hospital was a very valuable asset to have. I understood what the requirements were to truly have very special operating rooms and structure to support those operating rooms. It also showed me how wasteful we were, which they really didn't want to show me. What I ended up finding in the process was that although we had several really very spectacular operating rooms in Bethesda, as you would expect, we also had operating rooms that were not being utilized other than just being storage areas. And there was no way that overnight they could be turned back into operating rooms. It would take several days to get the junk out of there and the place sanitized again and brought into specs that you would be willing to use it as an operating room for major operations. So the fact was that we had some excess capacity in that area in Bethesda, which was pretty troubling and played into the hand of people who said, why can't Bethesda and Reed be merged?

EM: So even Bethesda had excess capacity?

SA: Yes. Now, we didn't have excess capacity in all areas. One of the things we had not enough capacity in was parking. If you wanted to bring more capability into Bethesda, you were going to have to build another parking garage.

EM: Right.

SA: So those types of things—I forget all of the areas that we had that happened, but we had several sort of misadventures associated with Bethesda during my time as Vice Chief, and Hal Koenig and I would have to sit down and talk about it. He was a very different kind of personality type guy, and so we had some difference on some of the issues. I think Jim thought I should probably have made a move to get rid of him, but I wasn't there. He had some great attributes that were important, he just wasn't as warm and fuzzy as a leader at the hospital. But technically he was great.

EM: So he had problems with some of his subordinates?

SA: Yeah. There were some issues there, but technically he was very, very good.

EM: He was head of BuMed?

SA: BuMed, yes.

EM: Yeah. At one time it was called Naval Medical Command. It went back and forth.

SA: Yeah. I think it was BuMed still, at that time.

EM: All right. Kendell Pease, ChInfo [Chief of Information].

SA: He was a good ChInfo.

EM: According to Jim Hinkle, he introduced the new Vice Chief, you, at the morning briefing by showing a picture of you clicking your heels as you left Seventh Fleet.

SA: Have you ever seen that photo?

EM: No. [Laughter]

SA: It's around someplace. I will have to look around and see if I still have one floating around here.

EM: He finds it hard to believe that you were excited about going to the Pentagon and leaving the fleet.

SA: Yeah, well, see, that's the way he interpreted that. So when he sent me this, I sent him back an email and I said, let me tell you the real story of why that picture was taken or how that picture was taken. It was traditional on the Seventh Fleet staff that when a staff member left to his next assignment, if he was in port on board the ship on *Blue Ridge,* we would have farewells down in the staff spaces. Then when it was time for him to depart, we'd all go up to the quarterdeck. He would render honors to the OOD, quarterdeck honors, and then when he got down the brow and got onto the pier, he would turn around and we would render a salute to him, and he would jump up and click his heels and off he would go. So that was sort of the traditional departure ceremony.

EM: It wasn't something that you did the first time?

SA: No. Here I am headed back to the helo deck, and everybody that's not in flight gear and stuff like that has to stay, stop earlier before they get back to the flight deck. There's a spot there where we used to say our farewells when people were leaving, departing by helo, and then they would head back and get in the helo and go.

So basically I started to head back for the helo, and I just turned around and gave the jump-up and click-your-heels and off I went. So there's this photo of me holding onto my flight helmet, or the helmet they had for getting on board the helo, and I'm in my life jacket and whites and up in the air with my—the old click-your-heels. I tried to remind everybody that if they look at my hardhat or the helmet, I'm on my way down and not on my way up, so I was really higher than I really was. [Laughter]

EM: This—actually this is an aside—but have you given any thought to retiring your papers and your other memorabilia from your Navy life?

SA: I haven't. I've been thinking lately.

EM: Well, Joe Prueher specifically said, "Ed, where can I put this stuff?" So I contacted the Naval War College. They have the Naval Historical Collection, which I think is probably the best place for your documents and the rest of it, photographs and all of that. For objects, things that you think would be worthy of historical retention, probably the Naval History and Heritage Command's curator function. I know people in both places if you ever decide. You know, it's valuable and will tell a good story, along with your oral history.

SA: I didn't keep much paper, which is amazing to me: as people worked for me, I would drive them all crazy because I lived with stacks of paper.

EM: No diaries?

SA: I didn't. Boy, and I wish I had, and I don't know why I didn't do that.

EM: Well, it's just a thought.

SA: But I've got lots of pieces of this and that.

EM: Sure.

SA: I gave a couple of things out to my old NROTC unit; I had a flag from *San Jose.* During the time I had *San Jose* we had the bicentennial, whatever it was, the year that the Navy—when we flew the "Don't Tread on Me" flag.

EM: '76?

SA: Yeah, it probably was. So I gave that flag to the NROTC unit. But I've got tons of plaques that I'm probably going to cut up and throw away.

EM: Well go ahead, because they don't like plaques.

SA: Yeah, I figured they didn't.

EM: They've got more plaques then they can deal with. All right, one last question on this list by Jim Hinkle. Rear Admiral (lower half) [Leonard F.] Picotte, he sued you for not getting promoted?

SA: Yeah. I had forgotten that until Jim reminded me, and I can't remember all of the details. I guess I must have been president of that board. But for some reason, he had it in his mind that I had specifically—somebody had to whisper in his ear, or he whispered in his own ear or something, that he had run afoul of me someplace, and it just never had happened. I knew of him, and I knew very little about him until all of a sudden this thing popped up on the horizon, and he thought I had deep-sixed him someplace.

EM: Did he have evidence?

SA: No.

EM: So where did the suit go?

SA: It just died. It was just one of those blips. Like I say, I hadn't even remembered it until Jim mentioned it there. I probably need to talk to Jim sometime and ask him what else happened, but I know it went away. And you saw the rest of his comment there, that the community probably wasn't as supportive as they could have been for him.

But the promotion boards, that's a whole new—part of the job is, you do sit on—well, for the Vice Chief you only sit on very few of those boards. But during the period that I was back there in Washington, I sat on several different boards at different stations in life, starting back when I was a commander and supporting one of the promotion boards.

I remember one of those boards as a little fly on the wall. It was a flag board, and there will be no names here, but the subject came up of this one individual, that his community was really pushing for him. He was sort of in the crunch zone and, you know, you are now down to just the last few numbers that can be selected, and his community was really pushing for him, saying we really need to have this guy as a flag officer. Actually I think it was from a one-star to a two-star—we need him in the mix for the next big jobs coming up. It was going back and forth, and you could see that it was right on the margin. And the Vice Chief at that time was the head of the board, and so everybody had been talking. Everybody had been talking, and then all of a sudden the Vice Chief starts to talk, and the room goes dead quiet. He says, "There's just not another job for that guy." [Laughter]

EM: That's the end of that. [Laughter]

SA: I said, whoa. So that's how the crunch gets solved every now and then.

EM: It comes down to he was the decider.

SA: Yeah.

EM: Okay. The last thing, you had mentioned earlier that you and Joe Prueher have different views about the job of VCNO.

SA: I think we do. And Joe and I have really never sat down and talked about it. I guess we ought to. One, Joe's a wonderful guy and a wonderful friend, so it has nothing to do with that. It has to do with Joe has a much bigger picture than I do. Joe really has that type of mind that looks at the big picture and can really make sense out of it. What I consider my strength is, I'm a nuts and bolts sort of guy. I like fiddling with cars and things. So I like to take things and tune it up a little bit, and Joe's a guy that can say, you are wasting your time tuning things up. You need to be thinking there.

Joe is also a very neat person. I would guess that the Vice Chief's office looked much different when Joe was there than when I was there. My desk was an absolute—the Japanese friends that I knew from Japan that would come into town every now and then, they would always want to come over and see the Vice Chief, you know, their old buddy from Seventh Fleet days. And they would always want to have their picture taken by my desk, because I had these stacks of papers. And the Japanese just couldn't imagine sitting at a desk with all of that paper on it. Because they are very neat.

I had somebody at one time in my career, when I had *Coral Sea,* they came in. It was in overhaul, and they were going to redo the captain's in-port cabin. It had a nice little working area off to the side, a little office space. This lady had been contracted to come in and redo the space. So she asked me different things, and what I did with paper and all of this stuff. She says, "Oh, you are a vertical guy." [Laughter]

I said, "You've got it." So I would say that Joe probably did not enjoy the pieces of some of the job that I enjoyed. I enjoyed some of the little nitty-gritty things that happened that you fussed around with and tried to do. I think those things were probably not as enjoyable for Joe as they were for me. So that's where that comment comes from. I would guess that if somebody had a picture of my tenure in the Vice Chief's office with me behind the desk and a picture of Joe behind the desk, they would be dramatically different, and Joe was probably much better at the job than I was.

EM: Well, I don't know about that. Okay.

SA: You asked about dealings with the press, and as I've mentioned it's always a two-edged sword. You've got to be very careful. You are not sure exactly how it's going to come. So I was always very cautious about interviews and things.

EM: Did you ever get burned, or something that came back to—

SA: No, not really. I forget if we discussed this during—yeah, we did talk about it during the Gulf War, when Sam Donaldson interviewed me. That was really my first big press exposure, because it went national TV. It gave me an appreciation for how talented—he could always see himself as the camera saw him. I didn't realize that until I watched the program, and I remembered what we were doing and how we were walking and talking and everything, and I remembered different things that all of a sudden he would just turn and look, and then when I saw the show, it was obvious to me that he could see what the camera was seeing. It was just like he was looking through the camera.

EM: I'm sure it's a skill you have to develop.

SA: Yeah, it's a skill. It was really interesting to me to see that what you see on TV, they know what it's going to look like. They know how to do that. I always thought it was the cameraman that did it.

EM: They just follow the directions.

SA: Sam—because I'd seen him move the guy [cameraman], so it was really eye-opening and he was very good, the interview went well. He didn't throw anything in there that was totally unexpected. So I was happy. After I retired I did another program for, I think PBS, and they were looking at things on my retirement from the Navy, and then the Scott Speicher recovery operation. What was going on there—the Rebecca Hansen stuff—I guess maybe that one wasn't the PBS one. I did one with PBS; I think maybe that was titled "Navy Blue." Then one with *60 Minutes.*

EM: *60 Minutes,* really?

SA: Yes. That may have been post-Boorda—probably was. But anyhow, those were three big ones that I did that had the potential to make pretty big headlines if there was a mistake, and they went well. They treated me well. They were fair.

EM: Did your PAOs give you briefings on how to respond?

SA: The one out in the war zone there with Sam, Schwarzkopf told me, "I want to make sure you leave Sam with the impression that we can make an amphibious landing." We keep wanting to sell that.

EM: He told you that?

SA: Yeah. He said, "So, you know Sam is going to be there, and I don't want him to walk away thinking that we are not going to do an amphibious landing. We are probably not." We already knew, we'd already discussed all of that. "But," he said, "I don't want that to get out. I want that sold that we can do it and are prepared to do it."

EM: Well, they bought it.

SA: And they bought it. That was about all of the direction I got there. The ones that I did after I retired, I always called my old friend Kendell Pease, who was ChInfo during the time I was Vice Chief. He knew the press really well. He also knew messaging very well. So he said, "You've got to stay clear of answering these kinds of questions. Don't volunteer more information than they ask for," basically, was one of the primary things. "Answer the question in as few words as you possibly can. Don't expand on them unless they ask a follow-on question." It was pretty good advice.

EM: Yeah. All right. So you are at the retirement time.

SA: Retirement time, yeah.

EM:1995. What day and month was that?

SA: I retired in June officially. I really had pretty much wrapped up work in May, but the official retirement day was in June, which marked 38 years. In those days we had already instituted a transitioning course for officers who were retiring from service, so I took part in the transition course.

EM: How long did that last?

SA: I think it was a week, a week course. An old supply flag officer ran the course and was very good.

EM: What can you learn in a week, though?

SA: They talked about financials, what do you need to do financially, what you are going to get, and are you going to receive it. Whether you are going to take the reduction in pay so you can have a follow-on benefit for your wife in case you depart the scene early.

EM: Right.

SA: So they gave you some of the financial ramifications of what was going to be available, and how you could manage your retired pay. And medical—if you are going to go on a medical disability, what you have to do with VA [Veteran's Administration]. They brought people in to coach you through how you did VA forms. Then they talked a bit about attire. How you dress. Ties, shoes—

EM: How you dress?

SA: Yeah.

EM: Out of uniform?

SA: Yes. How you dress out of uniform. And you had to come out of uniform, and they would sort of grade you. This is not what you would wear to such and such a place, or you think about this. It was an interesting course, and they talked a lot about your expectations and how you interfaced with interviews, and what kind of decisions do you need to make before you branch out into this effort. I mean, are you looking to be a consultant, or are you looking to be on boards, or are you looking to work full time? Here are the pluses and minuses of these, and then you need to make up your own plus and minus list of who you are and what you view for the future. So it was a quick course, but it got your mind out of the day-to-day business that you were involved in and got you thinking in a whole new—

EM: More long-term?

SA: Long-term. At first I thought it would not be useful, but it was.

EM: Going back a bit. Your actual day of retirement, where was it?

SA: I didn't have a retirement ceremony.

EM: You just left?

SA: I just left. But I did have an award ceremony, and I made that my retirement ceremony. So they presented me with the Distinguished Service Medal or something, Defense Distinguished Service Medal, I think.

EM: In the Pentagon?

SA: It was held up at Fort McNair—or Fort Myer, in the theater up at Fort Myer. The Secretary of Defense and Mike Boorda were both there and presented me with the award. I invited my mom, who came out from California, and I had some cousins from Ohio that came. So it looked like a retirement ceremony, but it wasn't. It was an award ceremony. I didn't want a retirement ceremony. They kept pushing, and so I finally agreed that I would do an award ceremony.

EM: Why did you not want a retirement ceremony?

SA: For some reason I just didn't. I don't know, and it's sort of funny, because I remember a couple of years earlier there was a Navy captain, really well known in the aviation community and fighter pilot. I got wind of the fact that he was going to retire without a retirement ceremony. So I got in contact with him and I asked the same question you did. He said, "You know, there's nobody around that really knows who I am or what I'm doing."

I said, "No, no. There's a bunch of us that know who you are, and a lot of people who want to come and say goodbye to you. So you are going to have a retirement ceremony, and I'm going to be the speaker, and we will take care of all the details." We did it there at the Navy Yard on the parade field, in front of the CNO's house. It worked out great. I think he enjoyed it, and it all went well.

But I thought, accept the award and take time to give a couple of messages out to the group, and let the attaché corps in D.C. know, of course. I was pretty close—especially the ones from the coalition forces. Some of the guys had been out in theater with me commanding ships from the other navies, so I wanted them to be there at the award ceremony. So that worked out. It satisfied my need for a retirement ceremony, yet it wasn't a retirement ceremony.

EM: Right. So, you went through your transition training, and now you went to your home in Fairfax?

SA: We had a home there in Vienna at that time. We just built this new home, so we moved into it as I retired.

EM: Did you take any time off to vacation?

SA: No. [Laughter]

EM: Right back in the saddle.

SA: Back in the saddle. John Lehman's brother, who was very active in the Washington scene, sought me out and said, "Let me help you write a resume," and stuff like this. "It will be productive as you go about your business." A couple of companies wanted to come and talk to me. So we were doing that during the sort of dead period right before I actually left the building. A week or so afterward he talked to me about jobs, and a couple of offers were made. Then General Bob [Robert W.] RisCassi, Army retired, four-star, very well thought of in the Army—I had met him when I was Seventh Fleet and especially after the war, I came back and he was U.S. Forces Korea—so Bob and I got acquainted there, and so he said, hey, you know, I'm over here working for Loral, and we need a Navy guy.

EM: What's Loral?

SA: Loral was a defense contractor. The CEO was Bernard Schwartz, Bernie Schwartz from New York. The chief operating officer/president was Frank Lanza. And so Loral was big in space, and then they had a whole potpourri of businesses that went from space to electronics. They were all missiles, lasers. They had arrayed a whole bunch of business areas. Bob said, "We think you would like to come to work with us." The other guy in the Washington office, the head of the Washington office at that time, was General Jimmie [V.] Adams; I think his last job was PacAF [Pacific Air Forces] at that time. He was a retired four-star, Air Force. So Bob said, "come on over and talk to Jimmie and I."

I went over and talked to them, and they said, "Bud Edney joined us for a very short time, but he wanted to get back to California. So he's left, and we are looking for a Navy guy of the same sort of horsepower we are, a retired four-star. We work out of this office here in Crystal City for Bernard Schwartz and Frank Lanza of Loral Corporation." I hadn't had any real direct dealing with Loral. "Frank Lanza would like to interview you for the job. So we will fly you up on the shuttle. There will be a car and driver there waiting for you to take you into New York City, and you go up and talk to Frank."

So, you know, here's probably the kind of salary that they are talking about. "We would like to have you with us. We are enjoying the work."

And so I said okay, I will go up and talk to Frank. So I flew up to New York City, and sure enough, driver there waiting for me and into town we go. Go up high in this building looking out over the city. Frank Lanza is Italian, short, squatty sort of guy. You know, very direct. Looks like a Mafia boss. [Laughter] And he said, oh, so you think you would like to work for us? I said yeah, it sounds pretty interesting. I think I would. He says fine.

EM: That was it?

SA: That was it. [Laughter]

EM: Short interview.

SA: No handshake, no nothing. So I come back to D.C. and I check in with RisCassi and Jimmie Adams and say I think I'm hired. They say yeah, you're hired. I say well, what's my salary? "We think it is this. And you'll have to fly up next week and get all of the paper; they've got the insurance for you, and they will give you your badges and all that."

EM: Before you get into that, was this what you were thinking you would want to do when you were retired?

SA: I knew I wanted to work full time.

EM: Right.

SA: I wasn't exactly excited about working in defense particularly, but again, I realized that that was sort of where I was going to go. I was worried about, was I going to be expected to play on my contacts within the fence to do the business. So I talked to Jimmie and Bob, and both of them had a little bit different versions.

Jimmie says, "You can be one of the guys that goes into the building to facilitate stuff. Personally, I don't like to do that. Now, Bob RisCassi has no compunction about walking in, and he's got access to the Army staff, and he will sit down with people and talk. But our job is not to sell the products. Think of us as the joint staff to Frank Lanza. Frank Lanza is the chairman of the joint staff, and we are the service reps. We are the CNO and the chief of staff of the Air Force, chief of staff of the Army. Frank will take us and bring us up to New York for every business development meeting that they do with the businesses. When they bring them in for their quarterly reviews and year-end plans and all that, we are sitting at the table with Frank." Many times Bernard Schwartz was there as well. "Our job is basically, we are the bullshit meter. We are in charge of the bullshit meters. If the business area is saying that they can sell this product at this level to the services, our job is to advise Frank and Bernie, not going to happen."

EM: Forget it. Yeah.

SA: That is not something that the service is going to go after at that level, or after at all. And point out areas where maybe they are not being as aggressive as they should be. "So," he says, "our job is to advise the front office, how solid are the plans of the operating units. Our job is not business development. They don't expect us to go into the building and sell anything."

EM: Sell products. Yeah.

SA: "But they do want us to be able to point individuals to the right direction. If they are trying to make an approach, you know, don't try to sell this to this part of the Army or Navy. If you are talking to this guy about this products, that's not going to help you at all. You ought to be talking to somebody that resides over here in this shop." So the deal was to point the business areas to the right place to do their business development, but the most important job was to be the keeper of the flame of, are these guys being realistic about what they can do in the business area relative to what we know about our services.

EM: Is that how it played out?

SA: It was exactly that. We traveled with Frank to every business review, everyone that came into New York to sit down and talk to him. We were on the shuttle constantly, back and forth, day trips. We would hop on the shuttle early in the morning and be back in the mid-evening. Then we traveled to the West Coast and all of the business areas, which were scattered all over God's green earth.

EM: What aspects did you deal with? What specific products?

SA: We dealt with Army—a lot of Army missiles for the Army. Navy, we had helo integration. We had submarine communications and command and control. We had radios, military radios. You name it, we touched every service with something. So I went to work in July with them, and by March we were absorbed by Lockheed Martin. Out of the clear blue sky, all of a sudden we find out that we've merged with Lockheed Martin. So this is the first introduction to how people get wealth, why you see these mergers and things, sometimes being thought of as very exciting events by people in business. Because part of the merger process is that stock options that you had as part of your compensation were made whole in one instant. So rather than sitting and waiting for this stock option to mature to option time, all of a sudden it was it was cash on the barrel head.

EM: That's nice.

SA: You say, wow, how did this happen? Well, part of the merger process, we've got to clean out these options. As we became acquainted with Lockheed Martin, the three of us, we had staff under us. We had guys more junior to us that were also helpful, and we had a pretty small shop. I guess there were probably 10 or so of us in the whole shop, together with secretaries and all.

How are we going to be integrated into Lockheed Martin? Frank tried to sell Norm [Norman R.] Augustine, who was head of Lockheed Martin—or Lockheed in those days, and then Lockheed Martin—on how we'd done it. He said, "No, no. We don't—we are not organized that way. These guys, we are not going to put them up here at headquarters with the chairman and the president. We are organized entirely different." And let me go back to Loral. Lockheed Martin was organized in a pyramid, almost exactly like the Pentagon in the military, but the pyramid was bigger and not as responsive. [Laughter] Which I never failed to mention to the guys as I got better acquainted with them.

EM: Not as responsive as the military bureaucracy?

SA: Yeah. And that was really surprising.

EM: Right.

SA: They always kept waving the BS flag at me; I said, "Look, if there was ever an issue that needed to be decided in the Pentagon, really needed to be decided, I can guarantee you I could get a paper signed by the Secretary of Defense by the end of that day. Now, it had to be an important issue, but that could be done. And I've done it." Impossible to do in Lockheed Martin. Absolutely impossible to do in Lockheed Martin. You're stepping over somebody's toes and no, no, you can't go that far. We've got to carry it to the next step. So I used to rail at them a little bit on this.

But Loral was a flat organization. You had the chairman, you had the president and chief operating officer, and then you had these operating units. I forget now what the number was, but it exceeded 15. So you had a flat structure. Two guys, 15-plus operating units. The operating units didn't go through a middleman to the president and then on up. So Frank Lanza was absolutely brilliant. I never ceased to be amazed at his depth of knowledge of each business area. I mean, we would go in for a business quarterly review and he knew the numbers better than the president of the organization.

EM: So he had to know each one of those 15 areas?

SA: He needed to know.

EM: And manage them?

SA: And he managed them. He didn't just accept what they were doing. I've seen him go from dealing with a budget number that was off by a significant margin that the president of the company couldn't figure out, how did that get there. Frank could tell him how it got there. We were out in Sunnyvale, at the space guys out in Sunnyvale, and we got to talking about jitter on a satellite where they were having trouble with communications. The president is waving his hands, and, you know, "This is our problem. We've got this jitter problem, and we don't know exactly what to do with it."

EM: Jitter problems? Jitter means what, vibration?

SA: Vibration, instability of the antenna. So Frank says, "Let's take a 15-minute break, and get me the chief engineer." Brings in the chief engineer for the program. He and Frank sit down and start discussing exactly what this problem was. When did it develop? What are the symptoms? What do you think is the problem? Is there a fix? Isn't there a fix? Let's, no I've seen that before. No, that's not the way to go. You need to be thinking about this, and hey, why don't you contact this guy someplace.

I mean, it was just amazing. He knew every piece of the business, so that was a great—I couldn't have picked a better place to start out in industry than to have that job. Because you got to be there, sitting there when all of the decisions were made, and all of the discussions and the pluses and minuses, and why you were doing it and why you weren't doing it.

EM: When Loral was bought out by Lockheed Martin or merged with them, did Lanza go away or was he brought into the structure with Lockheed Martin?

SA: So what happened was, Lockheed Martin was built on this pyramid, and basically you had the chairman and the president, and then you had the executive vice presidents of operating areas. A space guy, an electronics guy, a aeronautics guy, a commercial guy. We had five vice presidents. So you had those guys sitting on top of their business areas, and under each of those executive vice presidents, then you had the operating units that belonged in that sector of business. And so most of the stuff that Loral brought to the table went into the electronics sector of Lockheed Martin. At that time the executive vice president was Tom [Thomas] Cochran.

Each of the executive vice presidents of their areas had a Washington representative down in Crystal City. The office in Crystal City was manned by a guy who was on the staff of the president, out in Bethesda, of Lockheed Martin. He reported in at the staff level at headquarters, and he ran the infrastructure of the Washington office. And inside the Washington office was a guy who was the Washington rep for the electronics sector, for the aeronautics sector, for each of the sectors. Then they had a team of people that worked under them that were available to the operating units to do their work in Washington. You know, hey, I need to get this paper over to this guy, and this type of stuff, what a liaison office normally does.

So when Frank brought Loral into Lockheed Martin, what Lockheed Martin wanted to do was pick off some of these businesses and make sure that they went in to the right area, but most of them ended up coming into the electronics sector. So for a while, Frank wanted to just bring in the major piece and sort of run it. They wanted him to be the executive vice president to run those particular areas. So he was down here. Frank didn't like that at all. So he said that's not the way, and there were some of the businesses that Lockheed Martin found were pretty redundant. They knew this, so they were getting ready to sell these businesses off, because they were redundant, to some of the stuff that they already had. So he had enough of the pyramid structure and said, "I'll buy them."

EM: I'll what?

SA: "I will buy them," the units that they wanted to shed, and "I'll form L-3, L-3 Communications." So Frank split away and formed L-3, and he came to me and said, "Look, I'm going to take Bob RisCassi and Jimmie Adams back with me into L-3, and I will take you if you don't agree with what I'm going to say. But what I want is, we brought a bunch of businesses into Lockheed Martin that are involved with the Navy, and I would like to have you—I would wish for you to stay behind. I've already talked to the guy in Lockheed Martin that is going to make this decision, and he's willing to do it because they are going to switch out a guy in the Washington office. So I want you to head up—they are going to offer you the job of heading up the Washington office of the electronics sector. And all of the Navy business that we are leaving behind is going to be in that sector. So I'm asking, would you do that? Are you willing to do that?"

It just so happened we all resided in the same building, so I ended up walking down the passageway and taking over the head of the Washington office for the electronics sector of Lockheed Martin.

EM: Somewhere here I've got the title. You were President of Missile and Fire Control?

SA: This was later.

EM: Oh, later, okay.

SA: So this all happened. I went to work for Loral in July of '95. So by March of '96, I'm now Lockheed Martin and head of the electronics sector, Washington office, electronics sector for Lockheed Martin. The businesses that we brought over I was familiar with, and was able to help meld those businesses in with the business areas of Lockheed Martin that I hadn't been involved with before. It was not the same.

I mean, it was not the same job, although I was still sort of the guy—when Tom Cochran was having business development meetings, I always had a seat at the table to be, again, doing sort of what I did for Frank Lanza, but we are now another tier down in the structure. I was sort of the sanity guy to say, okay, from my Navy experience, and now from what we know from the Washington office—because now we had people who talked to people on the Hill, and we had people who were representatives of the companies out in the field. So it was sort of the same job, but just had a little bit of different feel to it. That went on until about spring of '99, and then one of the things that we had done from our Loral days, we had a missile business out in Dallas.

There was a missile business in Dallas that mirrored the missile business here in Orlando, which was under Lockheed Martin. It had originally been a Martin Marietta business, back in the Patriot days—built Patriot missiles and stuff like this. So then when Martin Marietta and Lockheed merged, that became part of the Lockheed Martin family. I had been familiar with the Dallas plant from my Loral days. That was a big Army business out there. Tom Cochran, and by this time I guess it was Bob [Robert B.] Coutts was now the executive vice president for the electronics sector, said Jim [James] Berry, who had been on his staff up in Bethesda, was now president out in Dallas, and the president here in Orlando they were going to move to a job up on the Aegis cruiser business in New Jersey.

EM: New Jersey, yeah. The conning tower in the cornfield?

SA: Yeah, and so they said, "Look, we want you to go and take over as president of Lockheed Martin Missiles in Orlando, and Jim Berry is going to be president out in Dallas. Jim and I then talked, because we were in agreement. Our job was to build one organization that was in the missiles business, missiles and fire-control business, because there was so much overlap. Although Dallas tended to stay on the big side of the missiles and we stayed on the small side of the missiles, there was still an awful lot of technical information that was the same and needed to flow, and we had some redundant manufacturing.

So the job was that I would come down here and take over as president, and Jim and I asked, "How long is it going to take you guys to get this all put together?" They said it would probably take two years, three at the max to get this done. So I said, "That works with me fine, because two or three years down there—"

I really wasn't that interested in coming to Orlando. But I said then I would retire. Jim was younger than I and had a lot more growth potential in Lockheed Martin than I did. I was going to age out pretty rapidly.

EM: They had an age ceiling?

SA: They had an age ceiling for certain jobs. In other words, I couldn't aspire to be executive vice president. I would be too old to be an executive vice president. And so I said that works on my time schedule: go down there three years or so. We will get this job done.

Well, we went to work at it, and it wasn't as easy as we'd thought. These two organizations had competed against each other forever. Each of them had a win-loss record. You know, it was like going into a football locker room. Orlando 6, Dallas 5. Both businesses were very profitable. In fact, the combination of us, we were as big as the aero sector of the airplane guy. We brought in the best margins of anybody in the corporation. So everybody was depending on us not to break anything in the process of doing this. We weren't given any relaxation on the goals and measurement standards. We were still expected to deliver financially. So we went to the business of trying to organize the two business areas and still be competitive in the open market with our adversaries in Raytheon and ATK [Alliant Techsystems Inc.] and other people that we competed against. It was a good experience. I would say I was a little reluctant to come down. Was not sure I was ready to be the president of an operating unit. That was another skill set that I hadn't quite figured out—from the financial standpoint more than the leadership. I wasn't worried about the leadership, but I was more concerned about the financial side of the house. So it was a very positive, as it turned out. I enjoyed it. It was a lot of work, but a lot of rewards as well.

EM: What do you think you accomplished in that position?

SA: I think I was able to—because of the merger and bringing in a lot of the Loral stuff, there was a lot of nervousness about the future of the organization. Here they saw they could end up being swallowed by Dallas eventually, and there would be nothing left behind here. So I think I was able to project to them that Orlando is always going to here. We are not going to fold completely into Dallas.

If the business is going to fold, we are going to be operating from two locations. "Now," I said, "I can't guarantee where the leadership will finally settle, whether the president will end up staying here in Orlando or whether you will stay in Dallas. But as long as Jim Berry is out in Dallas, he will stay in Dallas. He will not come back to Orlando. He will spend time here, and I don't think you are going to see any change once we get this merger complete. Our job here is, one, not to lose any of our current business."

The mentality of the two units was—and this is an oversimplification—Dallas had legacy programs and had worked with the Army on these legacy programs for years and years and years, Patriot being part of that scheme. And so their business area, or their business people were much more oriented to continuing the line, upgrading the line. Taking the next step for the next improvement to the system. The Orlando mentality was finding new business. We had some legacy stuff, but the business development people here were much more attuned to, okay, we've got people working the programs that can protect the legacy. Now we've got to find new areas where we can export our technology. So Orlando tended to be much more aggressive in pushing the envelope about, hey, we've got this new technology. We think you ought to look at it. We were trying to steal business away from Raytheon in the laser business and weapons systems. So we developed several new weapons systems over here.

EM: That was successful? You were bringing in new customers?

SA: It was successful. JASSM, the JASSM missile for the Air Force was—

EM: Joint?

SA: Joint air-to-surface standoff missile—what was it? Strike missile, I guess it was, and a very exotic missile, and we developed it here. We have a facility up in Alabama where the production was. We were heavily involved with the Hellfire. We developed many more versions of the Hellfire, and we are still selling Hellfire missiles today.

EM: Yeah.

SA: A different variety, but—

EM: Is JASSM still in the inventory?

SA: JASSM is now migrated to the JASSM-ER [joint air-to-surface standoff missile–extended range], the extended-range JASSM. We had a targeting pod that we developed that competed against the Raytheon targeting pod. We thought for a while that our first customer would be the Navy. The Navy wouldn't do it, but we did it with the Air Force instead, an advanced targeting pod. It's now a big moneymaker for Lockheed Martin. For a while, of course we were losing our shirts on the development, but we've recovered all of that.

EM: What about helicopters?

SA: Helicopters. The Apache helicopter. All the fire-control systems on the Apaches were developed out here. The night vision stuff and the targeting business. So it's a very technically capable unit here. Great engineers. We did very little production here in Orlando. Very heavily engineer staff here in Orlando. Some production, some exotic production stuff here. We've got a plant up in Ocala that does a lot of production, and a plant up in—

EM: So Lockheed Martin still has that division with Dallas and Orlando?

SA: Well, we still have Orlando Missiles and Fire Control, and we have Dallas Missiles and Fire Control. The missiles and fire-control business president resides in Dallas. The current president was one of my chief guys who worked for me here. So the president didn't stay here.

EM: Are they still competing?

SA: No. Now there's a complete division of labor, so everybody knows who their customer is. There's no Dallas going in one door and Orlando exits the other door, which was what we were having when we first started. So we were able to consolidate a lot. We consolidated. They had a manufacturing facility out closer to Dallas than ours up here in Alabama. Both of those units are still in production activities, but there is a very clean line of what is produced in one place and what is produced in another place, so there's no overlap of capabilities. You do this business here and you do this business there. So we made some efficiencies there.

EM: You were working full time on this job.

SA: Oh yeah. It was a full time job, yeah.

EM: When did you retire?

SA: So then I retired in '05. My desire there was, I ended up being 10 years with industry when I thought I would probably only do maybe five or so.

EM: Was it satisfying?

SA: It was. I had a ball. I really enjoyed it. I had a lot of fun. Met a lot of nice people. I was very proud of the stuff that we were building and the customers that used it. It took me back to where I started. When I came out of high school and headed off to college, what was I going to do? Well, I had hoped to someday be an aero engineer and build airplanes or something.

EM: Got close to it?

SA: I ended up building a lot of stuff that went on airplanes and realized that, boy, I sure made the right choice to go to the military first and industry second. I have a great admiration for the people who work in this industry. I mean, it's a scary industry to work in. You can be the top engineer running one of the very best programs, and the minute the government decides that program has come to an end, you're out of a job. There may be not another place for you in that business.

EM: No security.

SA: There's no security. On the other hand, if you want to stay in the organization and progress up to various levels of management and leadership—I don't know how many people I retired here in Orlando who had spent 30 to 40 years and had never moved out of the same area in the plant they started in. I just can't imagine going to work for 30 or 40 years and walking in the same door every morning and walking out the same door every night.

EM: I did.

SA: Well, you know what I'm talking about.

EM: I know, it's totally different. It's a lifestyle.

SA: I remember I was talking to the top guy who ran the top business area in my organization here, a guy by the name of Tom [Thomas] Gosselin. Wonderful guy. Smart guy. Still pretty young. Tom retired. I said, "Tom, what a great career you've had here."

He said, "You know, from my office, when I retired, I could look out and see the desk that I started at."

I said, "Tom that's just—I mean, I just can't—"

EM: Not for you?

SA: It's just something that I can't even identify with in any shape or form. I said, "How did you do it?"

He said, "Oh, it was wonderful."

EM: Yeah. Me too.

SA: It was wonderful.

EM: So you retired in 2005 again, the second time. What activities were you involved in after that?

SA: I did some boards, small boards, not big boards with big corporations, but I worked on the board for the Spectrum Group out of Washington, D.C., for a few years. On the Sun Trust Bank board here in central Florida. Was on the PG School board.

EM: Trustees?

SA: Trustees, I think they called us. Naval Aviation Museum over in Pensacola.

EM: You are president of the board of the Naval Aviation Museum?

SA: No, I was one of the directors and on the advisory board now.

EM: What projects were you involved in in that capacity?

SA: We were just sort of the oversight, like a board of governors for a corporation, to check and make sure that the business was run as a business.

EM: So periodically you would get reports from the staff about the status of the museum?

SA: We would have a least two board meetings a year over in Pensacola. For a while I was doing a lot of studies. I was on the defense science board for one study on logistics. I did a study for John Douglas, convened a study on what went wrong with the presidential helo program. So I was part of a team that had executives from Raytheon and Northrop Grumman. We called in all of the players from the business areas—and the business area that had done the presidential helo was one that I was familiar with, both from my Loral days and again a little bit from my Lockheed Martin days.

Well, a lot from my Lockheed Martin days in Washington and up in New York. They had put together the systems for the first of the Navy H-60 helo programs, and so they were the equipment integrator. Sikorsky built the airplane. Lockheed Martin installed the smarts [electronic gear]. So they had won the contract for the presidential helo, and of course that program blew up along the way. I was part of a team that did the study on what went wrong on both sides of the coin, both from the government side and from the industry side.

EM: What were your findings?

SA: Both at fault. Again, when we were talking about the LCS and Mike Boorda's idea of a green helo, or a green ship manufactured to maritime standards but not NavSea standards, the same mistake was made with the presidential helo. The Navy signed off on a contract that said green helo. Somebody build the helo, and then find an integrator to integrate all of the stuff that we want on the helo. Too many fingers in the pie.

You had the Navy, Marine Corps, who had certain things that they wanted the helo to have that the green helo didn't, wouldn't normally come with. You had the White House, who had specific things that they wanted on that helo that very few people could know what those things were. So you had a big security issue there on how to bob and weave and get this installed, and people not find out what was there. You had the manufacturer, Lockheed Martin, who was too easy to acquiesce to changing requirements, you know. They took the message that the customer is always right too seriously.

It was one of the things that I had a lot of trouble with my whole time in industry, was the fact that industry was too easy to roll over instead of pushing back at the customer when the customer was asking for something that was going to be a lot more expensive than the customer was willing to pay for.

EM: But you were always afraid that you were going to lose the Navy as your customer?

SA: But you were afraid you were going to lose the customer, so there was plenty of blame on both sides of the coin on that one. It was really sort of sad to see how far they had progressed, and then all of a sudden—but that's sort of the same thing with the A-12 program. I'm sure if I would have looked at the A-12 program, I would have found the same things.

EM: Yeah. How did the logistics excellence award come to be named for you? Was there a guy who had worked for you?

SA: No. Well, he had worked in the building at the same time I worked, and then he ended up with the OP-04 job, I think two times after me, Bill [William] J. Hancock [Jr.]. And it was surprise. I mean, nobody had cleared that with me. I was as surprised as the next guy.

EM: Hancock suggested creating this award?

SA: That's where it started.

EM: Right. Would you have been opposed to it?

SA: Oh, no. No. I was just surprised. [Laughter]

EM: Yeah, it's quite an honor.

SA: It is, and I think I've only missed one presentation since it was instituted—maybe two.

EM: Yearly presentation?

SA: It's a yearly presentation. I go up to Washington to participate. So yeah, it's quite an honor.

EM: And your hometown, which brings you full circle back to Jackson, Ohio, named a park in your honor.

SA: They've got a park named in my honor and a little picture of me at the park.

EM: Is there a statue of you on a horse, or—

SA: No. There's a billboard or sign board there, with a color photograph of me and a list of commands and places that I'd been and done.

EM: Did you have a ceremony for that?

SA: Little, short ceremony for that. It's nice to be remembered.

EM: Well, a long and distinguished career.

SA: It was. It was a lot of fun. It was something that I would certainly have regretted not doing, and I'm just absolutely pleased as punch that somebody let me do it.

EM: Your dad would have been proud, too.

SA: I think so, yeah.

EM: Any last thoughts? We will go over some broader insights tomorrow, but is there anything you wanted to add?

SA: No. Industry and the military both were very fulfilling opportunities. I don't think I could have done the reverse. I don't think I could have gone into industry for a while and then, if I had been there too long, I couldn't have come in the military anyhow. But you saw a lot of industry people come back into the building as political appointees for different jobs, and people like Bill Perry who came and had lots of industry, and Norm Augustine came into the building on the defense side to support lots of stuff.

But I think I was just extraordinarily fortunate that I was able to have two careers. The military career was much more than I ever—you go in there and you think, well, I'll do four years, and then I will be off doing something else. And at the end of four years, well, boy, that looks attractive. I can go to PG School. Yeah, that sounds good. I will go to PG School. Oh, I get to go to the developmental squadron and test new equipment. Yeah, that sounds good. I will go do that for a while. Oh, I get to go to sea on a ship and play like I'm a navigator? Yeah, that's sounds fun. Everything—you know, as time went on, everyone kept saying, how long do you think you are going to stay in the Navy? I said, as long as it's fun I'm going to stay. And it was, until the end.

EM: Fantastic.

SA: Yeah.

EM: And I've thoroughly enjoyed our interaction over these many months.

SA: Yeah, it's been fun. It's been fun to sort of revisit and go back. Now I've got to figure out—since you forced me to go back and figure out how I got as far as getting into the Navy to begin with—I've got to get that all written down for my sister, so she can add her things in it, so we can leave it back for our kids.

EM: Yes. This will be a great—not a memento, but a remembrance for your entire family.

SA: Yeah. I hope—I think the kids, because each one of them comes into the picture, of course, at a different time, and as they've matured, they went off and did things. And some of them were close to different events in my life, and so I think it will be interesting for them to see it all tied together.

EM: Thank you, sir.

SA: Okay.

[End of 20 January 2016 interview]

Launched in 1969, the U.S. Naval Institute's award-winning oral history program is among the oldest in the country. Used in combination with documentary sources, oral histories offer a richer understanding of naval history through candid recollections and explanations rarely entered into contemporary records. In addition, they help depict the atmosphere of a particular event or era in a manner not available in official documents.

The nonprofit Naval Institute accomplishes its history projects through contributed funds and gratefully accepts tax-deductible gifts of all sizes for this purpose. This support allows the Institute to preserve the life experiences of today's service men and women so they may enlighten and inspire future generations.

For information about opportunities to underwrite Naval Institute oral history projects, please contact the Naval Institute Foundation at 291 Wood Road, Annapolis, Maryland 21402; by phone at (410) 295-1054; or by e-mail at foundation@usni.org.

Index to the Oral History of Adm. Stanley R. Arthur, USN (Ret.)

www.ingramcontent.com/pod-product-compliance
Lightning Source LLC
LaVergne TN
LVHW080924210926
848485LV00004B/721